An Advanced Course in Nuclear Engineering

Series editors
Yoshiaki Oka, Tokyo, Japan
Haruki Madarame, Tokyo, Japan
Mitsuru Uesaka, Tokyo, Japan

Recently the global warming issues are advocated, so the nuclear power is paid to attention again as a low-carbon and ecology energy source; So-called "The Nuclear Power Renaissance1" has risen up The main nuclear power industrial countries are not staying only in the domestic sales but progressing global sales strategies of nuclear power plants, and not the few nations backup them. On such a background, the main & developing nuclear power industrial countries work positively on the human resources education of nuclear engineering Inside that nuclear power industry globally develops, the required human resources education should stand on the same viewpoint, however, no English textbook that becomes a global standard has existed up to now, and the publication was expected for a long time While seeing such a worldwide demand and aiming to become a global standard, the Japanese nuclear power specialists concentrated their wisdom, and the Japanese version "Nuclear engineering textbook series" were published in the cooperation of the University of Tokyo, Japan Atomic Energy Agency and other domestic organization related to nuclear engineering. Moreover, the covered range is various and includes not only the nuclear reactor but also the nuclear fuel cycle, the radiation use, and the maintenance, etc. This was possible only in Japan that has steadily advanced the research and development of all fields of nuclear engineering. The English version "Nuclear engineering textbook series", the English translation of the Japanese version, can be expected to be useful as a globally standard textbook for the he human resources education of nuclear engineering in the all domestic and foreign universities that have the department of nuclear engineering, such as the University of Tokyo, and all research laboratories and companies related to the nuclear engineering.

More information about this series at http://www.springer.com/series/10746

Naoto Kasahara
Editor

Fast Reactor System Design

Springer

Editor
Naoto Kasahara
The University of Tokyo
Tokyo, Japan

Original Japanese edition, *Genshiryoku Kyokasho: Kousokuro Sisutemu Sekkei* edited by Naoto KasaharaCopyright © 2014 Published by Ohmsha, Ltd. 3-1 Kanda Nishikicho, Chiyoda-ku, Tokyo, Japan

ISSN 2195-3708 ISSN 2195-3716 (electronic)
An Advanced Course in Nuclear Engineering
ISBN 978-981-10-2820-5 ISBN 978-981-10-2821-2 (eBook)
DOI 10.1007/978-981-10-2821-2

Library of Congress Control Number: 2017930399

Printed on acid-free paper

This Springer imprint is published by Springer Nature
The registered company is Springer Nature Singapore Pte Ltd.
The registered company address is: 152 Beach Road, #22-01/04 Gateway East, Singapore 189721, Singapore

Series Aims and Scopes

Since the beginning of the twenty-first century, expectations have been rising in Japan and abroad for nuclear power generation because the need for a safe and stable energy supply is increasing against the background of global environmental issues and the depletion of oil-based energy sources. The situation calls for the development of human resources with advanced knowledge and techniques of nuclear energy. However, many traditional textbooks on nuclear engineering have fallen out of print, and many of them were written about 30 years ago when full-scale R&D and use of nuclear energy began in Japan and therefore hardly reflect recent advances.

Anyone who tries to solve and counter practical problems and unknown application problems encountered in using nuclear energy will face the need to return to the basics. This requires the capability to envision various images of the subject. The systematic acquisition of knowledge of nuclear energy is fundamental to nuclear R&D and the development of the nuclear industry. This series is designed to serve this purpose.

With advances in the utilization of nuclear energy, there have emerged many challenges that cannot be solved by the traditional nuclear engineering approach, which addresses the relationship between society and nuclear energy in an individualized and fragmented manner. Traditional education that uses scientific and technological approaches is not enough to help understand and handle these problems appropriately. In an effort to reduce risks and uncertainties associated with the use of nuclear energy, the University of Tokyo has started the world's first systematic education and research that address the humanities and sociological aspects of nuclear energy. Some of the books in this series are used in these courses.

We have published this series as a set of standard books for systematically studying nuclear energy from basics to practice, with the aim of nurturing experts and engineers who can act from a global perspective. While being aware of the situation surrounding nuclear energy, students need to not only familiarize themselves with basic knowledge but also acquire practical expertise, including flexible knowledge and firsthand techniques that allow them to understand field practices without feeling any gap.

This series consists of three courses on nuclear energy: the basic course, the frontier course, and the practical course. The content is based on the education at the Nuclear Professional School and the Department of Nuclear Engineering and Management, both established by the University of Tokyo with the aim of developing high-level human resources specializing in nuclear energy to cope with the new century of nuclear energy. The books were written mainly by faculty members of the University of Tokyo and researchers at the Japan Atomic Energy Agency and in related industries.

We would like to extend our gratitude to all those who have kindly taken time to contribute to or cooperate in the creation and publication of this book series.

Department of Nuclear Engineering and Management
Nuclear Professional School
School of Engineering
The University of Tokyo
Tokyo, Japan

Yoshiaki Oka
Haruki Madarame
Mitsuru Uesaka

Preface

The fast reactor (FR) is a type of new reactor currently under research and development. In addition to the high level of safety required by nuclear power plants, creativity is also required when designing an FR system to ensure reliability and economy. FR research first began some 60 years ago, and the principles have largely been established. The path forward requires the development of a commercial system that integrates the experience and engineering know-how gained thus far.

While many conventional textbooks on FRs introduce the general principles and elemental technologies of FRs, discussion of design is limited mainly to the reactor core.

The aim of this book is to instill in young researchers and engineers of tomorrow who will be charged with commercializing this new type of reactor the ability to come up with creative designs that lead to the development of new components and systems for improved plant reliability and economy. Toward this end, this volume describes the methods of integrating the power plant in its entirety — from the reactor core to all of the various systems and components — and teaches the thinking that forms the background of these methods. This background includes the various organizational and management issues that are encountered as projects move forward and will be explored in great detail based on actual design and construction experience with Japan's prototype FR, Monju.

From this perspective, the main points of each chapter are listed below:

Chapter 1 describes a thinking process that is effective in creating a new reactor system while filling in gaps in knowledge and data between it and commercial reactors. This process analyzes the functional requirements by system and then hierarchically develops methods to realize these functions from the mechanism level to the structure level. These structures are then integrated into a plant system. This description, however, is simply a logical way to systematically understand the process. The actual design procedure becomes a more complex process of trial and error. This chapter summarizes the process of designing a new type of reactor, and it provides a logical framework for the subsequent chapters.

Chapter 2 describes the purpose of the FR and gives the history of research and development. It provides the reader with an understanding of how the requirements during the long period of research and development expanded from release from the uranium resource constraint to the reduction of environmental burden caused by high-level radioactive waste. This chapter also presents a general overview of the experience various nations have gained and how their circumstances have changed. This knowledge forms the premise for the determination of the plans and required functions that provide the starting points for system design.

Chapter 3 introduces the required functions of the FR, namely, to breed fuel and reduce the toxicity of high-level radioactive waste. It also describes the mechanism of how this is achieved using plutonium and fast neutrons. The use of fast neutrons requires the use of liquid metal sodium as the core coolant. Sodium is also used to transport heat from the reactor core to the steam turbine, which results in a unique design condition under which the entire plant becomes a high-temperature, low-pressure system.

Chapter 4 discusses the assurance of safety, which is particularly important for nuclear power plants. This becomes a design constraint and requires additional functions. This chapter first presents an overview of the basics of safety assurance for all nuclear reactors and then discusses the characteristics of the FR and the basic policies for assuring safety. It specifically describes the handling of core fuel possessing nuclear characteristics differing from that used in light water reactors, safety measures for plants using sodium, and severe accident measures.

The assurance of structural integrity, as discussed in Chap. 5, is a prerequisite for ensuring safety, and it forms the primary constraint for achieving reliability and economy. This chapter first introduces the mindset for ensuring the structural integrity of nuclear reactors in general and then discusses the load characteristics under the design conditions unique to FRs, such as high temperature and low pressure, and their related damage/failure modes. It also discusses thermal load and seismic design as measures for maintaining the structural integrity against the damage/failure modes.

Chapter 6 discusses the method of forming the plant concept. It explains the method employed in designing mechanisms to meet the required functions described in Chap. 3 within the constraints listed in Chaps. 4 and 5 and integrates the mechanisms into a system. At this stage, the outlines and major parameters of each plant subsystem, such as the core, the reactor structure, cooling system, and the fuel handling system, are specified as primary mechanisms. From this, the type of plant that characterizes the big picture formed by the integration of each system will be determined.

Chapter 7 develops the reactor core mechanism determined in Chap. 6 and explains the thinking behind and method of determining the specific structure of the reactor core/fuel within the constraints described in Chaps. 4 and 5. First, the reactor core design is broken down into nuclear, thermal, and flow design, and each design process is introduced while touching upon the fundamentals of reactor physics and thermal hydraulic engineering. Next, for fuel design, methods that meet the need to increase burnup within the constraints of fuel safety and cladding tube structural integrity are discussed together with the fundamentals of fuel material engineering that form the foundation of these methods.

Chapter 8 develops each of the mechanisms (systems) discussed in Chap. 6 that form the plant and explains the thinking behind and method of building the specifications and layout of the structures (components) that comprise each subsystem within the constraints discussed in Chaps. 4 and 5. Design samples of the different subsystems introduced in Chap. 6, such as the reactor structure and cooling system, are also discussed together with their background and rationale.

Maintenance, as discussed in Chap. 9, is important to ensure the structural integrity and reliability of the plant over its lifespan by compensating for any uncertainties that require attention when a new reactor is designed. Maintenance is also important for the implementation of improvements to the plant design through operational experience. This chapter first introduces the general thinking behind the maintenance of nuclear power plants and then explains the maintenance typical of FRs that is characterized by high-temperature low-pressure conditions due to the use of sodium and its respective reactions with air and water. Finally, this chapter discusses the specifics of FR maintenance, operation, inspection, and repair.

Chapter 10 specifically discusses the design experience of Japan's prototype FR, Monju, as an example of the FR system design discussed in previous chapters. Any actual project will be subject to a number of non-technological constraints by individuals and society. Consequently, the project will often not proceed as described in the previous chapters, and this will necessitate a certain amount of trial and error and judgment. The goal of this chapter is to convey actual experience to the next generation of engineers.

This book is based on lectures and practical exercises presented in classes at the Graduate School of Engineering, the University of Tokyo, and it incorporates opinions and requests from students. It is hoped that the students, engineers, researchers, and others who use this book will contribute to the world by applying their high-level problem-solving ability to comprehensively resolve the myriad issues faced in the development of new reactors.

Tokyo, Japan
August 2016

Naoto Kasahara

Contents

Editor, Authors, and Collaborators

Author and Editor

Naoto Kasahara (The University of Tokyo) [Chaps. 1 and 5, Sect. 8.3.3]

Authors

Masakazu Ichimiya (University of Fukui) [Chap. 3]
Hiroyuki Ohshima (Japan Atomic Energy Agency) [Sects. 5.2.2, 7.5.5]
Yoshio Kani (Tokai University) [Chap. 4]
Tadao Takahashi (FBR Senior Research Laboratory) [Chap. 10]
Toshikazu Takeda (University of Fukui) [Sect. 7.5.4]
Satoru Nakai (Japan Atomic Energy Agency) [Chap. 9]
Nobuo Nakae (Nuclear Regulation Authority) [Chap. 7]
Kiyohiko Maeda (Japan Atomic Energy Agency) [Chaps. 6 and 8]
Tsutomu Yanagisawa (Japan Atomic Energy Agency) [Chap. 2]

Collaborators

Takashi Nagata (MFBR Systems Inc.)
Shinji Yoshikawa (Japan Atomic Energy Agency)

Chapter 1
Designing a New Reactor

Naoto Kasahara

Fast reactors (FRs) are a new type of reactor under research and development that are unlike commercialized and standardized light water reactors (LWRs). Data and knowledge required for their design, however, are insufficient; manufacturing techniques have not yet been established, and operational experience is limited. When designing an FR, therefore, these uncertainties must be taken into consideration.

One important goal for a new type of reactor is to bring it to commercialization by improving related technologies to the degree possible through design, construction, and operation.

Therefore, unlike typical designs, the goal of the new reactor design is not limited to the achievement of its initially planned performance, but also includes contributions to the establishment of technological systems spanning the entire plant life cycle.

This chapter provides a diagram depicting the thinking process (thinking process diagram) underlying the creative design methodology needed for a new type of reactor. This process analyzes the functions required for a system and then hierarchically develops methods for realization from the functional level to the structure level; these structures are then integrated into a system. Although the procedure of actually designing a large system is a more complex process of trial and error, the thinking process diagram provides an effective method for logically understanding the system as a whole. The flow of the thinking process for new reactor design is, therefore, based on this diagram. This diagram is also used to provide a logical framework for subsequent chapters.

N. Kasahara (✉)
The University of Tokyo, Tokyo, Japan
e-mail: kasahara@n.t.u-tokyo.ac.jp

N. Kasahara (ed.), *Fast Reactor System Design*, An Advanced Course in Nuclear Engineering 8, DOI 10.1007/978-981-10-2821-2_1

1.1 Thinking Process and Design Methods for a New Reactor

1.1.1 New Technology Creation and Design Methodology

This section provides an overview of the thinking designers employ when creating a new technology such as a new type of reactor. It also provides an overview of the design methodology based on this thinking by referring to the latest study of design [1].

Generally, as shown in Fig. 1.1, the flow of the thinking process (design) in a creative design begins with the determination of the functional requirements. Mechanisms fulfilling those functions are then devised, and a structure enabling the realization of these mechanisms is determined. Although the actual design procedure involves a repeated process of trial and error, Fig. 1.1 is helpful in understanding the logic of this thinking process.

In addition, the actual design process for large, complex systems such as a nuclear reactor proceeds step by step through the phases of planning, conceptual design, basic design, and detailed design (Fig. 1.2) [2]. These steps are chronological; and as the process progresses through the steps, the design is reviewed and refined.

Figures 1.1 and 1.2 present two different views of the design flow; however, they are related to a certain degree. The conceptual design step, which comes first chronologically, is where the overall functions and approximate mechanisms are temporarily determined. The basic design step includes the design of everything from functions to structures. The detailed design step is where the details of the mechanisms and structures are refined for manufacturing. Since the purpose of this textbook is to provide an overview of system design as a whole, the discussion focuses mainly on the steps from planning through basic design. It is organized along the logical flow of a design based on the thinking process shown in Fig. 1.1.

Further analysis of the thinking process shown in Fig. 1.1 is illustrated in Fig. 1.3. The first half of the thinking process, the region of functions, is where the desired functions are considered. Here, the required broad functions from the planning stage are analyzed, and functional configuration is determined. These are further analyzed and broken down into functional elements. It is necessary to reduce the functions to a small level so that they can be placed with their respective mechanisms.

The second half of the thinking process, the region of mechanisms/structure, is where the end products that realize each function are described. If functional elements are sufficiently refined in the region of functions, then it can be mapped to the mechanical elements that will bring them to realization.

Typically there are multiple mechanical elements that are candidates for realizing a particular function. Therefore, all possible mechanical elements for a function should be listed, as shown in Fig. 1.4. Once the constraints and priorities that are important for nuclear reactors, such as safety, have been considered, then the

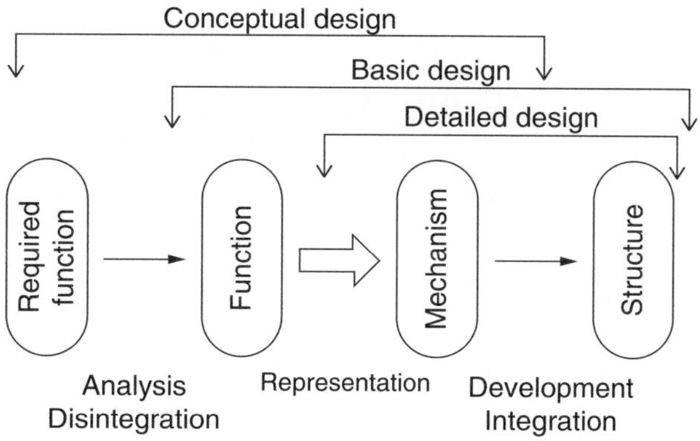

Fig. 1.1 Design thinking process

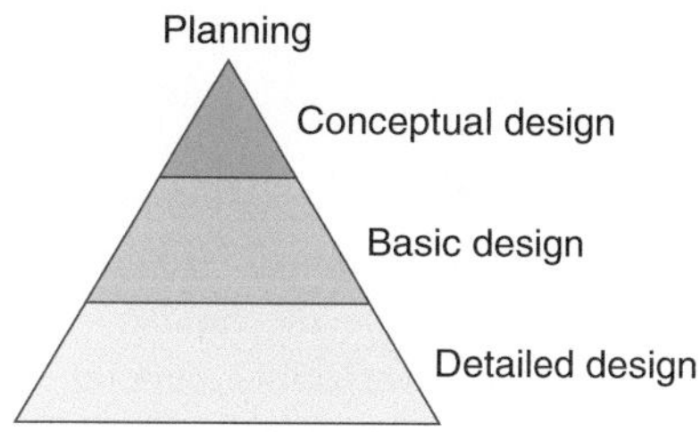

Fig. 1.2 Design steps

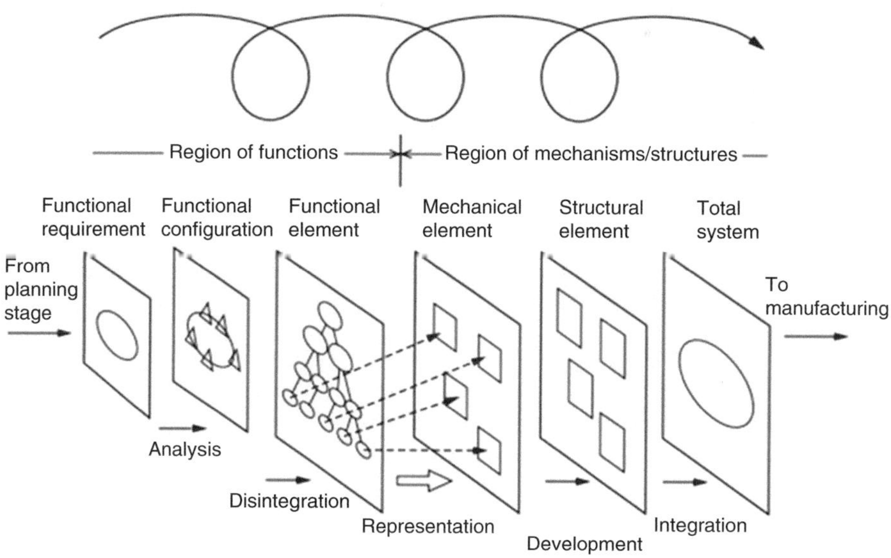

Fig. 1.3 Analysis of the design thinking process

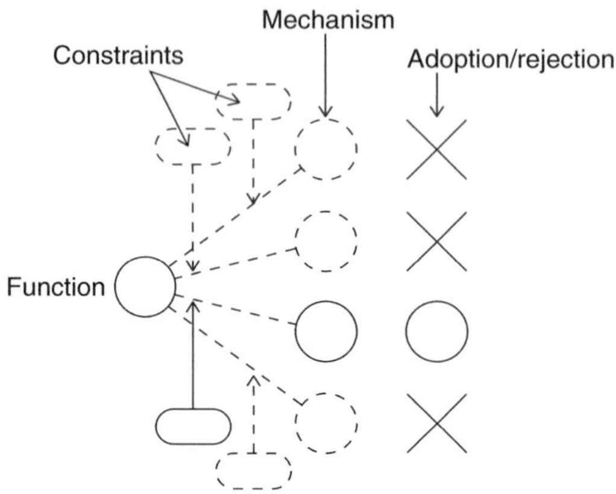

Fig. 1.4 Choosing mechanical elements under constraints

appropriate element can be chosen from the list. The mechanism elements that are not selected still have value for future designs since the selected mechanism elements may change if the constraints change. When designing a new reactor in particular, there are many mechanisms that can realize one function; and since the potential realization of a function depends on the results of research and development [2], multiple mechanisms are considered with one being selected. Furthermore, since constraints depend upon the current level of technology and the economy, they become less rigid as time passes. For example, improvements in manufacturing technologies or materials may relax the constraints themselves, and the required tolerances may be rationalized because of improved accuracy in predicting phenomena due to advances in analysis technology or because more data has reduced the level of uncertainty. Also, authorized construction budgets may change in response to the state of the energy supply or the environment. Therefore, both a reference and an alternative for each mechanism that is to realize a particular function should be prepared to enable the selection of a design that corresponds to the constraints. For example, multiple types of steam generators are often reviewed together because they are heavily dependent on material development. In one plant overseas, different types of steam generators were designed, manufactured, and installed [3]. If this type of design experience and data is retained, it can be used when technological innovation enables a relaxation in constraints.

The next step in the design thinking process is to determine, within structural integrity and other constraints, the size, weight, and materials for the selected mechanical elements so as to realize their translation into structural elements. These structural elements are then combined and integrated into a total system. In the real world, this process does not progress linearly; rather, it takes the spiral path of trial and error (Fig. 1.3).

One way of illustrating the design thinking process as described thus far is the thinking process diagram [1] shown in Fig. 1.5. This diagram is nearly symmetrical between function and mechanism since the initial functions in the function region

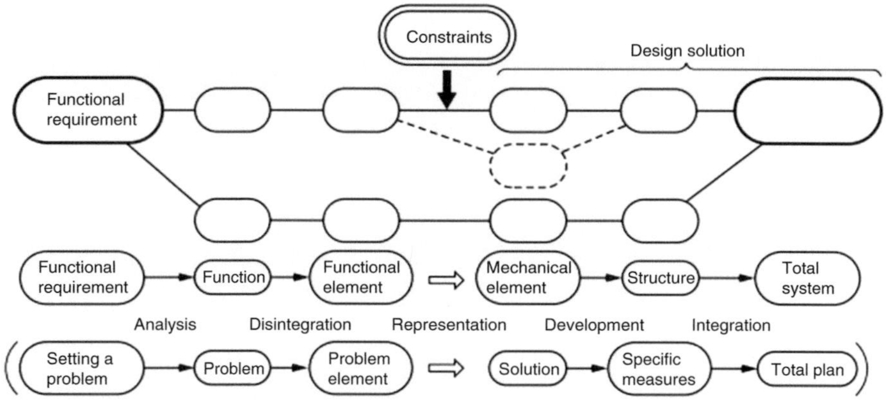

Fig. 1.5 Design process by the thinking process diagram

are hierarchically reduced and mapped one to one with mechanisms, and then the whole structure (system) is integrated in the mechanism/structure region. At this point, there are usually multiple mechanisms and structures that fulfill the functions; and there are constraints for determining them. This type of development diagram is useful not only for hardware design but also for the creation of project protocols and the development of software.

1.1.2 Thinking Process on New Reactor Design

The design flow for new type reactors is illustrated in Fig. 1.6 using the thinking process diagram for creative designs discussed thus far. The actual design does not flow in one direction from planning to the total system; rather it flows forward and backward through trial and error as shown in Fig. 1.3. In the conceptual design stage in particular, it is common to first temporarily create an approximate overall picture, determine the individual component structures, and then subject the integrated structures to hypothesis testing and verification. The entire process is then repeated. The content of this textbook thus flows according to the thinking process diagram from planning in Chap. 2. The overall function and plan systems are described in Chap. 3, mechanisms and subsystems are shown in Chap. 6, and mechanism elements and structures are explained in Chap. 7 (reactor core) and Chap. 8 (components). Chapters 2, 3, 4, 5, 6, 7 and 8 are outlined below.

Chapter 2: Purpose and History of Fast Reactors

The starting point for a design is planning. Determination of the purpose of the design must be based on the needs of society, and the identification of its functional requirements must be based on the current level of technology and know-how. Large-scale projects such as the development of a new type of reactor require a

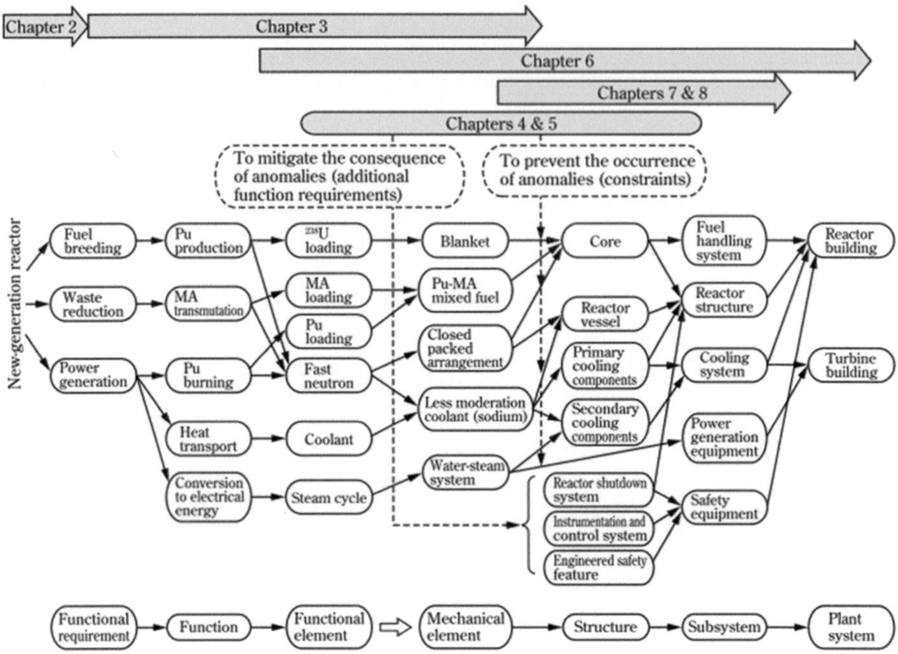

Fig. 1.6 Design flow of a new reactor as organized by the thinking process diagram

great deal of time from planning to construction and operation. During this time, social needs can change and technological innovations can occur. For example, the expectations for FRs now include not only the initial release from the uranium resource constraint but also the reduction of the environmental burden from high-level radioactive waste (HLW). In recent years, consequently, the primary functional requirements of FRs have been fuel breeding, reduction of waste, and power generation. To meet these needs, the design must include the realization of a core with characteristics sufficiently flexible to meet the functional requirements as well as accommodate any changes in the future. The history of development and changes in the significance of FRs are discussed in Chap. 2.

Chapter 3: Plant Concepts and Mechanisms

Next, the above-described functional requirements are analyzed into functional configurations. Furthermore, they are reduced into functional elements so that the functions can be realized. The fuel breeding requirement is met by the conversion of uranium 238 into plutonium 239 using fast neutron fission reactions. The requirement for decreased waste is met by preparing nuclear spallation of minor actinides (MAs) with long half-lives using fast neutrons. Its functional element is a blanket loaded with uranium 238 that is placed around the fuel so that the fast neutrons are irradiated efficiently. The fuel is a mixed load of Pu and MA. For power generation, sodium coolant, which does not moderate the fast neutrons, is used to transport the heat created from the combustion of Pu to water to produce steam to drive the turbines that generate electricity. These functional elements and their respective principles are discussed in Chap. 3.

Chapter 4: Policy of Safety Assurance (Design Constraints and Additional Functional Requirements)
Constraints must be clearly identified when replacing functions with mechanisms. This is especially true when designing a nuclear reactor because safety must be assured to protect both people and the environment from the effects of radiation. Systems and components responsible for the basic safety functions of shutdown, cooldown, and containment are required to prevent anomalies. These constraints are vital. Since so little is known of a new type of reactor, an essential constraint is the need for designs with large safety margins to prevent accidents. Additionally, functions that can mitigate the consequences of anomalies if they occur in the nuclear reactor are required. Based on this requirement, safety equipment such as a reactor shutdown system, instrumentation and control systems, and engineered safety features must be added.

Chapter 5: Mind-Set Required to Ensure Structural Integrity (Design Constraints)
The assurance of structural integrity is a prerequisite for the prevention of anomalies, which forms the basis of safety assurance and is an indispensable condition for the maintenance of plant reliability. Therefore, conservative designs with safety margins are needed. Such required designs bring various constraints to the design of the core and components, such as operating temperature, materials, and form. In addition, new mechanisms may need to be added to the design to mitigate load, such as thermal stress mitigators and seismic isolators.

In addition to safety, the cost of constructing a plant is also a constraint. Since structural design is closely related to the cost of plant construction, design solutions must be sought that meet both the constraints of construction cost and structural integrity assurance.

Chapter 6: System Conceptual Design (From Function to Mechanism)
First, consider the mechanisms and general structure for a core that can demonstrate the expected performance and can load heat-generating Pu, fuel breeding 238 U, and MA for the reduction of waste. Next, consider the reactor structures needed to support and cool the core. A cooling system must be designed to transport the heat from the reactor to the steam turbines. Additionally, a water-steam system for the generation of power and the power generation equipment itself must be considered.

Before these various mechanisms can be built as part of the plant's systems and structures, they must first meet the constraint of anomaly prevention. As the structures are evaluated for their accordance with safety design and structural design policies, they must also be optimized within their respective limits.

Furthermore, a reactor shutdown system, instrumentation and control systems, and engineered safety features are needed as functions to mitigate consequences from anomalies, as discussed in Chap. 4.

Last, a picture of the plant system must be drawn up to verify consistency when the major plant specifications and mechanisms are brought together. Consequently, even if certain mechanisms and structures have not yet been determined, an integrated design that conceptually combines the entire system must be created; and major plant parameters, such as inlet and outlet temperatures of components that balance heat, must be determined. Furthermore, the layout must be designed if the respective systems are to be housed in the reactor or turbine buildings.

Chapter 7: Core and Fuel Design (From Mechanism to Structure)
Determination of the actual structure of the reactor core/fuel (i.e., specifications) is based on the plant concept (primary parameters such as the type of nuclear reactor, reactor core thermal power, type of coolant and fuel, and core inlet/outlet temperatures) that is determined by system conceptual design within given constraints (safety assurance, preservation of structural integrity). The composition of the Pu-MA mixed fuel is determined by the design of the reactor physics that sustains criticality and by the design of the thermal hydraulics that obtain the required linear power density. The fuel cladding tube is to be designed to maintain structural integrity under fast neutron irradiations to prevent anomalies from occurring. The resulting fuel pins (combination of fuel and cladding) are closely packed in a wrapper tube with wire spacers to maintain the dense arrangement for the effective use of neutrons and sufficient space for cooling. The fuel assemblies form the core in combination with the blanket assemblies that control plutonium breeding, the neutron reflectors that support the achievement of criticality, and the control rods for shutting down the reactor that are all closely arranged.

Chapter 8: Plant Component Design (From Mechanism to Structure)
The major mechanisms determined by the system conceptual design are built as plant component structures as they meet safety assurance and structural integrity constraints. The reactor structure is designed with the spatial support needed to meet the conditions of the reactor physics and with the required channels for core cooling. The cooling system is designed as a major system consisting of the primary and secondary systems, heat exchangers, piping, and pumps that indirectly transport heat from the reactor vessel to the water-steam system via sodium coolant. Additionally, the fuel handling system is designed specifically for FRs so that fuel is exchanged and stored without contacting the atmosphere while the decay heat of the removed fuel is cooled.

1.1.3 Actual Design and Constraints

The thinking process diagram is an effective method for logically overviewing and organizing a whole system design; however, the procedure of actually designing a large-scale system such as a new type of reactor is significantly more complicated and includes trial and error. For this reason additional information is given here.

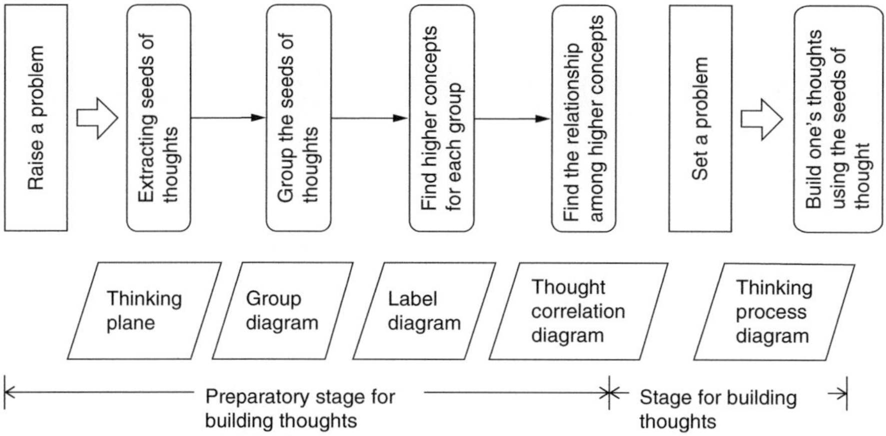

Fig. 1.7 Process of thinking development

When creating a new system, ideas are initially conjured up in piecemeal fragments. The actual design work starts with the organization and development of these fragments. This work is done at the preparation stage prior to using the thinking process diagram. A methodology exists that supports this preparatory work, and the reference [1] is described at the end of this chapter for those interested in learning more about it. A very simple illustration of this way of thinking is shown in Fig. 1.7. First, the seeds of the fragmentary thoughts are extracted in notes. Next, any similar thoughts are grouped together and then labeled with the characteristics for the group. These labels show the higher concepts of each of the groups. Through this process, the number and types of idea sources are developed. Next, relationships are found among the labels (higher concepts). Through this process the fragmentary thoughts become systemized, and the thinking process diagram becomes ready for use.

Before entering the thinking development stage shown in Fig. 1.7, one typically refers to examples of existing designs. As listed at the end of this chapter, there are excellent materials and databases [3, 4] available for reference in the world. The chronological flow of an actual design typically proceeds by consulting existing examples and creating ideas in the thinking process diagram. The feasibility of these ideas are then explored, and plans are made to build them out step by step.

A few more words must be added regarding the importance of constraints when evaluating the feasibility of an idea. The greatest constraint for a nuclear reactor is safety. To prevent the occurrence of anomalies in the fundamental safety functions of shutdown, cooldown, and containment, the individual components must be highly reliable, and their diversity, multiplicity, and independence must be assured. These requirements constrain the design. As discussed above, new safety functions that exceed the constraints are needed to moderate the consequences of anomalies. To verify that these constraints are met, safety evaluations that assume the occurrence of various accidents are conducted.

The next prerequisite for preventing anomalies is the structural integrity of components during normal operation. The structural integrity of the safety equipment and other components is required also in case of an accident. These are major constraints on the structural design, and thus evaluations are conducted to verify that structural integrity is assured under operating conditions, such as temperature and pressure, and external forces such as earthquakes.

Once the reactor demonstration stage has been completed, the next important constraint is economic performance. There are many factors that comprise economic performance; however, since this textbook is about plant design, it uses a narrow definition and focuses discussion mainly on the cost of plant construction. As previously discussed, the cost of plant construction greatly influences the constraints of structural design, thus making it necessary to rationalize the structures in ways that reduce costs, such as simplifying and reducing the size of the core and components. The safety of any rationalized structure must still be assured.

On the other hand, economic performance includes various other factors, such as the costs of plant operation, fuel manufacturing, reprocessing, and waste processing. One of the major characteristics of FRs is a fuel cycle that burns up uranium 238. Consequently, the proportion of the cycle cost is often greater than the cost of plant construction when evaluating the overall cost of an FR. Therefore, a decrease in the cycle cost (i.e., a decrease in the number of reprocessing cycles) by an increase in fuel burnup may contribute more to a reduction in the total cost of power generation than cuts in the cost of construction. The absolute value of the targeted economic performance will change relative to any competing power sources.

Furthermore, social acceptance, in a broad sense, becomes a constraint. The use of plutonium involves pros and cons in terms of reducing the burden on the environment and nuclear nonproliferation; however, the balance of these may change depending on social circumstances and temporal and spatial perspectives.

1.2 Research and Development

1.2.1 Roles of Reactors in Research and Development

Compared to commercial reactors, the data and knowledge needed to design a new type of reactor are insufficient. Manufacturing technologies have not yet been established, and there has been little operating experience. All of these large uncertainties must be considered when a new type of reactor is designed.

The development of new technology in any engineering system such as a new type of reactor requires trial and error experience. When developing a nuclear reactor, utmost care must be given to safe operation.

Therefore, research and development is carried out as nonlethal experience is gained by proceeding incrementally in complexity and scale. The design and construction stages of the world's FRs are listed below [3].

Experimental reactor: Reactor used in basic research for theory verification. Also known as a research reactor. Also used in fuel/material irradiation tests.

Prototype reactor: Reactor used to establish technologies and to evaluate the feasibility of large reactors. Provides feedback on experimental reactor experience. Also provides technological and economic data needed for the construction of demonstration reactors.

Demonstration reactor: Reactor used to verify large-scale plants. Used to demonstrate reliability and economic competitiveness for commercial use. A design that passes this stage is viewed as complete, thus allowing plants to be constructed.

Figures 1.8, 1.9, and 1.10 show the different FR types and plant parameters that have been developed in various countries at different times. The closed circle symbol indicates an experimental reactor; the closed square symbol indicates a prototype or demonstration reactor; and the closed triangle symbol indicates a commercial-size reactor.

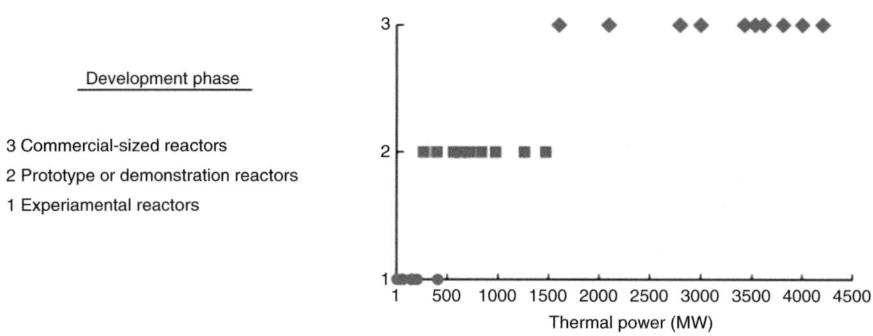

Fig. 1.8 Development phase vs. power of fast reactors

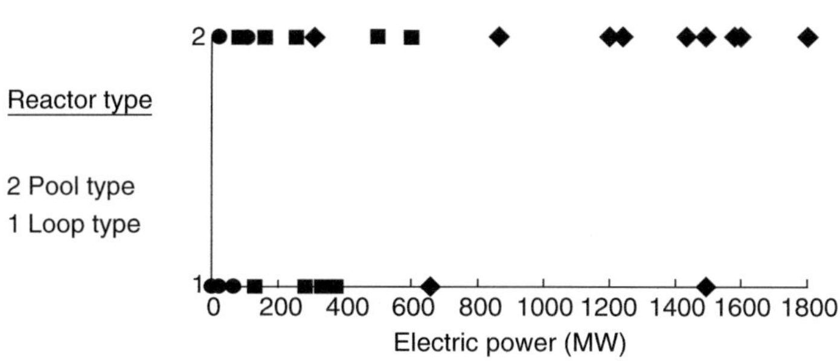

Fig. 1.9 Reactor type vs. power of fast reactors

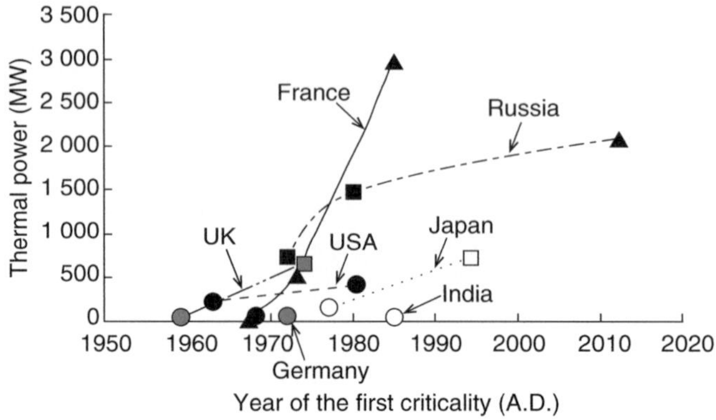

Fig. 1.10 Development steps for fast reactors in various countries

The design of a reactor presents different perspectives according to stage. For example, experimental reactors often do not include power generation equipment, and operation tests that are independent of external power systems can be run because they are not aimed at demonstrating the power generation function. Experimental and prototype reactors are often of the convenient loop type since each component is independent and connected by piping, thus enabling easier access, repair, and maintenance. The loop-type reactor is suited for specialized work in design and production since its components and systems tend to have comparatively simple constraints relative to each other. On the other hand, prototype and demonstration reactors are often of the pool type in which the individual element components can be installed in one vessel.

1.2.2 Basic and Mock-Up Tests

When designing a new type of reactor, design margins based on major uncertainties are overcome, data and knowledge for the rational design of future commercial reactors are accumulated, and research and development that promote technological innovation is conducted. In particular, research and development needed for the selection of mechanisms from the many candidates that meet the functional requirements is begun from the early stages according to their level of importance.

During research and development, the below-listed tests are prioritized according to the stage of development and design.

1. Basic tests to clarify phenomena and to obtain fundamental data for the design

In the initial stage prior to the construction of an experimental reactor, critical experiments to identify nuclear characteristics were conducted in various countries. In the prototype reactor stage, many thermal hydraulic tests were conducted to clarify the thermal and hydraulic characteristics of sodium so that new components for the handling of sodium could be designed.

2. Tests to obtain data that greatly influence construction cost

Economic performance is important in the prototype and demonstration stages, as well; therefore, tests to obtain data which influence plant construction cost are vital. For example, data on the material strength characteristics of the fuel cladding tube are among the most important because they influence not only the operating temperature and cost of plant construction but also the cost of the nuclear fuel cycle. Furthermore, the structural material of the sodium components is a factor that limits their size and lifespan and thus has a large influence on the cost of construction.

The influence that materials have on a plant is generally strong; therefore, new materials may be developed. When they are, the period of development may be significant and there may be large unknowns; therefore, testing must be started in the early stages of the project. Alternative designs must also be prepared in case sufficient data cannot be obtained during the period.

3. Mock-up tests to validate design

Mock-up tests to validate sodium component designs must be conducted at large-scale sodium research and development facilities because there has been so little experience with them in other fields. Full-size tests are generally preferred because of the difficulty of establishing a scaling rule for all of the physical phenomena.

1.2.3 Using Numerical Simulation

Certain full-size mock-up tests cannot be conducted. These include behavioral and destruction strength confirmation tests during a reactor core accident as well as verification of the performance of the safety facilities based on those tests. This is particularly true for large-scale reactors that have passed the demonstration stage. In these cases, numerical analysis techniques are used, thanks to the remarkable developments in computers and computer science.

The development of numerical simulation programs is based on phenomena identified from basic tests and data. Validity is then confirmed with midsize tests that are similar to the actual phenomena of interest. Research, development, and maintenance of the methods for verifying and validating numerical simulation programs are progressing. Universal analysis codes for commercial use as well as analysis codes for developers to understand content at the source level may become necessary for approval and licensing.

1.3 Design, Construction and Operation

1.3.1 Design Standards and Design

As a design moves from the basic design stage to the detailed design stage, the different types of standards for approval and licensing must be organized. These standards must be organized and consolidated into a knowledge base to enable the application of the research and development results and data to design. Design standards and guidelines are organized from these needs. Some representative examples are the safety design review guide and the elevated temperature structural design guide.

Furthermore, numerical simulation programs must be thoroughly verified to pass approval and licensing. Representative examples include the reactor physics analysis code, thermal hydraulics analysis code, structure analysis code, and safety analysis code.

These codes are products of design, and they will be superseded and developed for the design of a new reactor in the next stage.

1.3.2 Construction and Operation Experience

The ultimate goal for the design of a new type of reactor is the confirmation of plant safety and reliability through actual construction and operation. Added to this goal for a demonstration reactor is the confirmation of its economic competitiveness.

Since construction and manufacturing technologies are major constraints in the design of new reactors, research and development is required to relax the constraints of the principal technologies. In particular, the development of the manufacturing technologies and construction of the manufacturing facilities for the structures unique to FRs, such as sodium components, must be incorporated into the plan from the design stage. There is a tendency for components to be extremely thin-walled, large-diameter cylindrical structures to withstand thermal stress at low pressure; however, these structures sag under their own weight when positioned horizontally, and thus technological innovations in manufacturing are needed. These include the imposition of strict dimensional error limits of thin-walled components.

Performance tests are conducted after plant construction to confirm designs for newly developed components or when phenomena cannot be verified easily using small-scale test equipment. Data must be acquired that verifies the nuclear, thermal hydraulic, and other characteristics predicted by numerical simulations. Nuclear characteristic data, in particular, that are unique to new plants that have not been activated can be collected during construction. It is also necessary to confirm the performance of newly developed components and measurement systems that are unique to sodium. In this case, consideration must be given at the time of design to include measurement equipment under development in a system that is separate from the alarm system.

The way a plant is operated greatly influences its reliability and economic performance; thus, optimization is required through operational experience. One operating characteristic of FRs is the large thermal transient stress that accompanies reactor startup/shutdown. To limit this, determination of the upper limit to the rate of temperature change during startup and shutdown operation must take into account the balance of influence on the plant's economic performance. Furthermore, consideration must be given to the operation of auxiliary systems that are designed to allow filling/preheating of inert gases to prevent sodium from oxidizing and freezing and handling/cleaning fuel in the inert gases.

The development of maintenance technologies as discussed in Chap. 9 requires experience through actual operation maintenance. When designing a plant, particular care must be given to the ease of sodium component maintenance, radiation control, and waste management. To allow personnel access to sodium components for maintenance, unique operations, such as the draining of the activated primary sodium, the replacement of nitrogen with air, and repairs within an inert gas environment, are needed.

1.3.3 Development Organizations and Systems

The development of a new reactor proceeds in stages over a long period of time with enormous investments of labor and capital; therefore, it is very important to build suitable development organizations and systems.

The period from the design of a new reactor to its operation may be longer than the lifetime of an engineer or organization; therefore, systems must remain flexible in response to each stage of research and development. Also, for technology to be passed along, the retention and education of staff must be ongoing. Project leaders must be able to see the big picture relative to the technology and society.

In the initial stages of a project, the government often tends to take the lead because of the great risk involved, risks that may determine success or failure. National or public research institutions often implement the research, development, and operation.

At the prototype reactor stage, since the range of technology development widens and technology begins to be transferred to the private sector, the government may be leading the project; however, the private sector participates in its implementation. Power companies, in particular, play a large role in plant operation.

At the commercialization stage, the private sector takes over development and operation since efficiency and economic competitiveness are desired. At this stage, the plant is scaled up and its impact on society becomes greater; therefore, it becomes subject to the same regulations that existing commercial reactors are.

During the transition through these stages, special care must be given to the balance between research and development and safe operation and the smooth transfer of technology from the public to the private sector.

Furthermore, a great amount of time and effort is needed to gain public understanding, from the municipality in which the construction site is located, surrounding towns and villages, and from the residents, of the significance of the new reactor (how it addresses particular social issues), its technological difference from existing reactors, and the safety of its characteristic phenomena.

As discussed above, because the scale and time needed to develop a new type of reactor may be too much for one country, various countries may join in an international effort [4]. Learning from experience is important in the development of an engineering system; therefore, it is of significant importance that countries share their experience in new reactor development with the world when the number of years of such experience is so limited. Representative examples, both past and present, of this type of international cooperation are discussed in Chap. 2.

Further Readings[1]

M. Kanda, K. Umeda, S. Miyake, T. Shimizu, M. Ichimiya, K. Yamashita, H. Mochizuki, T. Yonomoto, Y. Oka, *Genshiryoku puranto kogaku* (Nuclear Plant Engineering) (in Japanese) (Ohmsha, Ltd., 2009)

References[2]

1. Y. Hatamura, *Gijutsu no sozo to sekkei* (Design and creation of technology) (in Japanese), Iwanami Shoten (2006)
2. M. Hori (supervising editor), *Kosokuro kogaku* (Fast reactor technology) (in Japanese) (Power Reactor and Nuclear Fuel Development Corporation, 1993)
3. IAEA-TECDOC-1531, *Fast Reactor Database 2006 Update*. (IAEA, Vienna, 2006)
4. IAEA-TECDOC-1691, *Status of Fast Reactor Research and Technology Development*. (IAEA, Vienna, 2012)

[1]References useful for those who want to learn more in detail are listed below in random order.

[2]The following literatures were referred to in this chapter.

Chapter 2
Purpose and History of Fast Reactors

Tsutomu Yanagisawa

2.1 Continual Use of Nuclear Energy

It is estimated that the world's population will reach 10 billion in the last half of this century, and there is concern about drastic changes in the global environment due to serious global warming and abnormal weather. Considering these conditions, it is essential to realize structural reform to shift from mass consumption to a recycling-oriented society to achieve the slow growth required for sustainability. Under these circumstances, we should apply ourselves to the development of nuclear energy because of its potential to provide resource saving and contribute to the creation of a recycling-oriented society as a sustainable, clean, and economically feasible energy source.

The FR is expected to be a key successor to the LWR as a next-generation reactor that supports the realization of a recycling-oriented society. The effective use of fast neutrons in FRs to generate more neutrons than LWRs allows the development of nuclear energy as a more diversified technology.

An extremely serious drawback of nuclear energy is that the radioactive waste produced by it cannot be disposed of without processing. Consequently, expectations have been placed on the full development of FR technologies to respond to the need to establish a system capable of addressing settlement issues such as spent fuel management, decommissioning, and waste disposal.

T. Yanagisawa (✉)
Japan Atomic Energy Agency, Ibaraki, Japan
e-mail: yanagisawa@syskon.jp

N. Kasahara (ed.), *Fast Reactor System Design*, An Advanced Course in Nuclear Engineering 8, DOI 10.1007/978-981-10-2821-2_2

2.1.1 Release from the Restrictions of Uranium Resources

With the continued use of fissionable uranium isotopes such as ^{235}U, potential uranium reserves are not significantly larger than oil and coal reserves that are expected to be exhausted in the near future.

In addition to achieving criticality for heat, breeder reactors aim to produce more fuel than they consume. Since the development and use of nuclear energy began, there have been concerns about the finiteness and uneven distribution of uranium resources. Since the exclusive use of fissionable nuclides, particularly ^{235}U, would deplete uranium resources, appropriate countermeasures were thought necessary. To address this problem, a method of using ^{238}U, which is contained in large quantities in natural uranium, as a fuel through conversion into ^{239}Pu has been examined. One of the two to three neutrons released during fission is used for the next fission reaction to maintain the chain reaction. Therefore, if more than one neutron is absorbed by ^{238}U to convert it into ^{239}Pu, it is possible to produce more fuel than is consumed (i.e., breeding) while producing electricity by nuclear energy.

Fission caused by fast neutrons releases more neutrons than fission by thermal neutrons. Additionally, fast neutrons can cause fission of ^{238}U. Consequently, the use of fast neutrons is more advantageous compared with thermal neutrons in realizing breeding. Since some neutrons leak out of the core and are absorbed by structural material, etc. in an actual system, the value of neutron regeneration factor η (the number of neutrons produced when one neutron is absorbed by the fuel nuclide) must be greater than 2 to achieve breeding. η shows the largest value in higher neutron energy regions. In particular, since the η value of ^{239}Pu is close to 3, ^{239}Pu is the most suitable nuclide for breeding. Fast neutron reactors (fast reactors) designed mainly for breeding are called fast breeder reactors [1].

2.1.2 Reduction of Environmental Burden of High-Level Radioactive Waste

The feature of FRs from the viewpoint of the reduction of environmental burden is the repeated recycling of plutonium and minor actinides (MAs). MAs are actinides produced from plutonium in a reactor core, such as neptunium, americium, and curium, and the term "minor" is used because the amounts of these elements are limited. Compared with an LWR-based once-through fuel cycle in which spent fuel is directly disposed of without reprocessing, deployment of an FR cycle (nuclear fuel cycle based on FRs) could greatly reduce the amount of high-level radioactive waste generated and suppress the size and burden of radioactive waste repositories needed in the future. In addition, compared with the direct disposal of LWR spent fuel, the potential hazard of radioactive material contained in disposed waste generated from the FR cycle is markedly lower. In particular, as shown in Fig. 2.1, it has been estimated that recycling MAs, which together with plutonium

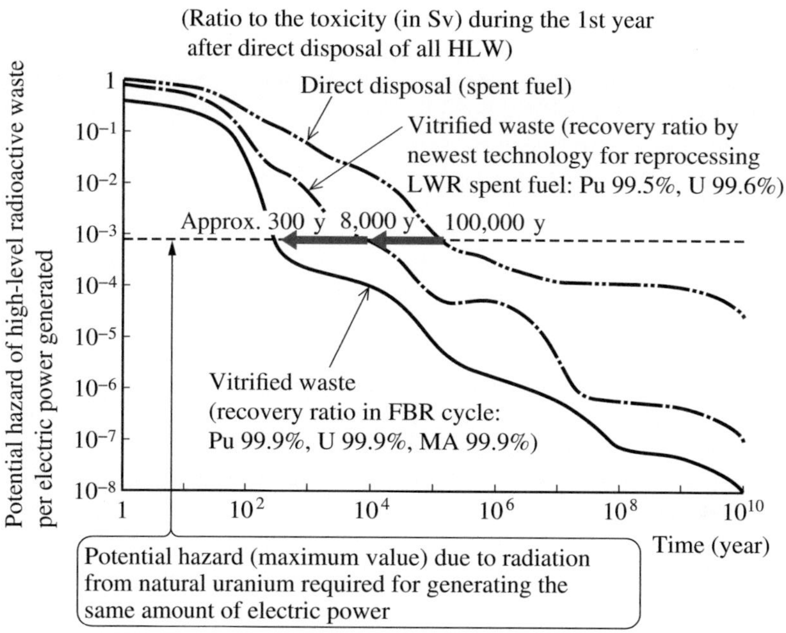

Fig. 2.1 Temporal change in potential hazard of high-level radioactive waste per electric power generated [4]

have significant potential hazard in FRs, can reduce the amount of time required for the potential hazard of high-level radioactive waste (HLW) to return to that of natural uranium required to produce the equivalent electricity to approximately 300 years, an amount of time that can be grasped on a human scale, from the approximately one hundred thousand years required in the case of direct HLW disposal [3].

2.1.3 Targets of FR Development

LWRs are taking root globally as the first step in the development of nuclear energy as a sustainable energy system. In the development of science and technology following LWRs, it is essential to select a concept based on a technology that is likely to bring innovation in a candidate for commercialization. The forthcoming era of nuclear energy is the fourth generation, in which FRs will play a key role. The targets of FRs should be superior to those of LWRs, as described below.

1. Safety
   FRs aim at ensuring safety based on new technical systems that are different from those of LWRs. These include innovative materials, nuclear reaction characteristics specific to FRs, and high-temperature, low-pressure plant systems

that address the high melting point of sodium coolant. Research and development is, therefore, based on experimental demonstration and simulation. Furthermore, based on the experience gained through the nuclear accident at the Tokyo Electric Power Company's (TEPCO's) Fukushima Daiichi Nuclear Power Station (Fukushima accident), measures are being taken in plant design to cope with severe accidents that are beyond the conventional design basis and which take the features of FRs into consideration.

2. Economic competitiveness
   For future commercialization, FR cost-effectiveness is aimed at surpassing that of existing LWRs. The cost of power generation includes capital, operation, and fuel costs. The cost of facility construction is higher, while fuel costs represent a smaller percentage of the total costs of nuclear power generation than in thermal power generation. For LWRs, fuel accounts for approximately 20 % of the total costs; however, the proportion of the fuel costs for FRs depends on the relationship with the nuclear fuel cycle, creating the potential for increase. The development target of FRs is to reduce the construction cost to below that of next-generation LWRs. In addition to innovative technologies, relevant factors for improved economic competitiveness include the reactor power level, standardization, and the number of reactors to be constructed.

3. Effective use of resources
   Valuable uranium resources will be recycled for effective use in FR cycles. Plutonium used in FRs comprises fissionable 239Pu and 241Pu and hard-to-fission 240Pu and 242Pu. In addition to plutonium, MAs, including americium, neptunium, and curium, are produced in an FR core. Technology development has only just begun to burn these transuranium elements exactly as planned.

   Other ideas that would provide efficient nuclear fuel cycle performance include a plant facility in which the nuclear power generation, reprocessing, and fuel manufacturing functions are integrated and a nuclear park concept in which a fuel facility is located adjacent to a number of FRs.

4. Environmental protection
   Nuclear power is based on science that addresses nuclear reactions focusing on nuclides in a microscopic manner. Consequently, radioactive waste disposal methods should be established on the same scientific basis. Burning and transmutation of long-lived transuranium elements and fission products that impose a heavy environmental burden are expected to be performed using nuclear reactions driven by fast neutrons in consideration of nuclide characteristics to reduce the burden on the final disposal. Cooperation with nuclear fuel technologies, such as reprocessing and fuel manufacturing, is essential to this development.

5. Nuclear nonproliferation
   A basic policy of Japan's nuclear fuel cycle is not to have surplus plutonium with no particular objective. To follow this policy, it becomes necessary to develop a

plutonium-burning technology to suppress plutonium production and efficiently consume plutonium. This technology can also contribute to the use of plutonium generated from the dismantling of nuclear weapons.

From the perspective of the establishment of an FR system, it is important to take measures inherent to the system itself separately from the institutional development of safeguard systems. To achieve technology that makes the use of plutonium as a weapon difficult, it is possible in plant design to aim for the use of fuel containing minor actinides or to make it difficult to produce high-purity plutonium equivalent to a weapon-grade plutonium that is generated in the blanket regions.

Furthermore, it is required in FR design to provide safeguards in hard-to-access areas with consideration of the separation from the air and the opacity of sodium. In plant design, efforts are made to develop techniques for the effective and secure monitoring of fuel transfer routes, etc. and the abovementioned nuclear park concept in which a fuel facility is integrated into FRs.

2.2 Development History of FRs

2.2.1 Changes in Coolant

1. FR core concept
   The main feature of an FR core is the reduction of useless neutron absorption through the use of fast neutrons produced by fission to maintain the chain reaction and, thus, to use the increased neutrons for conversion of fertile nuclides (e.g., 238 U) into fuel nuclides (e.g., 239 Pu). Consequently, as a coolant material having a lower neutron moderation effect, liquid metals, such as sodium, NaK (an alloy of sodium and potassium), lead, bismuth, and mercury, and gases, such as helium, carbon dioxide, and steam, were examined [5].

Figure 2.2 shows the relationship between neutron energy and the likelihood of fission (expressed by the unit of cross section: barn) of fuel nuclides. As shown in the figure, the fission cross sections of fuel nuclides, such as ^{235}U and ^{239}Pu, are approximately 500 barns (1 barn = 10^{-24} cm^2)in the thermal neutron region (below 1 eV), and a few barns in the fast neutron region (approximately 100 keV). Therefore, although useless absorption is reduced, it is necessary to increase the number of neutrons produced in an FR core by increasing the density of fuel nuclides by approximately five times that of a thermal neutron reactor to maintain the chain reaction by accounting for the leak of neutrons to the outside of the core.

For this reason, the dense arrangement of fuel nuclides in an FR core is inevitable. In addition, it becomes necessary to increase the power density of the core (thermal power per unit volume) to prevent a significant increase in the amount of fuel installed per reactor power. In addition to avoiding the use of light elements to minimize neutron moderation, therefore, a material that has good heat

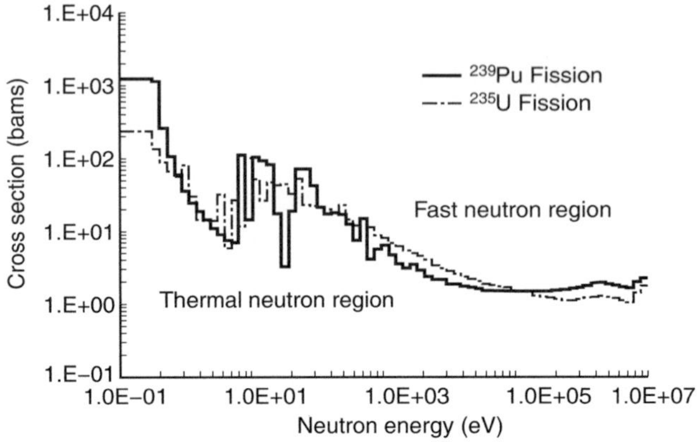

Fig. 2.2 Fission cross section of fuel nuclides

Table 2.1 Comparison of physical properties and pump power among liquid metals

	Sodium (Na)	NaK (Na:K = 25:75)	Lead (Pb)	Lead-bismuth (Pb:Bi = 45:55)	Mercury (Hg)
Melting point (°C)	98	−11	328	124	−38.9
Boiling point (°C)	882	784	1737	1670	357
Density (kg/m^3): ρ	826	757	10,360	10,087	12,945
Specific heat (kJ/kg°C): C_p	1.30	0.958	0.16	0.146	0.135
Thermal conductivity (W/m°C)	66.1	28.7	15.4	14.9	12.9
Coefficient of viscosity (mPaS): μ	0.232	0.194	1.74	1.33	0.97
$\rho^{0.8}\mu^{0.2}$	161	143	1821	1690	1937
ρC_p (MJ/m^3 °C)	1.07	0.72	1.66	1.47	1.75
Relative volume flow rate at a constant ΔT	1.00	1.50	0.65	0.73	0.61
Pump power/amount of heat transported (relative value)	1.0	2.8	3.4	4.3	3.1

Physical properties listed are those at 277 °C for mercury and at 527 °C for the other liquid metals (Reference: Heat Transfer Handbook (Japan Society of Mechanical Engineers))

conductivity, heat transfer characteristics, and a high heat removal capability is desired as the coolant of an FR. As a result, attention was first paid to liquid metal coolants.

2. Comparison among liquid metal coolants
   What physical properties are required for the coolant to achieve "high heat removal capability?" Table 2.1 lists the physical properties of the liquid metals that have already been used in FRs. As shown in the table, only NaK and

mercury are liquid at room temperature, and the other materials require some preheating equipment to enable their use as coolant. This property is thought to be the reason why mercury was used in both the US and the former Soviet Union in early days of the development of FRs. Another reason is thought to have been the fact that a very high-energy neutron spectrum can be obtained to allow for the collection of data for the development of nuclear weapons. However, mercury was not used by either country as the coolant for a fast power reactor after Clementine (US) and BR-2 (former Soviet Union) because, in part, the boiling point is too low (i.e., 357 °C). We compare various liquid metals in terms of heat removal capability. Here, we compare the pump power per unit amount of heat transported by assuming the same flow cross-sectional area and the amount of increase in coolant temperature.

Table 2.1 shows the pump power per unit amount of heat transported that was calculated from the physical properties of each liquid metal in the form of a relative value assuming that of sodium as 1. The table shows that the pump power for sodium is the lowest among all liquid metals, suggesting that sodium has the best heat removal capability. Another indicator of heat removal capability is the heat transfer rate that affects the fuel cladding temperature, etc. From this perspective also, sodium is preferable because it has approximately twice the heat transfer rate of other liquid metals.

In addition to the heat removal capability, other important factors in the selection of coolant include:

① Boiling point
② Neutron absorption cross section and moderation effect (the greater the mass number is, the more suitable it is as coolant)
③ Chemical activities (chemical reactions during leakage, toxicity)
④ Compatibility with structural material (corrosiveness)

The challenges facing NaK are a rather low boiling point and high chemical activity. Lead and lead-bismuth have the following advantages: i) the coolant void effect need not be taken into account because the boiling point is sufficiently high; and ii) application of a simplified heat transport system eliminates the need for a secondary cooling system because neither material reacts with water. However, disadvantages have been noted in terms of plant integrity and maintainability. These include the production of highly radioactive ^{210}Po (half-life: 138 days) through neutron absorption by bismuth and, for the case of lead, the high melting point (328 °C) and corrosion/erosion of structural material. It should be noted that lead-bismuth was used in nuclear submarines of the former Soviet Union.

As is described above, sodium was selected as the FR coolant worldwide through comprehensive evaluation of the heat removal capability, boiling point, compatibility with structural material, chemical activity, and the behavior during leakage.

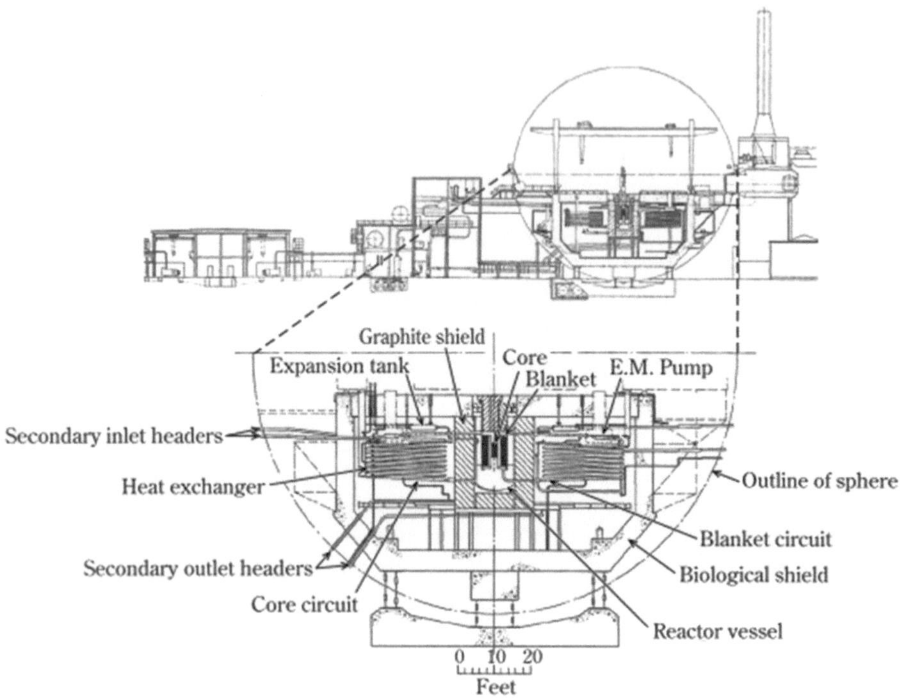

Fig. 2.3 Reactor structure of experimental reactor DFR [7]

3. DFR, a NaK-cooled experimental reactor

The FRs for which NaK was used as the coolant after EBR-I (US) include the UK's experimental reactor DFR (primary and secondary systems) and the former Soviet Union's experimental reactor BR-5 (secondary system). Figure 2.3 shows an overview and the reactor structure of the experimental reactor DFR, and Table 2.2 shows its main specifications. The DFR was designed and constructed to verify the applicability of an FR concept and first achieved criticality in 1959. As shown in Table 2.2, the coolant temperatures were approximately 200 °C lower than those of the subsequent sodium-cooled FRs. It is likely that the reason for the reduced operating temperature was to ensure the maintainability of the whole plant from the perspective of elevated temperature structural design. The same reason is likely to apply to the use of NaK as the coolant. The number of primary and secondary system loops is 24 and 12, respectively. This is due to the limited capacity of the pumps and heat exchangers in those days. Since commissioning, the DFR had been used as an irradiation bed for fuels and materials to develop mixed-oxide fuel and high Ni alloy cladding material; however, it was closed in March 1977, after completion of a series of safety tests, and decommissioning is in progress. As is described above, one advantage of NaK is that it is liquid at room temperature, allowing

Table 2.2 Main specifications of experimental reactor DFR [7]

Thermal power (MW)	60
Electric power (MW)	15
Reactor outlet/inlet temperature (°C)	350/230
Steam conditions (temp. (°C)/pressure (MPa))	270/1
Reactor type	Loop type
Number of loops (primary/secondary)	24/12
Coolant (primary/secondary)	NaK/NaK
Fuel/cladding material	U-7%Mo alloy/Nb alloy
Uranium enrichment (%)	75
Core diameter/height (cm)	53/53
Core power density (W/cc)	900
Fuel pin diameter/thickness (mm)	20/0.35
Fuel discharge burnup (MWd/t) (maximum/average)	3000/2500

Table 2.3 Comparison of heavy element density among fuel materials

	Density(g/cc)	Heavy element density(g/cc)
Metal $U_{0.75}Pu_{0.15}Zr_{0.1}$	16.0	14.4
Carbide $(U_{0.85}Pu_{0.15})C$	13.6	13.0
Nitride $(U_{0.85}Pu_{0.15})N$	14.3	13.5
Oxide $(U_{0.85}Pu_{0.15})O_2$	11.0	9.7

simplification of a preheating system designed to avoid freezing. However, due to the fact that NaK has a low boiling point, which makes it difficult to improve the thermal efficiency by increasing the reactor outlet temperature, and the fact that NaK has higher chemical activity than sodium, no FR using NaK was designed after the DFR.

2.2.2 *Changes in Core and Fuel*

1. Metal fuel used in early days
   Table 2.3 compares the heavy element density among metal, carbide, nitride, and oxide fuels. The table shows that metal fuel has the largest heavy element density and oxide fuel has the smallest. Since a fuel form with a larger heavy element density and containing no light element is advantageous in pursuing the benefit of the FR core, such as high breeding ratio, metal fuel alloy fuels were used in the early days. Table 2.4 shows a comparison of fuel types employed in early FRs in the USA and the UK. As an example of the early metal-fueled FR, the Enrico Fermi fast breeder reactor (EFFBR, hereinafter referred to as the "Fermi reactor," criticality achieved in 1963, closed in 1972) is explained below.

Table 2.4 Metal fuel materials employed in early fast reactors

	First criticality(year)	Fuel material
Clementine	1946	Metal Pu
EBR-I	1951	U-2% Zr alloy
DFR	1959	U-7% Mo alloy
EBR-II	1961	U-5% Fs[a] alloy
Fermi	1963	U-10% Mo alloy

[a]General term of metal FP after pyrometallurgical reprocessing. Composition: 2%Zr, 98% metal (Mo, Ru, Rh, Pb)

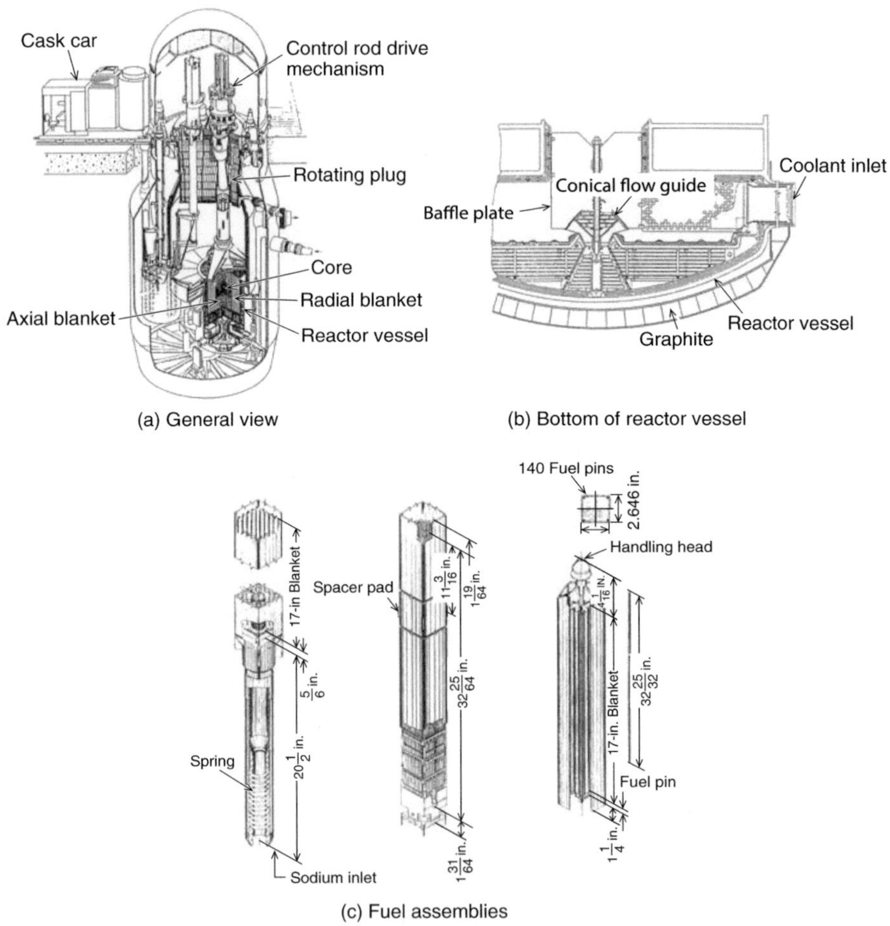

Fig. 2.4 Reactor structure of Fermi reactor [7, 13] (**a**) General view. (**b**) Bottom of reactor vessel. (**c**) Fuel assemblies

Figure 2.4 shows the structure of the Fermi reactor. The Fermi reactor is a nuclear power plant that was designed and constructed to evaluate the economic performance of FR power generation. Since 1962, vigorous efforts had been exerted to perform various types of operation and testing at a thermal power level up to 100 MW; however, several fuel subassemblies melted due to flow

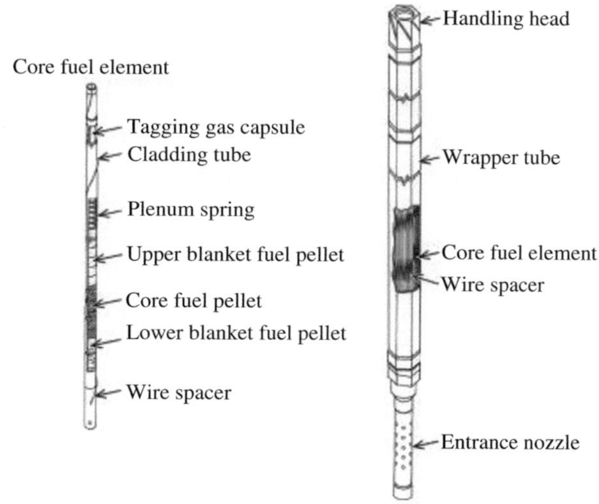

Fig. 2.5 Fuel assembly of Monju [9]

blockage during operation at a thermal power of 34 MW in October 1966. This was caused by the fall of the zirconium flow guide (current plate) shown in Fig. 2.4, which blocked the inlets of several fuel subassemblies. Ironically, the flow guide was installed at the lower part of the reactor vessel as a measure against core melt during a major accident. The structure of the fuel assembly is shown in Fig. 2.4c. It can be pointed out that the assembly inlet nozzle design is likely to be blocked by foreign material. In the subsequent assembly design, it became common to adopt a side entry method, as is found in the entrance nozzle of the fuel assembly of Monju (see Fig. 2.5), to prevent a complete blockage of the entrance by foreign material, etc.

Another feature of the fuel assembly of the Fermi reactor is that fuel rods (25.6 % enriched U-Mo alloy, zirconium cladding) are arranged in a tetragonal lattice. This is different from the arrangement in a hexagonal lattice, which is used in present FR fuel subassemblies and advantageous for highly dense layout. The reason is thought to be that there is a relatively large degree of freedom in designing the fuel volume ratio because the Fermi reactor used metal fuel. Another reason is to make use of experience with LWR fuels. However, the hexagonal arrangement has been adopted in all FR fuels designed after the Fermi reactor.

2. Sodium and oxide fuel play the main role
   The blockage of fuel subassemblies and the resulting fuel melt in the Fermi reactor raised the need to develop ceramic fuel types, such as oxide, carbide, and nitride. The prediction that metal fuel is not suitable for increased fuel burnup because of potential fuel failure due to significant swelling under irradiation also enhanced the need.

In particular, mixed-oxide (MOX) fuel, in which plutonium is enriched in uranium dioxide, was expected to be the main FR fuel-type candidate because the irradiation stability of oxide fuel was well proven in LWRs and the suitability for increased burnup was widely recognized.

3. Verification of Doppler effect in SEFOR
   Under these circumstances, the Southwest Experimental Fast Oxide Reactor (SEFOR) was constructed to achieve the following goals:

   1. To gain MOX-fueled FR operational experience
   2. To verify the Doppler effect that represents the level of negative feedback during performance testing of MOX fuel and power variation
   3. To demonstrate the role of the Doppler effect for the inherent safety of a MOX-fueled core
   4. To establish the safety design standards for future large commercial reactors
      SEFOR was designed, constructed, and operated by General Electric (GE) under commission from the US Atomic Energy Commission (20 MWth, criticality achieved in 1969, closed in 1972). Figure 2.6 shows the cross-sectional view of the SEFOR reactor structure. The SEFOR fuel assembly was designed to prevent a change in axial fuel length due to thermal expansion. The only test that was ever performed in an FR core to measure the Doppler coefficient of the whole core was successfully

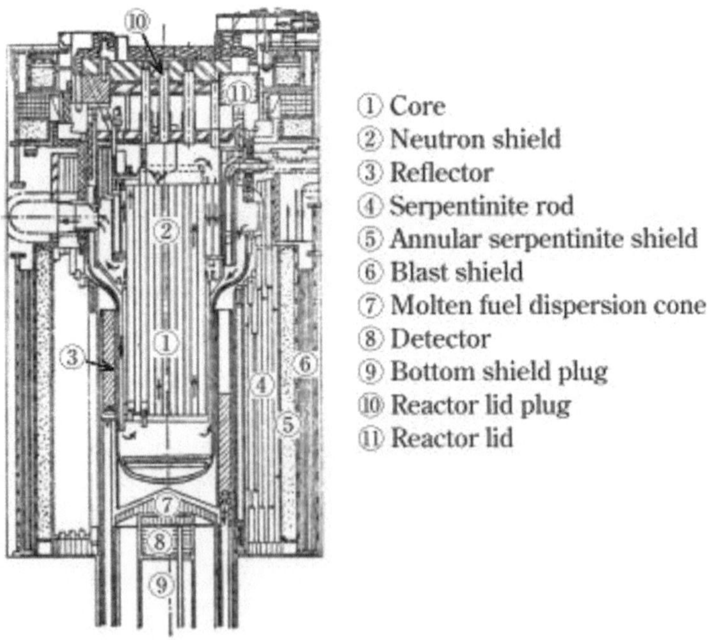

Fig. 2.6 Reactor structure of SEFOR [6]

conducted at SEFOR. With this success as a springboard, MOX fuel was recognized as the main candidate for the FR fuel type.

4. Representative sodium-cooled, oxide-fueled cores
   The main specifications among representative sodium-cooled, oxide-fueled FRs that have operational performance are compared in Table 2.5. Present FR fuel subassemblies have a configuration in which a fuel pin bundle is supported with a wrapper tube and coolant flow is distributed by the entrance nozzle with a side entry and the below-core structure according to the power of the assembly. As shown in Table 2.5, dense arrangement of fuel rods in a hexagonal lattice provides a fuel volume ratio of 31 to 48 %. In addition, a wire-wrapped spacer is generally used because this method makes it easier to maintain narrow gaps between fuel rods while limiting the pressure drop of coolant in an assembly. The wire-wrapped spacer method is predicted to become the mainstream from the perspective of manufacturability and cost. As for the irradiation data of MOX fuel, data for approximately 380,000 fuel rods have been obtained, and the peak burnup reached approximately 240,000 MWd/t in Western countries and Japan. These data suggest that MOX fuel has a high potential for enhanced economy by increasing burnup. On the other hand, based on previous irradiation experience, it has been predicted that the limiting factor for the lifetime of FR fuel, for which austenitic stainless steel, such as SUS316, is used as core material, is the swelling deformation (swelling of the cladding and wrapper tubes due to fast neutron irradiation) and the resulting bundle-duct interaction or duct-duct interaction. Therefore, in Japan, R&D for commercialization of a cladding tube made of oxide dispersion strengthened (ODS) ferritic steel with improved high-temperature structural strength and limited swelling is under way to achieve higher burnup.

5. EBR-II, metal-fueled FR with several achievements
   After the SEFOR experiment, metal fuel gave the leading role to oxide fuel. However, at the Argonne National Laboratory (ANL) experimental fast reactor EBR-II (USA, criticality achieved in 1963, closed in 1998), metal fuel was continuously used as driver fuel, and many fuel/material irradiation experiments were successfully performed. The EBR II was designed and constructed to demonstrate an integrated FR concept, a closed cycle system that combines a power reactor, dry reprocessing, and fuel fabrication facilities (ANL began to call the reactor the integral fast reactor (IFR) from approximately 1984). Figure 2.7 shows the EBR-II plant layout and Table 2.6 lists the main specifications. As shown in Tables 2.4 and 2.6, the driver fuel of the EBR-II consists of U-5%Fs alloy. In the early days, since the fuel failed due to rapid fuel swelling, the fuel burnup was limited to approximately 12,000 MWd/t. However, it was confirmed that swelling can be saturated by reducing the fuel packing density to 75 %.

Table 2.5 Comparison of main specifications among sodium-cooled oxide-fueled fast reactors[8])

	Experimental reactor				Prototype/Demonstration reactor				
	Rapsodie (France)	Joyo (Japan)	BOR-60 (Russia)	FFTF (US)	PFR (UK)	PHENIX (France)	Monju (Japan)	BN-600 (Russia)	SPX (France)
First criticality (Calendar year)	1967	1977	1968	1980	1974	1973	1994	1980	1985
Thermal/electrical power (MW)	40/0	140/0	55/12	400/0	650/250	563/255	714/280	1470/600	2990/1242
Reactor outlet/inlet temperature (°C)	515/400	500/350	530/330	503/360	560/399	560/395	529/397	535/365	545/395
Reactor type	Loop	Loop	Loop	Loop	Pool	Pool	Loop	Pool	Pool
Fuel material	PuO_2-UO_2	PuO_2-UO_2	PuO_2-UO_2	PuO_2-UO_2	PuO_2-UO_2	PuO_2-UO_2	PuO_2-UO_2	UO_2	PuO_2-UO_2
Equivalent core diameter/height (cm)	44.6/32	80/50	46/45	120.2/91.4	147/91	139/85	180/93	205/103	370/100
Fuel pin diameter (mm)	5.1	5.5	6.0	5.84	5.8/6.6/8.5	6.6	6.5	6.9	8.5
Fuel:sodium:structural material (%)	42.5:39.6:13.6	37:37:23	48:29:23	31:39:26	35:41:21	37:35:25	33.5:40:24.5	37.5:34:21.5	37:34:24
Burnup (actual peak, MWd/t)	102,000	86,900	176,000	155,000	200,000	150,000	–	97,000	90,000
Cladding tube material	316SS	316SS (20%CW)	Cr16Ni15	316SS (20%CW)	PE16 etc.	Cr17Ni13 Mo2.5 Mn1.5TiSi	mod316	Cr16Ni15 Mo2 +MnTiSi (CW)	Cr17Ni13 Mo2.5 Mn1.5TiSi
Fuel pin spacing method	Wire	Wire	Wire	Wire	Grid	Wire	Wire	Wire	Wire

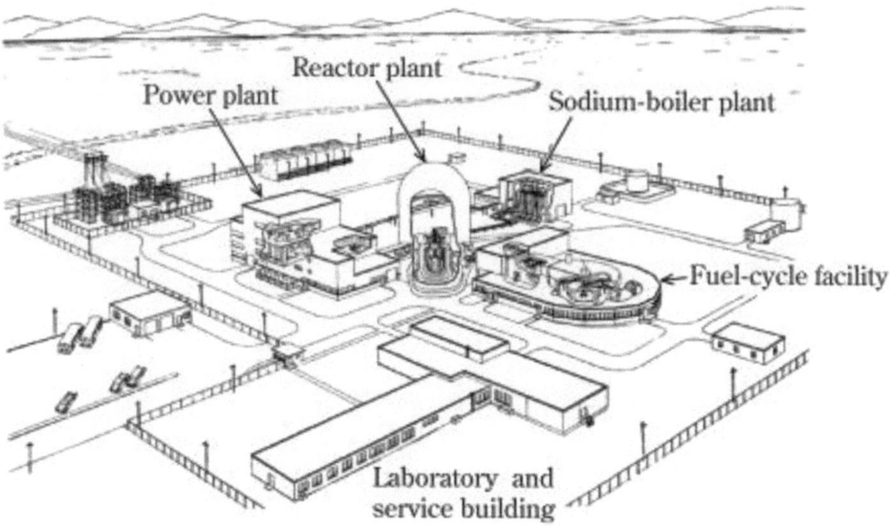

Fig. 2.7 Plant layout of experimental reactor EBR- II [7]

Table 2.6 Main specifications of experimental reactor EBR-II [7]

Thermal power (MW)	62.5
Electrical power (MW)	20
Reactor outlet/inlet temperature (°C)	473/371
Steam conditions (temp. (°C)/pressure (MPa))	433/8.79
Reactor type	Tank type
Number of loops (primary/secondary)	2/1
Coolant (primary/secondary)	Na/Na
Fuel/cladding materials	U-5%Fs alloy/316SS
Uranium enrichment (%)	67
Core diameter/height (cm)	69.7/34.3
Fuel pin diameter/thickness (mm)	4.42/0.305
Fuel discharge burnup (MWd/t) (maximum/average)	80,000/66,000

Through the change in design specifications of the driver fuel cladding tube, a peak burnup of 80,000 MWd/t was achieved. Various types of fuel irradiation tests were simultaneously performed at the EBR-II. The peak burnup achieved is 190,000 MWd/t for MOX fuel and 185,000 MWd/t for metal fuel (U-Pu-Zr alloy), respectively. The EBR-II was completely closed in 1998 according to the Democratic Party's policy on reprocessing strategy. However, basic research in the US on the recycle of transuranium elements using metal fuel continued afterward.

Table 2.7 Main specifications of experimental reactor FBTR[8])

Thermal power (MW)	40
Electrical power(MW)	13
Reactor outlet/inlet temperature (°C)	515/380
Steam conditions (Temp. (°C)/pressure (MPa))	490/16.7
Reactor type	Loop type
Number of loops	2/2
Coolant (primary/secondary)	Na/Na
Fuel/cladding materials	PuC-UC/316(20%CW)
Pu enrichment (%)	55
Core diameter/height (cm)	49.2/32
Fuel pin diameter/thickness (mm)	5.1/0.37
Fuel discharge burnup (target value, MWd/t) (maximum/average)	50,000/38,000

6. FBTR, a carbide-fueled experimental reactor in operation
   One of the FRs using fuel other than MOX fuel is the Indian experimental fast reactor FBTR that is in operation using carbide fuel (criticality achieved in 1985). The FBTR was constructed in technical cooperation with France. Its reactor structure is approximately the same as that of French experimental reactor Rapsodie except that the water-steam system was added to produce electricity. Table 2.7 lists the main specifications of the experimental reactor FBTR. Concerning the core fuel, India originally planned to use MOX fuel, as France had. Since neutron leakage occurs more easily in a small core, the Pu enrichment must be increased to a level significantly greater than the 45 % that is known as the metallurgical limit of MOX fuel, and it becomes necessary to use highly enriched uranium. However, since India, a non-NPT country, was not able to obtain highly enriched uranium, it was decided to use carbide fuel, which has no metallurgical limit, for the Pu enrichment.

The number of FBTR core fuel subassemblies was gradually increased from the initial 25, and a thermal power of 18 MW was achieved with 39 fuel subassemblies at the end of 2003. It is known that the maximum burnup achieved was 123,000 MWd/t. Furthermore, India is constructing the Prototype Fast Breeder Reactor (PFBR) (electric power of 500 MW, pool type), which is designed using a MOX-fueled core.

2.2.3 Major Accident/Trouble Experiences

1. Accidents of core/fuel

    1. Core melt in EBR-I
    Before the fuel melt that occurred at the Fermi reactor as described in paragraph 2–2-2, one major US accident related to the core and fuel of an early FR was the core melt that occurred at the EBR-I. The EBR-I was an experimental fast reactor with a thermal power of 1.2 MWt and electric power of 200 kW. It succeeded in the first ever nuclear power production prior to LWRs on December 20, 1951. Liquid metal NaK was used as coolant for both the primary and secondary systems. Enriched metallic uranium was originally used, while metallic plutonium was used afterward.

A core melt accident occurred at the EBR-I on November 29, 1955 during a reactivity kinetics test because of an inadvertent slow scram that was performed by the operator instead of a fast scram. During the test, circulation pumps were stopped for a short time to measure the reactivity coefficient for the increase in fuel temperature and, thereby, to observe the effect of the coolant flow velocity. In this condition, the temperature coefficient became positive, the reactor power increased, and 40 to 45 % of the core fuel volume melted. The release of radioactive material to the outside of the facility was minimal. A schematic view of the fuel melt is shown in Fig. 2.8.

The fuel material was enriched U-2%Zr alloy. The fuel pin cladding material was 304 stainless steel. There were no wire spacers. No wrapper tube that contains the fuel pins was used. Investigation revealed that fuel pins were bowed at high temperatures, and, thereby, a positive reactivity was added. Taking into account the investigation results, wire spacers were wrapped around the fuel pins to prevent the bowing, and the fuel pins were installed in a wrapper tube. These changes established the basic concept of FR fuel design.

The EBR-I was restarted after decontamination, in-core repair and improvements that took approximately 2 years, and closed in 1964.

Fig. 2.8 Core damage condition in EBR- I [12]

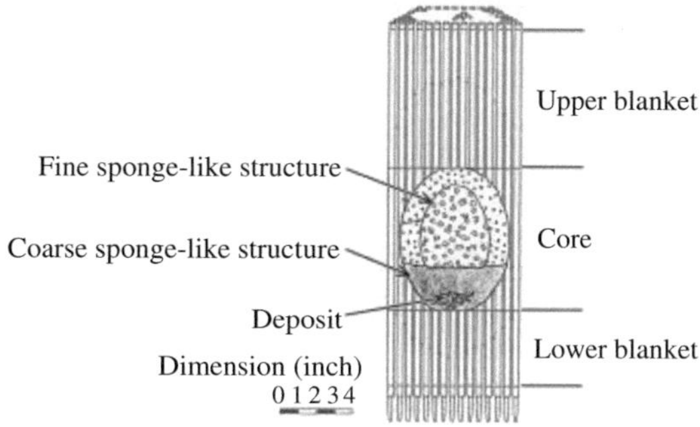

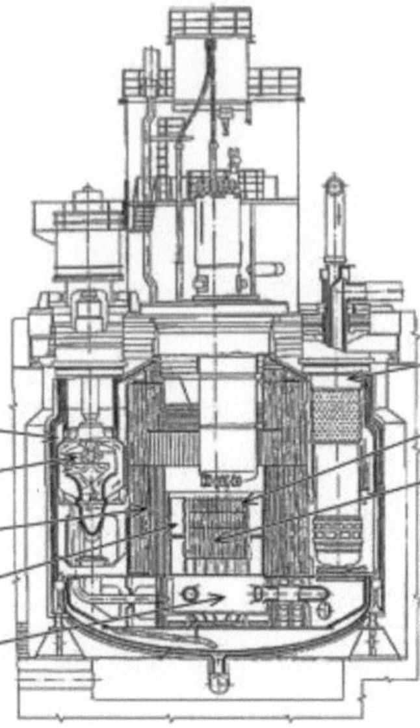

Fig. 2.9 Reactor structure of BN-600 [7]

2. Accidents concerning sodium

   1. Sodium leak

      Sodium leaks have occurred many times at FRs. Monju in Japan also experienced sodium leaks. The cases of the Russian BN-600 and the French Super Phenix are presented as examples of large sodium leak accidents. The BN-600 continued successful operation for many years though it often experienced sodium leaks. Super Phenix marked the maximum amount of sodium leaked.

      ① BN-600

      BN-600 is a large-scale prototype reactor with an electric power of 600 MWe. Since a part of the steam generated is also used for a heat source to supply hot water, the BN-600 is a practically used power plant that is indispensable for local residents. The reactor structure is shown in Fig. 2.9. The BN-600 started commercial operation in April 1980. Sodium leak occurred 27 times before November 1995, including 5 times from the primary system, 17 times from the secondary system, and 3 times from the sodium-receiving system. Fourteen of these leaks were followed by fire. The amount of sodium leaked during the primary sodium purification system leak accident that occurred in October 1993 was approximately 1 m3, and approximately 10 Ci of radioactive material was released to the air. The accident was caused by cracking at a T-type weld between the main pipe and a branch pipe due to differential thermal expansion.

      ② Super Phenix

      Super Phenix is a demonstration reactor with an electric power of 1240 MWe that started operation in January 1986. In March 1987, sodium

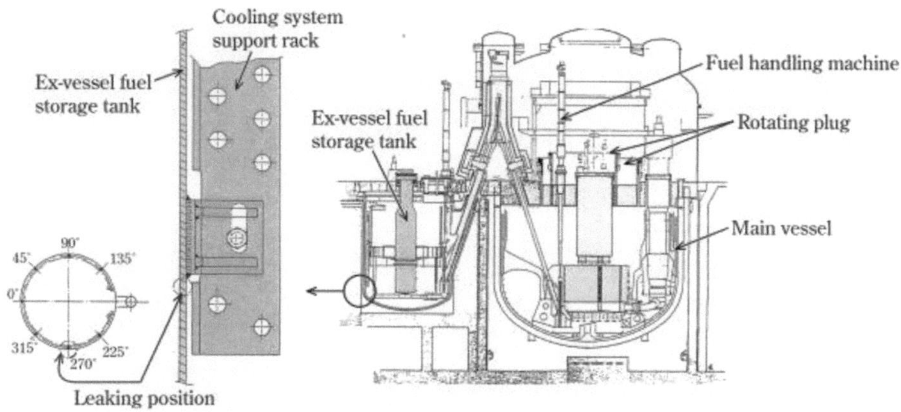

Fig. 2.10 Ex-vessel fuel storage tank of Super Phenix [15, 16]

leaked from the ex-vessel fuel storage tank (sodium capacity, 700 tons; diameter, approx. 9.5 m; height, 13 m). Since this storage tank is double walled, 20 m3 of leaked sodium was completely gathered in the gap between the outer and inner vessels. Since this gap was in an inert nitrogen gas atmosphere, the leaked sodium did not burn.

The leak position was located at a weld of the support plate that was welded at a height of approximately 3 m from the bottom of the storage tank inner shell, as shown in Fig. 2.10. It was estimated that hydrogen embrittlement of the weld progressed to cracking. After replacement of the storage tank with a newly installed, gas-filled fuel transfer tank, Super Phenix restarted in August 1994 and continued successful operation. However, it was decided to close the reactor in February 1998.

2. Water leak in steam generator
   Large-scale sodium-water reaction accidents in steam generator that activated a pressure relief system occurred at the Fermi reactor, the BN-350, the PFR, etc. All these accidents were addressed without affecting the safety of the reactors or the external environment, and each reactor restarted operation shortly after repair.

   ① Fermi reactor
   The structure of the steam generator for the Fermi reactor is shown in Fig. 2.11. During a sodium flow test (at 250 °C), 45 heat transfer tubes were ruptured in the steam generator, and a significant amount of water and steam leaked out into sodium. After the water-steam side was automatically separated, the system pressure and the reaction products were released to the pressure relief system, and the accident ceased.

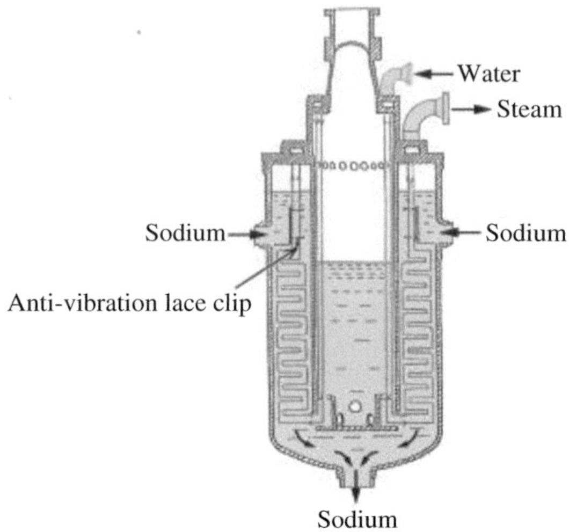

Fig. 2.11 Steam generator of Fermi reactor [14]

It was revealed that the rupture of the heat transfer tubes was due to vibration induced by sodium flow that caused thinning of the tubes due to impact and wear, resulting in tube cracking by high pressure of the water side. In addition, it was clarified that the sensitivity of the water leak detectors was not sufficient. As a measure to prevent vibration, improvement was made by installing a buffer plate and anti-vibration bars at the front of the sodium inlet.

② BN-350

The BN-350 is the world's first prototype fast reactor. Construction was started in 1964 by the former Soviet Union in an industrial complex located on the coast of the Caspian Sea. The BN-350 is a unique FR that was commercially operated for the dual purposes of power production and seawater desalinization.

Each loop of the secondary system has two evaporators and two superheaters. The heat transfer tube of the evaporator has a bayonet-type configuration in which the tube is inserted into the outer tube that is closed by a cap (end plug), as shown in Fig. 2.12, while the heat transfer tube of the superheater is U-shaped. During operational testing, leakage occurred several times at welds of the end plugs and heat transfer tubes. All heat transfer tubes of the evaporator in which leakage occurred were replaced; however, significant leakage occurred in one evaporator in 1973, and 120 heat transfer tubes were ruptured. Although a significant sodium-water reaction occurred due to the leak of 800 kg of water, the reaction products were burnt and released to the air through activation of the safety systems: pressure relief plate and blowdown, and the sodium loops were not damaged. Based on these experiences, improvement was made on the purity control of sodium and water, early leak detection, and early blowdown, and a plugging technique of heat transfer tube was established as a repair technique, leading to stable operation at the design power level.

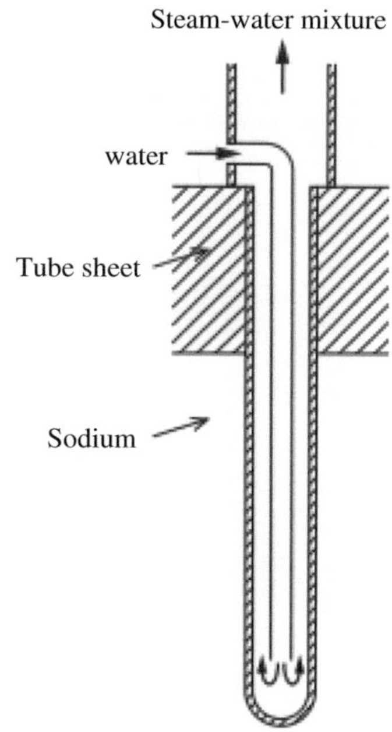

Fig. 2.12 Bayonet-type heat transfer tube of BN350

In addition, this accident was reported worldwide because an American satellite captured images of the flames at the accident site.

③ BN-600

Significant water leakage occurred three times in the steam generator of the BN-600. One module of the steam generator comprises an evaporator, superheater, and reheater. As shown in Fig. 2.13, one loop consists of eight modules, and, accordingly, 24 modules in total are installed for three loops.

In the case of the leakage that occurred in the initial operation before reaching the rated power, since operational responses and the separation of steam generator modules were delayed, the range of heat transfer tube rupture was increased to cause large-scale sodium-water reactions, which activated the pressure relief system, and the event ceased.

As the inspection during manufacturing was not sufficient, a small defect at the weld between heat transfer tube and tube plate grew and expanded during operation. Measures were taken to enhance the sensitivity of the hydrogen meter, to improve the operational method to reduce the cyclic thermal stress, and to replace the steam generator modules. After achievement of the rated power, high plant availability has been achieved by exerting the advantage of the module-type steam generator (separation of damaged modules) though small leaks occurred.

④ PFR

The PFR is a UK prototype reactor with an electric power of 280 MW that achieved full-power operation in February 1977. In February 1987, during

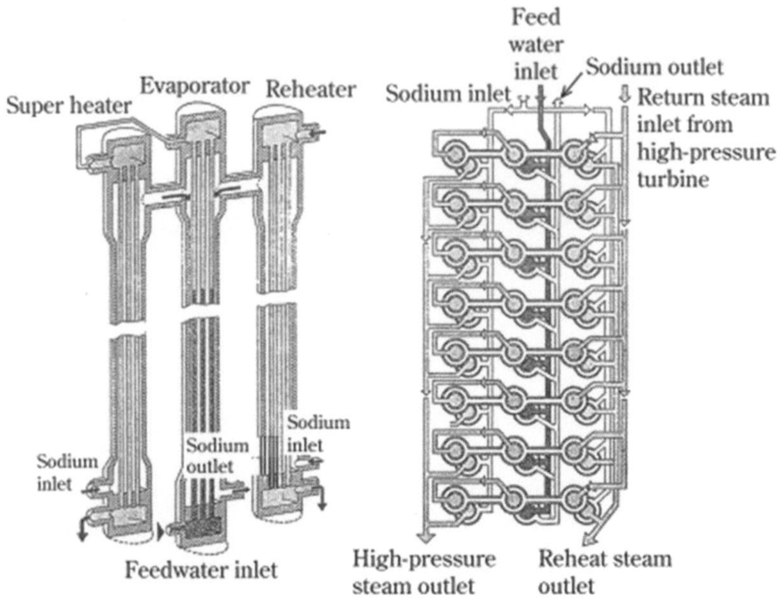

Fig. 2.13 Steam generators of BN-600 [17]

full-power operation, significant water leakage occurred after a small water leak in the superheater. The pressure relief plate was opened, the safety protection system was activated, and the reactor and turbine were automatically shut down. The configuration of the superheater is shown in Fig. 2.14a.

A bypass flow that flows through a gap at the lap joint of the inner cylinder at the sodium inlet to the tube bundle induced vibration of the innermost heat transfer tubes, and, thereby, more than a dozen heat transfer tubes were in contact with the cylinder and thinned, resulting in a small water leak from one tube. Since the hydrogen meter in sodium was removed because of malfunction and the hydrogen meter in the cover gas was out of order, the plant could not be shut down by detecting the increase in hydrogen concentration. As a result, the material strength of heat transfer tubes decreased due to high temperatures near the ruptured heat transfer tube, resulting in the rupture of a total of 39 heat transfer tubes. Since the superheater was not equipped with a system to rapidly exhaust steam, the number of ruptured heat transfer tubes increased to more than 70, as shown in Fig. 2.14b.

The following modifications were made: the material of heat transfer tube was changed from austenitic stainless steel to high chrome steel; heat transfer tubes with no welds were employed; the bypass flow was suppressed; and the number of heat transfer tube support plates was increased. Additionally, a system to rapidly exhaust steam was added. There was no influence on the reactor due to this accident, and operations restarted approximately 4 months later.

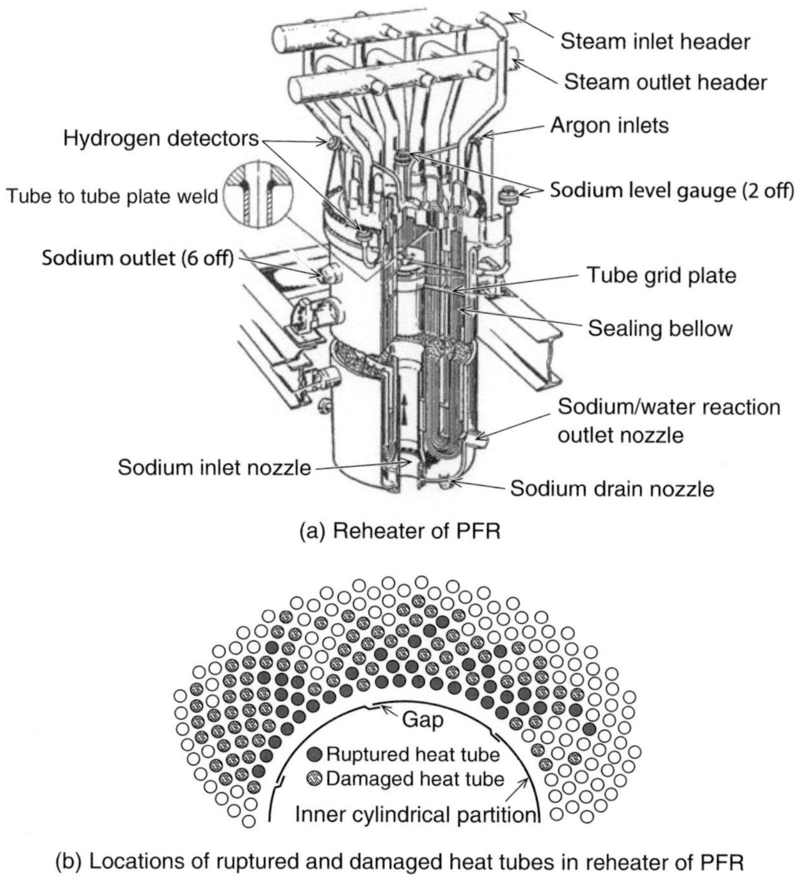

Fig. 2.14 Water leak from reheater of PFR [7, 18]. (**a**) Reheater of PFR. (**b**) Locations of ruptured and damaged heat tubes in reheater of PFR

3. Representative failures and troubles
   The cumulative operating experience of FRs in the world is approximately 400 reactor-years, while that of LWRs is approximately 13,000 reactor-years. Accordingly, it should be understood that the development of FR technology has just started. Representative troubles peculiar to FRs, including 413 cases from foreign reactors and 69 cases from the domestic experimental reactor "Joyo" and prototype reactor "Monju," are categorized in Fig. 2.15a and 2.15b, respectively, by cause and equipment.

Concerning the trouble cases that occurred in FRs, sticking or malfunction of components due to sodium accounts for 28 %, sodium leakage 20 %, fuel failure 4 %, water leak from steam generator heat transfer tube 3 %, and so on. That is, sodium-related troubles account for approximately a half of all troubles.

Sodium leaks were often caused by design factors, such as poor quality control during manufacturing and vibration, thermal deformation, and thermal stress during operation. In the case of sticking and malfunction, one important factor is that since

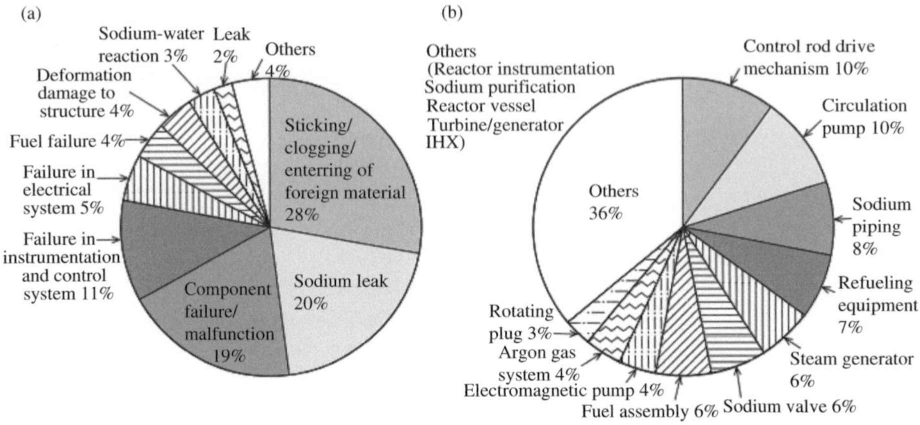

Fig. 2.15 Breakdown of troubles occurred in fast reactors [10]. (**a**) Breakdown of troubles by cause. (**b**) Breakdown of troubles by equipment

the operating temperature of sodium is approximately 200 °C higher than that of LWRs, the deposition behavior of sodium vapor may change depending on the temperature gradient in plant systems, such as cooling and gas systems.

To cope with these failures and troubles, the following measures have been taken: modification of equipment, enhancement of measurement and diagnostic techniques, and improvement in operational control methods. The findings from these experiences will be used to ensure the integrity of components and equipment and will be reflected in design. It is necessary to accumulate the operational experiences and enhance the component/equipment reliability database systems on a worldwide scale.

4. Monju sodium leak accident
   Improvements made after the sodium leak accident at Monju are described as a specific case of efforts exerted following troubles and accidents.

   1. Accident overview
      Monju achieved first criticality in April 1994. During a test at an electric power of 40 % on December 8, 1995, a sodium leak accident from loop C piping of the secondary cooling system occurred. This accident was caused by the failure of a temperature gauge that was inserted into the piping and resulted in the leak of approximately 0.7 tons of sodium at approximately 480 °C for 3 h and 40 min. The leaked sodium burned, dropped, and damaged air-conditioning ducts, etc. and finally accumulated on the steel-lined floor in the form of the compounds of sodium oxide. This state is shown in Fig. 2.16. It was revealed that the cause of the temperature gauge failure was high-cycle fatigue at the skirt of the temperature gauge protection tube due to flow-induced vibration. The vibration was found to be caused by symmetric vortex shedding in the downstream of the protection tube.

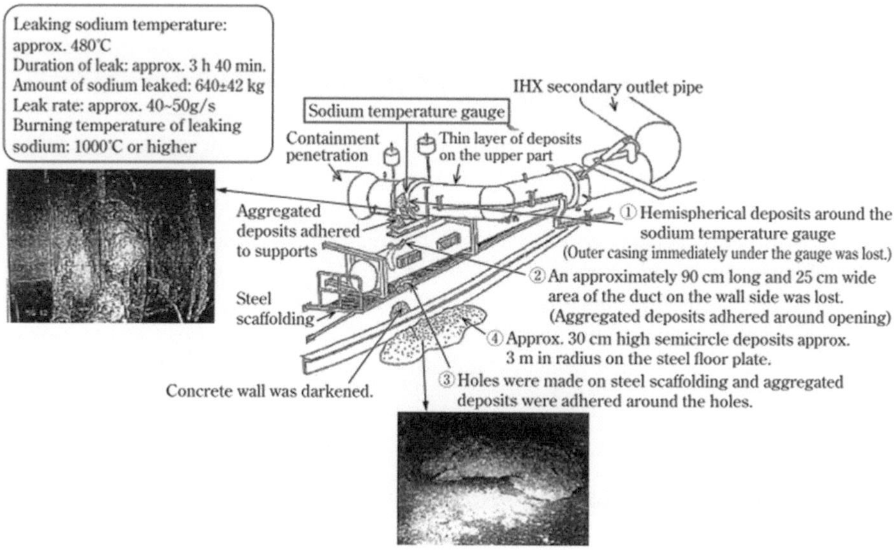

Fig. 2.16 State of sodium leak at Monju [11]

2. Improvements
   Postaccident activities included efforts to perform a comprehensive safety inspection of various aspects, including design, quality assurance, and personnel education, and to recover public trust through fundamental reform of the organizations that had responded inappropriately to society. Plant equipment improvements are described below.

The mechanism of the temperature gauge protection tube failure was confirmed by experiment and evaluated by computer simulation. Based on the results, "Guidelines for evaluation of flow-induced vibration of a cylindrical structure in a pipe" (1998) were newly established as Japan Society of Mechanical Engineers (JSME) standards. The number of protection tubes was reduced, and modification of the remaining protection tubes was undertaken to meet the guidelines by decreasing length and increasing diameter to increase rigidity.

Thinning was found on the steel liner on which the leaked sodium accumulated. Additionally, in a sodium leak and combustion reproducibility experiment, a different type of corrosion from that found after the accident was observed. Based on the experimental results, new findings were obtained on the mechanism of a molten salt-type corrosion that was caused by chemical reactions with moisture generated by the increase in concrete temperature, in addition to sodium and iron temperatures. In consideration of these findings, a heat insulator was newly placed on the concrete surface of the concerned rooms.

Analysis methods were also improved to estimate thermal displacement behavior, when the temperature of a local spot is increased by sodium dropped onto the steel liner, and to confirm integrity. Additionally, to reduce the amount of sodium leaked, an operating method was established to shut down the reactor immediately after identification of sodium leak, and the performance of the pipeline through which the sodium in piping can be rapidly drained to the underfloor tanks was enhanced. Pipelines to supply inert nitrogen gas into the relevant rooms were added to suppress combustion.

On the other hand, new efforts were exerted to develop a method to measure the sodium temperature from the outside of piping because the temperature gauge inserted into piping had failed in the above accident. A temperature gauge that can measure the change in density with the change in sodium temperature using ultrasonic waves was installed and testing was conducted. It is necessary to improve the reliability of measurement instrumentation in the future.

2.3 Present Development Situations of FRs

2.3.1 Situations by Country

In the initial phase of the development of FRs, since there were concerns about early depletion of natural resources, such as oil and uranium, due to an explosive increase in future energy demand, vigorous development of fast breeder reactors was undertaken in the US, the UK, France, and other countries. Later, development of fast breeder reactors from the perspective of resource availability was terminated in some countries, such as the USA and the UK, for the following reasons: i) the increase in proven reserves of natural resources and updating of energy supply and demand revealed a significant time margin for the depletion of resources and ii) FRs are inferior to LWRs in terms of economic competitiveness. On the other hand, Asian countries, particularly China and India, have recently instituted national programs to actively expand the use of nuclear energy. The expected nuclear power capacity of China and India in 2050 is approximately 5.5 and 6 times, respectively, that of Japan's current nuclear power capacity. As such, programs to commercialize the fast breeder reactor cycle are well under way as one of the main future sources of power in these countries [8].

In addition, research and development of the recycling of minor actinides using FRs has been conducted from the perspective of reduced environmental burden. This R&D is related to the establishment of solutions to issues associated with the disposal of spent fuel that accumulates as a result of continuous power production by nuclear reactors, including LWRs. At present, the significance of FR R&D somewhat differs by and is a reflection of the energy/nuclear power use situation in the respective countries; however, even after the Fukushima accident, programs for the development and use of FRs are maintained by some Asian countries and countries that continued R&D for many years, such as the US, France, and Russia.

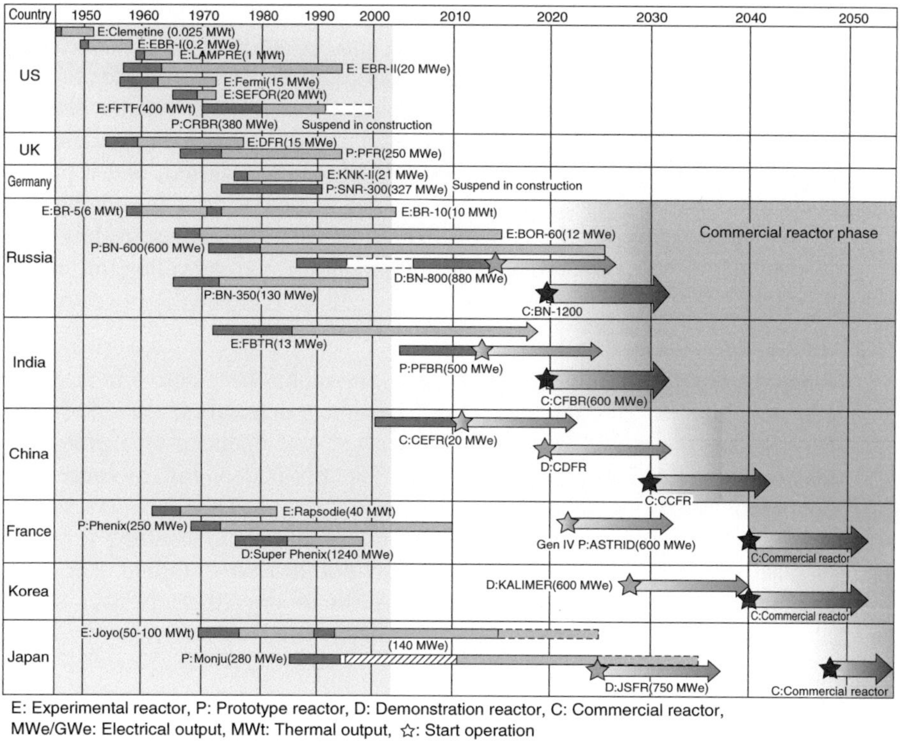

Fig. 2.17 Overview of fast reactor development programs in the world [4]

1. FR cycle development programs and situations in the world

   The achievements and situations of the main countries that are developing FRs are described below. The future development programs are shown in Fig. 2.17. These are current situations in 2015 and future plan may change. The purpose of this subsection is not to give new information, but to explain development process with time scale of FRs as new reactors.

   1. US

      The US has experience in the construction and operation of many experimental reactors (e.g., metal-fueled reactors, such as the EBR-I, the EBR-II, and the Fermi reactor, and MOX-fueled reactors, such as the FFTF) from the early 1940s to the early 1990s. With the change of administration in 1977, the then fast breeder reactor program was amended, and a new nuclear power policy focused on the postponement of FR commercialization was announced. At present, there is no program to construct a new FR; however, fundamental research has been continued to develop technologies for both directions: the eventual use of plutonium while reducing the volume of waste and the direct disposal of spent fuel.

2. France
   France has experience in the construction and operation of Rapsodie (experimental reactor), Phenix (prototype reactor), and Super Phenix (demonstration reactor) and is currently developing the ASTRID reactor that is positioned as a technology demonstration reactor aiming at the start of operation in the 2020s. The conceptual design stage started, and a policy decision regarding construction of the reactor will be made at the completion of the basic design at the end of 2017. France is playing a leading role worldwide in the research and development of the recycling of minor actinides.

3. Russia
   Since the era of the former Soviet Union, Russia has been active in research and development of FRs, including experimental reactors, the BR-5, the BR-10, and the BOR-60. Russia constructed and operated two prototype reactors: the BN-350 and the BN-600. The BN-600 is still in successful operation. Russia has constructed a demonstration reactor BN-800, where the first criticality achieved in 2014. The use of plutonium resulting from the dismantling of nuclear weapons in the BN-800 has been studied. The construction of a commercial reactor BN-1200 in the 2020s is also under planning.

4. India
   India has been operating a carbide-fueled experimental reactor FBTR and is constructing a MOX-fueled prototype reactor PFBR, which is expected to become critical in 2016. After that, it is planned that some reactors with similar capacity would be constructed for commercialization to achieve a nuclear fuel cycle with FRs. In addition, it is also planned for thorium to be installed in the blanket areas to convert it into 233 U that can be used as fuel in order to effectively use thorium, which is found in abundance in India. It is thought that India plans to proceed to a metal-fueled FR with a high breeding ratio on a long-term basis.

5. China
   China constructed an experimental reactor CEFR in technical cooperation with Russia. CEFR achieved first criticality in July 2010 and was first connected to the grid in July 2011 at a power level of 40 %. After that, CEFR has been in cold shutdown to wait for the permission of operation for the power ascension test. It is planned that a demonstration reactor CDFR would be deployed based on the BN-800 technologies in cooperation with Russia by the mid-2020s. In addition, the deployment of commercial reactors is planned to start around 2030. Furthermore, a metal-fueled FR with a high breeding ratio is also examined for the future.

6. South Korea
   The development of FRs is integrated into the national long-term research program. Research and development has been performed to construct a metal-fueled fast prototype reactor.

2. Japan's efforts and future perspectives

   In the sodium-cooled experimental fast reactor Joyo, the basic characteristics of a fast breeder reactor have been confirmed during approximately 71,000 hours of operation since first criticality in 1977, and various tests, including irradiation tests for the development of fuel and core materials, have been performed. Subsequently, modification was undertaken to enlarge the core for enhanced irradiation performance, improved cooling capacity, and increased thermal power (1.4-fold increase from 100 to 140 MW) using accumulated technical findings (MK-III project). The design and construction of prototype reactor Monju was based on the achievements of Joyo and adoption of the results of various full-scale mockup experiments. Monju achieved first criticality in 1994 and was first connected to the grid in 1995. In December of the same year, a sodium leak accident occurred. Following the accident, Monju was shut down for 14 years, during which measures against sodium leakage were enhanced and the responsible organizations were reformed. Commissioning tests were resumed in 2010. After completion of the core performance confirmation tests in the first stage, trouble in which the in-vessel fuel transfer machine was dropped occurred; however, recovery work was completed in 2012.

In 1999, in addition to the advantages of FRs from resource and environmental perspectives, the Feasibility Study (FS) on Commercialized Fast Breeder Reactor Cycle Systems started to establish a concept that excels in the "safety, reliability, and economic competitiveness" required for commercialization. As a result, "a combination system of the sodium-cooled reactor (with MOX fuel), advanced aqueous reprocessing, and simplified pelletizing fuel fabrication process" was selected as the main concept to be developed. Subsequently, the Fast Reactor Cycle Technology Development (FaCT) project started in 2006 with the aim of commercialization in approximately 2050. In this project, in addition to the improvement of safety and economic competitiveness, the combination of a core concept, reprocessing technology, and fuel fabrication process suitable for the effective recycling of minor actinides is included as a main research and development subject in order to achieve the reduction of environmental burden, for which expectations have risen A study on metal fuel cycle as a subconcept was also planned.

Internationally, Japan has actively participated in the Generation IV International Forum (GIF) and played key roles, including a leading position as the third chair country after the US and France from 2010. According to Japan's proposal at the GIF meeting in September 2010, efforts were started to promote the safety assurance strategy for sodium-cooled FRs and the international standardization of safety design criteria.

The seriousness of the Fukushima accident triggered by the Great East Japan Earthquake on March 11, 2011 was ranked at level 7 on the INES (International Nuclear Event Scale). This accident led to the shutdown of all of Japan's nuclear power plants. Being faced with the vulnerability of energy supply and demand, a new energy basic plan was developed in April 2014. The plan states that the promotion of the nuclear fuel cycle is one of the basic policies from the viewpoints of the efficient use of the resources and reduction of the volume and radiotoxicity of high-level radioactive waste. It states that Japan promotes the reprocessing of spent fuel and plutonium use in LWRs and further works on research and development of FRs, in part, in terms of appropriate control and use of plutonium.

A nuclear fuel cycle based on FRs is a recycle-based system that encompasses the treatment of spent fuel and the disposal of radioactive waste. High expectations are placed on Japan to take the lead in establishing such a system while promoting international cooperation as a country that has developed the use of nuclear energy exclusively for peaceful purposes. In particular, concerning the safety of the FR as a next-generation reactor, it is necessary to build international consensus by persistent efforts to ensure safety, such as the virtual elimination of the need of off-site emergency responses by enhancing robustness against severe accidents through the use of sodium properties and the prevention of recriticality in consideration of the lessons learned from the Fukushima accident.

2.3.2 International Cooperation

International forums on innovative reactors were organized, and research cooperation based on international harmonization has commenced. Since the development goals include the safety, economic competitiveness, effective use of resources, reduction of waste, and enhancement of nuclear nonproliferation that are described in paragraph 2-1-3 and have the same direction as those determined by Japan's Feasibility Study on Commercialized Fast Breeder Reactor Cycle Systems, active cooperation is under way.

1. Gen-IV (Generation IV International Forum)
   Taking the international workshop on the generation IV nuclear power generation system that was held by the US DOE in January 2000 as a starting point, the generation IV (Gen-IV) project started to promote a development program of next-generation nuclear technology under international framework aiming for commercialization in the first half of the twenty-first century. This project aims for a next-generation nuclear system that follows third-generation reactors, such as the ABWR and the EPR, and excels in sustainability (efficient use of resources, minimization of waste, and enhancement of nuclear nonproliferation) and economic competitiveness (advantage in life cycle cost) and safety/reliability (safe/reliable operation, elimination of the need for off-site emergency response), and study has been carried out to introduce the first unit in approximately 2030, as shown in Fig. 2.18.

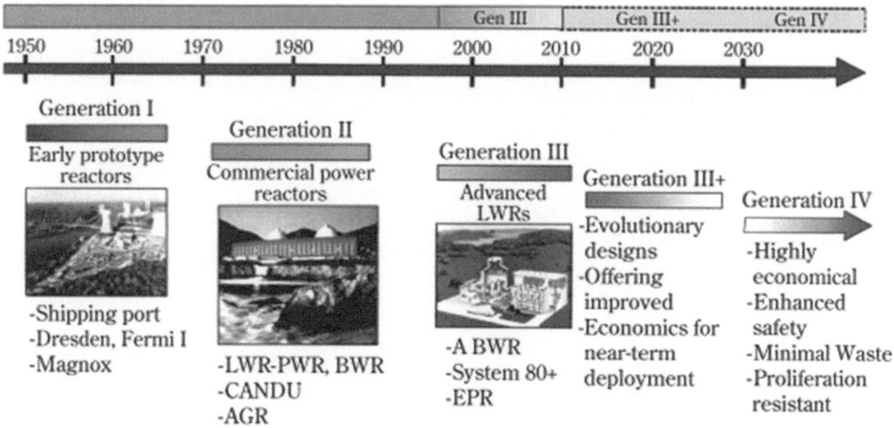

Fig. 2.18 Generation IV International Forum (GIF) [2]

To build a cooperative framework, the Generation IV International Forum (GIF) was launched. To date, twelve countries (Argentina, Brazil, Canada, China, France, Japan, South Korea, Russia, South Africa, Switzerland, UK, and US) and one organization (EU) have signed the GIF charter. The GIF selected six promising concepts (sodium-cooled FR, gas-cooled FR, lead-cooled FR, ultrahigh-temperature gas-cooled reactor, supercritical water-cooled reactor, and molten-salt reactor) and developed the respective technology development plans (December 2002). The GIF framework agreement was signed in February 2005. The GIF system agreement for joint research on sodium-cooled FRs was signed in February 2002, and Japan, France, South Korea, the US, and EU participated in the activity. Japan took the lead in discussion of sodium-cooled FRs and proposed an advanced loop-type sodium-cooled reactor as a reference concept. Specific projects for research and development of sodium-cooled FRs include the advanced fuel project, the component design and BOP project, and the GACID (Global Actinide Cycle International Demonstration) project. Agreements for these projects were signed, and cooperative research and development is under way. GACID is intended to perform engineering-scale irradiation tests of minor actinide (MA) bearing fuel with cooperation from three countries, Japan, the US, and France, to obtain a clearer view of the feasibility of practical recycling of MAs, as shown in Fig. 2.19.

2. INPRO (International Project on Innovative Nuclear Reactors and Fuel Cycles)
   It is necessary to innovate technologies and institutions to secure and make use of sustainable nuclear energy. INPRO is an international forum aiming for the establishment of indices for potential contribution to sustainable development and the innovation of technologies and institutions. The IAEA serves as the INPRO secretariat, in which 27 countries and one international organization have participated as of 2013, as shown in Fig. 2.20. In phase 1, from the start in 2001 to June 2006, focus was placed on the development of methods to assess

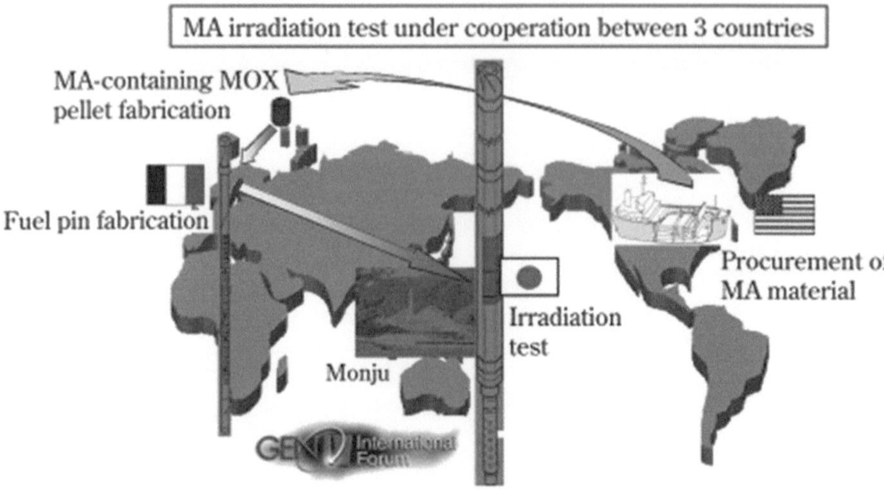

Fig. 2.19 Global Actinide Cycle International Demonstration (GACID) project [2]

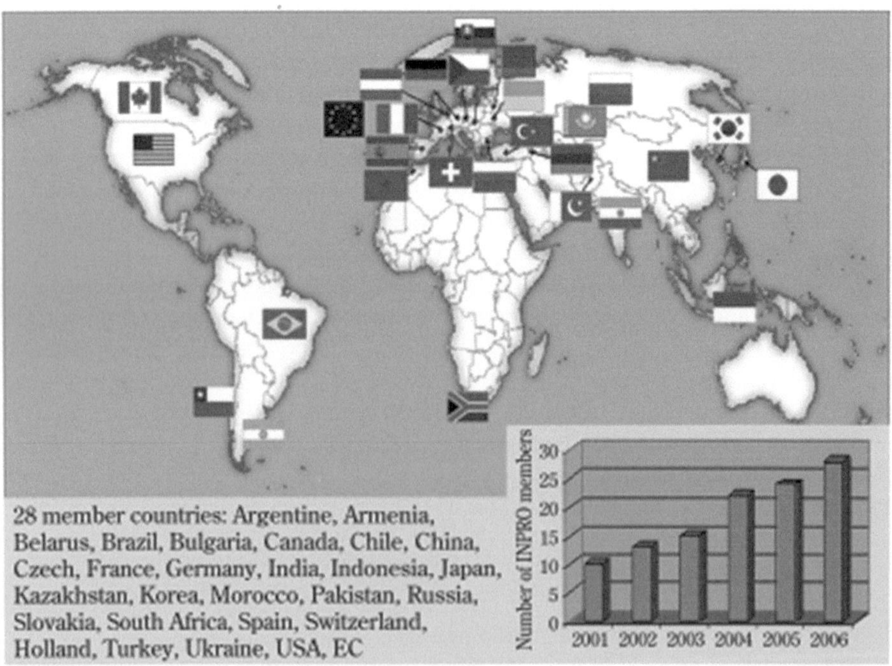

Fig. 2.20 International Project on Innovative Nuclear Reactors and Fuel Cycles (INPRO) [2]

innovative nuclear systems (INS). The INS assessment methods were comprehensively prepared by categorizing the requirements in seven fields (economic competitiveness, safety, environment, waste management, enhancement of nuclear nonproliferation, security, and infrastructure) for INS that should contribute to sustainable development into three steps (basics/user requirement/assessment criteria).

Subsequently, using these methods, application study was conducted. Japan participated in INPRO from April 2006 and performed the "International joint study on the closed fuel cycle system with FR" as part of the application study using the INPRO methods with the main countries that are developing FRs, including France, Russia, India, China, and South Korea. Through these joint studies, requirements to be provided for sustainable development were examined to share recognition of the FR cycle.

GIF is a forum for technology development by countries that already have the relevant technologies. On the other hand, INPRO is a forum through which countries and organizations, including user countries that are planning the use of nuclear power generation, can discuss technological and institutional innovation. Accordingly, INPRO is an international forum that attaches importance to assessment methods and institutions, which are expected by users.

Further Readings

J.G. Yevick, A. Amorosi, *Fast Reactor Technology: Plant Design* (M.I.T. Press, Cambridge, MA, 1966)

S. An, in *Kosoku zoshokuro* (Fast Breeder Reactor), (in Japanese), Dobunshoin, (1982)

IAEA-TECDOC-1180: Unusual Occurrences during LMFR Operation, IAEA Vienna (2000)

Y. Fujiie, in *Genshiryoku – Shizen ni manabi shizen o maneru* (Nuclear Energy – Learn from the Nature and Follow the Nature) (in Japanese), ERC Publishing Company (2005)

Japan Atomic Energy Agency, "Feasibility Study on Commercialized Fast Reactor Cycle System – Phase II Final Report" (in Japanese), JAEA-EVALUATION 2006–002 (2006)

Jean-François Sauvage; Phénix 35 years of history the heart of a reactor, CEA France (2009)

Georges Vendryes, translated by M.M. Uematsu: Superphénix Pourquoi ?,Nucléon (1997)

T. Yanagisawa, Monju (kosoku zoushoku genkeiro) nakama no shosoku (News from colleagues engaged in (the prototype fast breeder reactor) Monju) (in Japanese). **21**(4) (2011)

Gen IV International Forum, Annual Report (2014)

References

1. Edited by Atomic Energy Society of Japan (AESJ), in *Genshiryoku ga hiraku seiki* (Era developed by nuclear energy) (in Japanese), Atomic Energy Society of Japan (1988)
2. Edited by AESJ Editorial Committee, in *Keisuiro puranto – sono hanseiki no shinka no ayumi, kosokuro no hensen to genjo* (LWR plant – progress during half a century, and changes and present status of fast reactors) (in Japanese), Atomic Energy Society of Japan (1988)

3. Edited by Nuclear Power Association, in *Kakunenryo saikuru to kosokuro kaihatsu* (Nuclear Fuel Cycle and Fast Reactor Development) (in Japanese), NSA/COMMENTARIES:No.20, Nuclear Power Association of Japan Atomic Industrial Forum (2013)
4. Special Committee on Nuclear Fuel Cycle of Japan Electrical Manufacturers' Association, in *Kakunenryo saikuru gijyutsu no hitsuyousei to igi* (Need and Significance of Nuclear Fuel Cycle Technology) (in Japanese), (2013)
5. Alan E. Waltar, Donald R.Todd, Pavel V., Tsvetkov: Fast Spectrum Reactors, Springer (2012)
6. R. Miki, in *Kosoku zoshokuro* (Fast Breeder Reactor), (in Japanese), Nikkan Kogyo Shimbunsha (1972)
7. IAEA-TECDOC-1531: Fast Reactor Database2006Update、IAEA Vienna (2006)
8. IAEA-TECDOC-1691: Status of Fast Reactor Research and Technology Development, IAEA Vienna (2012)
9. "Kosoku zoshokuro, Monju, genshiro secchi kyoka shinseisho (Application for reactor establishment license: Prototype Fast Breeder Reactor, Monju)" (in Japanese) (1980)
10. Japan Atomic Energy Agency (JAEA), in *Monju no anzen kakuho ni mukete – Unten to ni oite sotei sareru jiko toraburu to no jirei to sono taiou* – (Toward ensuring safety of Monju – Cases of accidents and troubles assumed during operation, etc. and their countermeasures) (in Japanese), Tsuruga Head Office of JAEA (2008)
11. Website of Japan Atomic Energy Agency, in *Monju no ayumi – Niji reikyakukei jiko genin kyumei* (History of Monju – Investigation into the cause for the secondary cooling system sodium leak accident) (in Japanese) http://www.jaea.go.jp/04/monju/category05/mj_cause/cause.html
12. J.H. Kittel, M. Novick, R.F. Buchanan, The EBR-I meltdown-physical and metallurgical changes in the core. Nucl. Sci. Eng. **4**, 180–199 (1958)
13. Enrico Fermi atomic power plant technical information and hazards summary report, Vols. 1, 2 and 3, Power Reactor Develop. Co., Michigan (1962)
14. W. J. McCarthy: Enrico Fermi experience. Lecture, Nucl. Power Reactor Safety Course M.I.T (1969)
15. J. Branchu et al., Superphénix 1 primary handling system fabrication and testing. Nucl. Technol. **68**, 160–170 (1985)
16. How the Superphénix leak was pinpointed, Nuclear Europe, p. 49, 11/12 (1987)
17. Simon Rippon: BN600 status report, Nuclear Engineering International, pp. 551–555, 11/12 (1975)
18. C. M. Robertson, J. D. Walford: The Leak in Super Heater2-PFR, Feb. 1987, Fourth International Conference on Liquid Metal Engineering and Technology, Avignon. France (Oct. 1988)

Chapter 3
Plant Concepts and Mechanisms

Masakazu Ichimiya

Critical functional requirements for fast reactors are to breed fuel (generation of plutonium) and to reduce radiotoxicity of high-level waste and power generation. The relationship from these functional requirements to mechanism elements is shown in Fig. 3.1. This chapter describes the mechanism through which these requirements are achieved using plutonium and fast neutrons. To this end, the adoption of a reactor core utilizing fast neutrons and the selection of sodium as the coolant are discussed. This chapter also describes the mechanism involved in the transfer of heat from the reactor core to steam turbine, the unique design features of fast reactors, such as the primary and the secondary cooling systems between the reactor core and water-steam system, and the elevated-temperature and low-pressure systems.

3.1 Mechanism of Breeding

3.1.1 Fission and Plutonium Generation by Fast Neutrons

Breeder reactors are nuclear reactors which generate a greater amount of fissile material than that consumed. Breeder reactors are classified into two categories: fast neutron reactors which utilize Pu as a nuclear fuel and generate new Pu from ^{238}U and thermal neutron reactors which utilize ^{233}U as a nuclear fuel and generate new ^{233}U from ^{232}Th. All breeder reactors that have been constructed and operated in various countries are fast breeder reactors.

M. Ichimiya (✉)
University of Fukui, Fukui, Japan
e-mail: ichimiya@u-fukui.ac.jp

N. Kasahara (ed.), *Fast Reactor System Design*, An Advanced Course in Nuclear Engineering 8, DOI 10.1007/978-981-10-2821-2_3

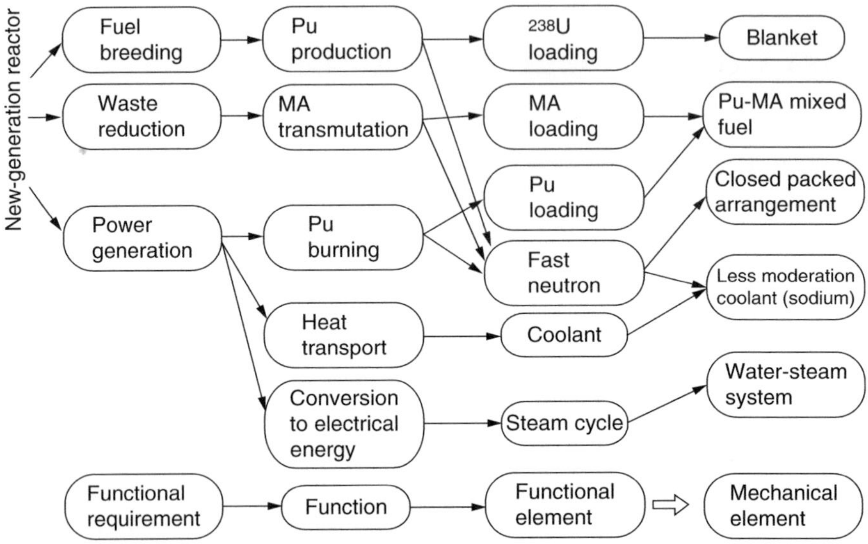

Fig. 3.1 Functional and mechanical elements needed to satisfy functional requirements

In fast breeders which use Pu fuel, fuel isotopes, such as ^{239}Pu, undergo fission. In parallel, ^{239}Pu ($T_{1/2} = 24{,}110$ year), a fuel isotope of the fast breeder, is generated from ^{238}U in fuel by neutron capture via the short-life isotopes, ^{239}U ($T_{1/2} = 23.5$ month) and ^{239}Np ($T_{1/2} = 2.35$ days).

In addition, ^{240}Pu ($T_{1/2} = 6560$ year) which does not cause nuclear fission is produced by neutron capture of ^{239}Pu, and ^{241}Pu ($T_{1/2} = 14.3$ year), a fissile, is produced by neutron capture of ^{240}Pu. Furthermore, ^{242}Pu ($T_{1/2} = 3.73 \times 10^5$ year) is produced by neutron capture of ^{241}Pu.

The isotopes that cause fission, such as ^{235}U and ^{239}Pu, are called fissile material, and the isotopes from which fissile is produced, such as ^{238}U, are called fertile material. The core of breeder reactors is usually designed with fertile material (called a blanket) surrounding the fuel of the fissile material so that the fertile material can efficiently capture neutrons leaking from the fuel region.

Why are fast neutrons, plutonium, and sodium suitable for breeder reactors?

Here, we refer to η (neutron regeneration factor), i.e., the number of neutrons emitted per absorption. The η value has the following relationship with other neutron parameters:

- One neutron is necessary for sustaining the chain reaction.
- The number of neutrons lost by leak and parasitic capture (L).
- The number of neutrons absorbed by fertile* material (C).

*Generation of fissile materials

$$^{238}\mathrm{U}(n,\gamma)^{239}\mathrm{U} \rightarrow {}^{239}\mathrm{Np} \rightarrow {}^{239}\mathrm{Pu}$$

Then,

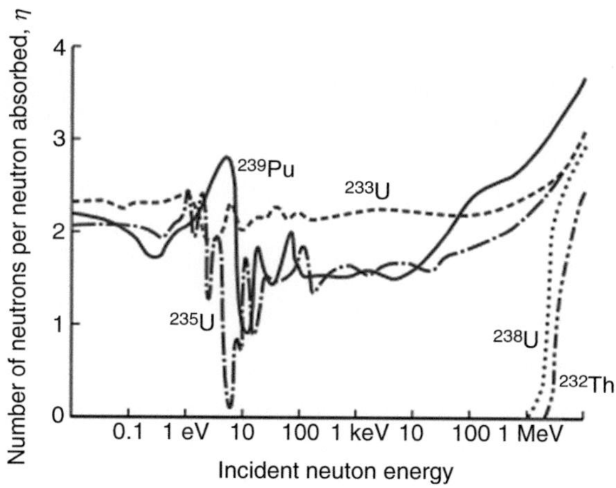

Fig. 3.2 Dependence of η value on neutron energy for candidate fuel isotopes

$$\eta = 1 + L + C$$

The number of C is

$$C \sim \eta - 1 - L$$

It is known from previous reactor core designs that the number of L is unlikely to be 0.2 or less*.[1] Breeding is achieved when C is greater than 1:

$$C > 1$$
$$\text{namely}, \eta > 2 + L$$

Considering the η value of candidate fissile materials over the neutron energy (Fig. 3.2), the material and the neutron energy region that can achieve breeding are summarized as follows:

- In the fast neutron domain, η can be greater than 2.2 for some materials, and η of ^{239}Pu is the largest.
- In the thermal neutron domain, η of ^{233}U is the largest, η>2.2.

Therefore, when designing a fast breeder reactor, ^{239}Pu is the most suitable as fuel.

In case of a thermal energy region, ^{233}U is the most suitable as fuel.

[1]While neutrons produced by fission have an average energy of 2 MeV, since those neutrons are easily moderated to 1 MeV or less by inelastic scattering with heavy nuclides, such as fertile material, it is considerably difficult to make C larger than 1 in the case of ^{235}U.

Breeder reactors constructed and operated around the world to date are fast neutron reactors with Pu as a fuel, and the fuel cycle is U-Pu cycle. Thus, this book focuses on the fast breeder.

In addition, however, naturally existing ^{232}Th can be the fuel of the breeder reactor because it is converted to ^{233}U, a fissile, via two β decay schemes after capturing one neutron. Because the η value of 2.3 for ^{233}U is relatively larger, a thermal reactor using ^{232}Th as a fertile material is regarded as a candidate concept for a breeder. A breeder reactor which consumes Th nuclear fuel in conjunction with a thermal reactor (it becomes the fuel cycle using the Th) and the related fuel cycle is being studied in India. A few comments are added below.

The following chain reaction is used in the Th cycle system:

$$^{232}\mathrm{Th} + n \rightarrow {}^{233}\mathrm{Th}\left(T_{1/2} = 22\,m\right) \rightarrow {}^{233}\mathrm{Pa}\left(T_{1/2} = 27\,d\right) \\ \rightarrow {}^{233}\mathrm{U}\left(T_{1/2} = 1.5 \times 10^5 y\right)$$

A fissile material ^{233}U is generated from ^{232}Th, a fertile material. Th does not have any isotopes undergoing fission and is available only as a fertile material which is converted to ^{233}U. The typical fuel cycle forms with Th include (1) a molten salt reactor using ^{233}U and Th as fluoride molten salt and (2) the use of a thermal reactor (light water reactor or heavy water reactor) and a fast reactor. (1) was studied at Oak Ridge National Laboratory; however, it was abandoned due to material problems, and the USA chose a U-Pu chain for its fast breeder reactor.

On the other hand, (2) is advocated by India. First, ^{233}U is produced from Th in the blanket of a fast reactor fueled with Pu and natural U. Then, the produced ^{233}U is mixed with natural U to form fuel for a thermal reactor. The fast reactor is driven by the Pu converted in the thermal reactor as well as Pu produced by breeding in the fast reactor. Three steps are needed to implement this fuel cycle strategy:

1. First, develop a thermal reactor technique through which Pu is produced from natural U to accumulate it.
2. Develop a fast reactor technique through which Pu is produced from U-Pu fuel and ^{233}U is produced from Th blanket to accumulate them.
3. Finally, mix ^{233}U with the fuel of the thermal reactor and use it.

3.1.2 *Mechanism of Fast Neutron Utilization*

The fast breeder reactor core setup is based on the abovementioned concept. This concept is shown in Fig. 3.3. In the core region, fissile materials, including ^{239}Pu, drive the reactor by fission while compensating for the decreased fuel by generating ^{239}Pu from ^{238}U by neutron capture. A blanket region is arranged in the surroundings of the core region. The blanket region contributes to the generation of ^{239}Pu from ^{238}U using neutrons leaking from the core and enables the breeding of the fissile material.

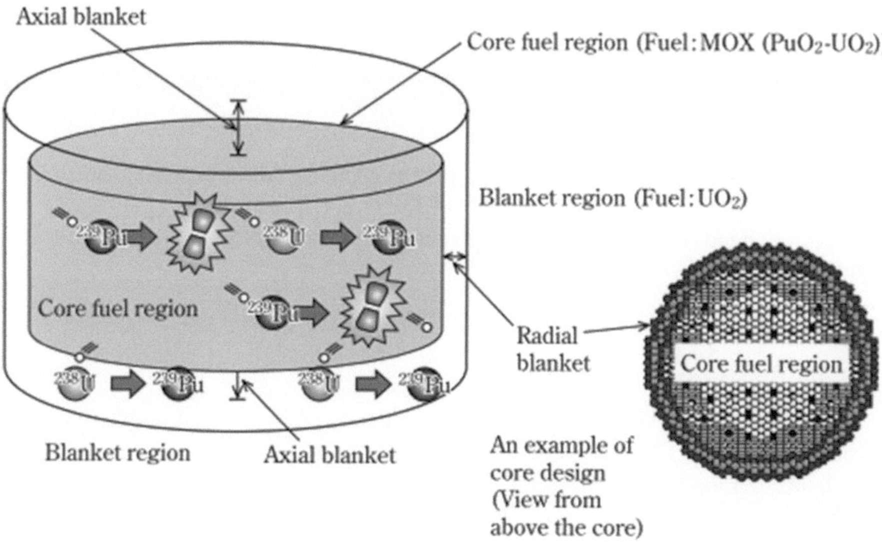

Fig. 3.3 Concept of the fast reactor core (MOX-fueled core)

The blanket region design is such that the blanket material, ^{238}U (called the fertile material), is placed in the upper and lower portions of a fuel pin (axial blanket) and along the outer periphery of the core (radial blanket). It is also possible to place the blanket material in the inside of the core.

Fast reactors are capable of flexibly coping with Pu demand through the arrangement of reflector materials instead of fertile materials in the blanket region. For example, it is possible to utilize the fast reactor as a Pu burner instead of a Pu breeder.

3.2 Mechanism for the Reduction of Toxicity in High-Level Radioactive Waste

A transuranium (TRU) is a nuclide whose atomic number is larger than that of uranium. Neptunium (Np), plutonium (Pu), americium (Am), curium (Cm), berkelium (Bk), and californium (Cf) are included in TRU. Every TRU nuclide is an artificial radioactive element which is generated, for example, from nuclear fuel irradiated in a reactor core in the course of repetition of neutron capture and β-decay. A spent fuel contains plutonium as well as a small amount of TRU elements other than plutonium. ^{237}Np ($T_{1/2} = 2.14 \times 10^6$ y), ^{241}Am ($T_{1/2} = 432.6$ y), ^{243}Am ($T_{1/2} = 7370$ y), ^{242}Cm ($T_{1/2} = 163$ d), and ^{244}Cm ($T_{1/2} = 18.1$ y) are called minor actinides*2(MA). These nuclides are long lived in radioactive waste and are generated in comparatively larger amounts among the TRU elements. Their characteristics should be taken into account during handling. For example, ^{241}Am and ^{243}Am are long-lived α-emitters, and ^{244}Cm is a long-lived decay heat source in spent fuel.

For this reason, research and development focus on the transmutation of MA utilizing a fast reactor with fuel bearing TRU extracted from LWR spent fuels. For example, instead of Pu, TRU-bearing fuel with the following composition is examined:

$$^{238}\mathrm{Pu}/^{239}\mathrm{Pu}/^{240}\mathrm{Pu}/^{241}\mathrm{Pu}/^{242}\mathrm{PU}/^{237}\mathrm{Np}/^{241}\mathrm{Am}/^{243}\mathrm{Am}/^{244}\mathrm{Cm}$$
$$= 1.1/54.1/32.1/4.3/3.9/0.5/2.0/1.0/1.0$$

In the case of this TRU composition, MA decreased by approximately 30–40 % by burning in a MOX-fueled core and approximately 20–40 % in a metal-fueled core, from fuel loading to extraction. The amount of decreased MA is called as MA transmutation, and this amount is the sum of the amount of MA incinerated by fission and the amount of net production of Pu from MA. The breeding ratio increased by 0.2–0.6 and by 0.4–0.8 in the case of the MOX-fueled core and the metal-fueled core, respectively; and the amount of fissile Pu loaded in the initial core decreased slightly, suggesting the enhancement of breeding performance. The sodium void reactivity increased, although within the allowable design limit, and the Doppler coefficient decreased, although with only a minor effect on safety. Thus, it was expected that design of a core containing the abovementioned TRU is technically feasible without serious loss of core performance.

An attempt is being made to transmute long-lived fission products (LLFP) using excess neutrons in a fast reactor core. The selected LLFP nuclides for the target of transmutation were ^{129}I and ^{99}Tc, with the reduction of environmental burden in mind. Three nuclides, ^{129}I, ^{99}Tc, and ^{135}Cs, are particularly important from the viewpoint of the amount generated and the amount of heat and migration; however, transmutation of ^{135}Cs could not be effectively achieved due to the fact that ^{135}Cs is generated from ^{133}Cs during the transmutation process. Concerning the loading positions of LLFP transmutation targets, two patterns are possible, namely, loading into the core and into the outer blanket region with special subassemblies containing Tc pins (Tc metal) and I pins (YI_3) surrounded by moderator pins (e.g., Zr-H). It was reported that self-generated amounts of ^{129}I and ^{99}Tc may be transmuted provided that the manufacture of the LLFP subassemblies is feasible. It was found in the former pattern that the volume ratio of LLFP (i.e., inventory) needs to be large to transmute the self-generated ^{129}I and ^{99}Tc, requiring a significant amount of time to accumulate sufficient LLFP for loading.

2*Fifteen elements from 89th (actinium) to 103rd (lawrencium) in the periodic table are generally called actinoid elements, and all these elements except actinium are called actinide elements. Furthermore, Np, Am, and Cm that are less abundant than U, Pu in spent fuel are called minor actinides.

3.3 Heat Transport for Core Cooling and Power Generation Using Liquid Metal Sodium

3.3.1 Liquid Metal Coolant Suitable for the Use of Fast Neutrons

As mentioned earlier, the fast breeder reactor can breed plutonium by the fission reaction in a fast neutron domain. Thus, there is no moderator, and the neutron energy spectrum is hard in a fast reactor core. The fission cross sections of ^{235}U and ^{239}Pu are two orders of magnitude smaller in the high-energy region than in the lower-energy regions, as shown in Fig. 3.4. Thus, in order to sustain nuclear fission and to obtain the energy equivalent to or greater than that of a thermal neutron reactor, it is necessary to raise both fuel enrichment and neutron flux density. Pu enrichment depends on design variations and is generally set at approximately 20 %. As a result, the core power density is high. Therefore, core cooling with high efficiency is required for a fast reactor core. A typical core power density of about 300 kW/l for a fast reactor core is much larger than about 100 kW/l for PWR and about 50 kW/l for BWR; however, it should be noted that the core power density also depends on reactor core design.

Considering the above features, it is concluded that the following coolant conditions are required for fast breeder reactors:

1. Neutrons should not be slowed down, and neutron capture should be insignificant.
2. Heat removal capability should be high because of high core power density.

We understand that gas, or liquid with a large atomic number, is desirable from condition 1. In addition, liquid metal whose cooling capability is high attracts

Fig. 3.4 Fission cross section of the fuel candidate material

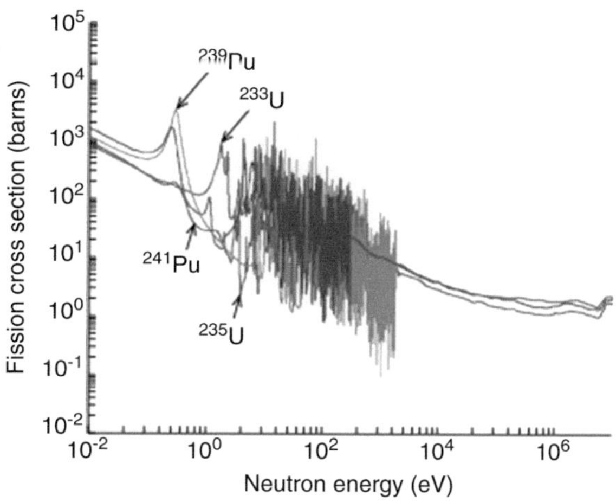

attention from condition 2. As a result, helium gas and liquid metals, such as Na, NaK, Hg, and Pb (Pb-Bi), became candidates. And Na has been used for fast reactor coolant following experience in design, construction, and operation in the world, as was illustrated in Chap. 1.

3.3.2 Sodium-Cooled Fast Reactor Plant System: Heat Transport from Reactor Core to Steam Turbine

Sodium-cooled fast reactors are comprised of:

1. A primary cooling system that removes heat from the core
2. An intermediate system (also referred to as "secondary cooling system") that receives heat from the primary sodium via the intermediate heat exchanger (hereinafter referred to as "IHX")
3. A water-steam system that generates steam to drive the turbine at the steam generator using the secondary sodium as a heat source

The main components of the primary cooling system include a core, a reactor vessel that contains the core, a main circulation pump that circulates the primary sodium, and the IHX, through which heat is transferred from the primary system to the secondary sodium. Depending on the arrangement of these components, two types of the plant configuration, the loop type and the pool type (also referred to as "tank type"), are available, and the hybrid-type configuration is also available as an intermediate type between the two. The loop-type and pool-type fast reactors have already been built.

The sodium-cooled fast reactor differs significantly from the LWR in that it has a secondary system containing nonradioactive sodium regardless of whether the loop type or pool type is employed (see Fig. 3.5). This secondary system containing nonradioactive sodium is located between the radioactive primary sodium and steam generator and is provided in consideration of the following factors: (1) direct effect on the reactor core of hydrogen and reaction products that might be generated due to reaction between sodium and water in case of a failure of heat transfer tube in the steam generator should be avoided and (2) experience with a steam generator using sodium was insufficient. In an effort to develop a sodium-cooled fast reactor aiming at commercialization, a design study on a plant configuration consisting of a primary system and water-steam system through removal of the secondary system is under way. This configuration can make the entire system more compact; however, it requires remarkable improvement in the reliability of the steam generator's heat transfer tubes.

In a loop-type reactor, the primary main circulation pump and IHX are installed outside the reactor vessel and connected by piping. This configuration allows a relatively simple component structure since the reactor vessel, primary main circulation pump, and IHX are individually installed and the independence of

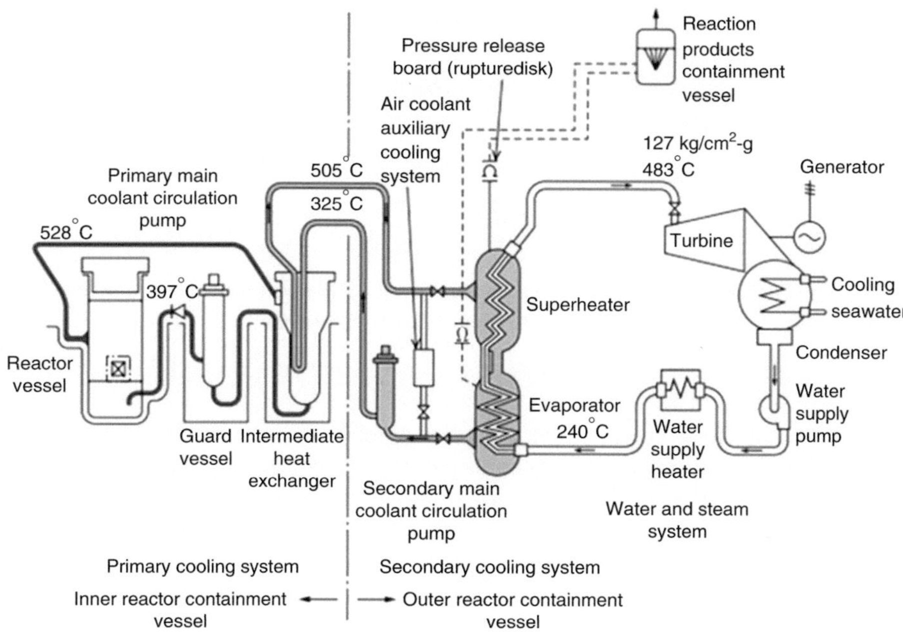

Fig. 3.5 Installation of the intermediate (secondary) cooling system in the sodium-cooled fast reactor system

each component facilitates access during maintenance and repair. On the other hand, a pool-type reactor allows a more compact primary system because the primary main circulation pump and IHX, as well as the core, are contained in a large single tank referred to as the primary tank. Both the experimental fast reactor Joyo and the prototype fast reactor Monju (hereinafter referred to as "Monju") in Japan are loop-type reactors. Joyo, Japan's first sodium-cooled fast reactor, achieved first criticality in 1977. Because Joyo is not designed to produce electricity, it has no water-steam system or power-generating equipment. Monju is a midsized prototype fast reactor developed prior to a demonstration reactor. It generates 714 MWt of thermal power and approximately 280 MWe of electrical power and is Japan's first sodium-cooled fast breeder reactor containing power-generating equipment.

The loop-type reactor includes two configurations, one in which the primary main circulation pump is located on the downstream side of the IHX and the other in which it is located on the upstream side of the IHX (i.e., between the reactor vessel outlet and the IHX). In the latter case, the IHX is located on the discharge side of the primary main circulation pump. The coolant flows out of the reactor vessel outlet and into the primary main circulation pump at a high temperature, where the coolant pressure is increased, and exchanges heat at the IHX before returning to the reactor vessel.

In an actual pool-type reactor, the primary main circulation pump is located in cold plenum, downstream from the IHX.

3.3.3 Plant System Suitable for Liquid Metal Coolant with a High Boiling Point

Here we consider the plant concepts peculiar to a sodium-cooled fast breeder reactor which are different from those of a light water reactor.

1. Sodium has been widely used as a coolant because it has a high boiling point, which enables a sufficient subcool temperature margin in the cooling system design even in a low-pressure condition, and also superior characteristics in thermal conductivity. The high boiling point of sodium (i.e., approx. 880 °C at atmospheric pressures) makes it possible to create a high-temperature (500~550 °C) and low-pressure system. This represents a significant departure from the LWR, which features a high-pressure design, allowing the thin-wall structural design of fast reactor components. In addition, since the higher temperatures allow a superheated steam cycle, the thermal efficiency of fast reactor is approximately 40 %, higher than that of the LWR. Sodium systems maintained at low pressures enable thin-wall structure, as shown in Table 3.1. Design accommodation is necessary to meet seismic requirement.

2. The sodium-cooled fast reactor is capable of operation at nearly atmospheric pressure since there is no need for pressurization to maintain the liquid phase of sodium coolant that has a high boiling point (approx. 880 °C at atmospheric pressure). In addition, sodium does not boil at reduced pressure, even when assuming a pipe failure, due to its low vapor pressure (approx. 9 mmHg). As a result, in the event of a leak of coolant from the reactor coolant boundary, a rapid injection of coolant, which is required for the LWR, is unnecessary. Instead, all that is required is to maintain a sufficient amount of the primary coolant for circulation to cool the core, and deacy heat removal. Consequently, a safety vessel, called the guard

Table 3.1 Comparison of design between FR and LWR

	Sodium-cooled fast reactor	Light water reactor
Operation temperature, °C	500 ~ 550	280 ~ 330[a]
Operation pressure, MPa	Ordinary pressure	7 ~ 16
Thermal efficiency, %	~ 40	32 ~ 35
Temperature difference between core inlet and outlet, °C	~ 150	5 ~ 30
Thickness of reactor vessel, mm	25 ~ 50	100 ~ 250

BWR: 286 °C (saturated steam)
[a]PWR: 320 °C

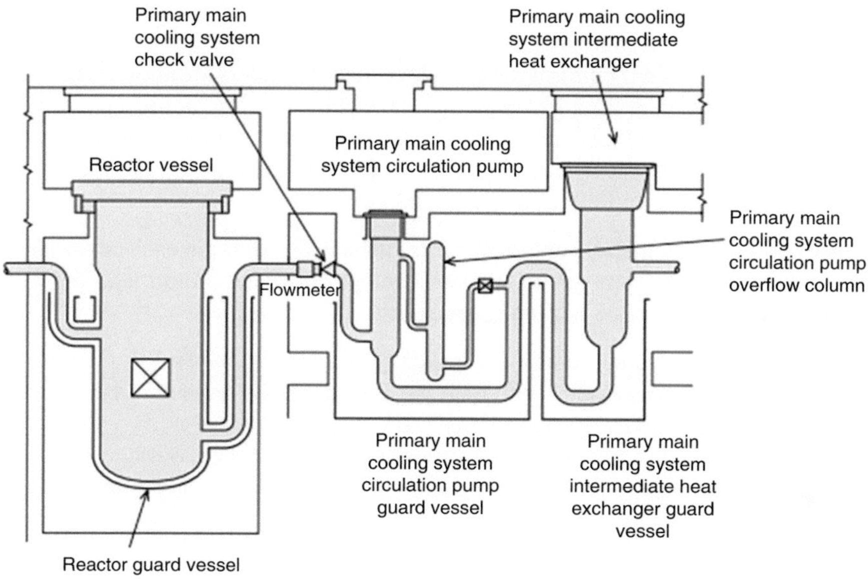

Fig. 3.6 Concept for measures against coolant leak in sodium-cooled fast reactors

vessel (see paragraph 3-5-2), has been designed to surround the reactor vessel, primary main circulation pump, IHX, and connecting piping around these components. The concept for measures against coolant leak is shown in Fig. 3.6.

3. The secondary system containing nonradioactive sodium is located between the radioactive primary sodium and steam generator. Steam generator design requires a leak protection system to prevent sodium-water reaction in case of heat transfer tube failure. This is due to the possibility that following the failure of a heat transfer tube, high-pressure steam jets would leak out into sodium to produce a high-temperature reaction jet, which may cause wear damage called "wastage" to the neighboring intact tube walls by corrosion/erosion and, thereby, expand the damaged region. In case of large-scale damage, the pressure in the sodium system is likely to rise rapidly due to the effect of hydrogen, etc., which are produced by chemical reactions. The steam generator leak protection system generally consists of three elements, the leak detection system, the steam blowdown system, and the pressure relief system. The water leakage detection system which monitors hydrogen concentration in the sodium and the cover gas is provided to detect small leaks. Intermediate water leakage is detected by cover gas pressure monitoring. A rupture disk is installed in the cover gas region of each evaporator and superheater to cope with large-scale water leak (see Fig. 3.5). When the pressure buildup exceeds a certain limit due to the large

leak, the rupture disk bursts and discharges the sodium-water reaction products to prevent the steam generator itself from being destroyed and to protect IHX heat transfer tubes, which form the reactor coolant boundary, from damage. The reaction products flow through the rupture disk to the reaction product container. In addition, the steam generator is isolated from both the main secondary circuit and water-steam circuit by a rupture disk burst signal and blows out the inside water to the outside rapidly through the discharge valve.

4. Since sodium is chemically active, the free surface of sodium in each component is covered with an inert gas to prevent contact between sodium and air. The following are features of sodium components:

   1. All sodium pipes and vessels are basically welded structures.
   2. Sodium surface is covered with inert gas (cover gas) to prevent direct contact with oxygen in air.
   3. Spent fuel handling facilities are different from those for LWRs.
   4. Measures against sodium fire are required because sodium burns in air.
   5. The components and piping in the primary main cooling system that contain radioactive materials are designed to be enclosed in steel-lined gastight cells filled with nitrogen to prevent, to the degree possible, chemical reaction between sodium and oxygen in case of sodium leak.

5. Sodium void reactivity may be positive at the center region of the core.*[3]

Thus, system design should consider prevention of sodium boiling and gas entrainment.

A dipped plate is installed to suppress sodium surface waving, which might cause gas entrainment. A gas-vent line is attached to the main piping to extract gas dissolving in sodium when the piping is filled with the primary sodium. A gas-vent hole is provided to the in-vessel structure to prevent gas retention.

6. Since the neutron migration length M is long, fewer control rods are necessary,*[4] and the requirement for distribution homogeneity is less strict, as shown in Fig. 3.7. The sodium-cooled fast reactor, in particular the loop-type reactor, is advantageous in that neutron detectors can be placed outside the reactor vessel, as shown in Fig. 3.8.

---

[3]The following are influencing factors. These values are dependent on reactor core design. Usually factors (1) and (2) are dominant.

1. Spectrum hardening → increase in η value of 239Pu → change toward positive
2. Increase in neutron leakage →change toward negative
3. Decrease in neutron capture → change toward positive
4. Change in self-shielding effect

[4]The range where a control rod exerts influence is within two or three times the neutron migration length M from the relevant control rod position.

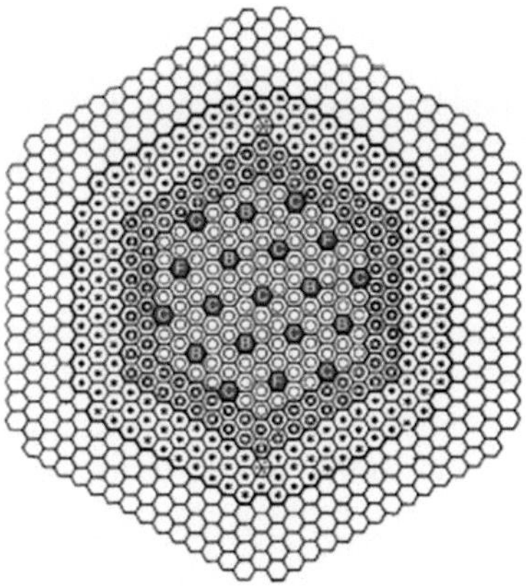

Core component		Symbol	Number
Core fuel assembly	Inner core		108
	Outer core		90
Blanket fuel assembly		*	172
Control rod assembly	Fine control rod	F	3
	Coarse control rod	C	10
	Backup control rod	B	6
Neutron source assembly			2
Neutron shield assembly			316
Surveillance assembly			8

Fig. 3.7 Control rod positions in the Monju core

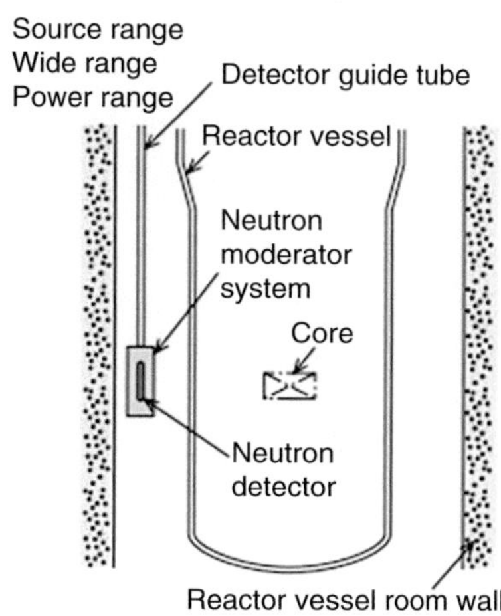

Fig. 3.8 Position of the neutron detector in Monju

– Example of M and M/D:

M: migration length of neutrons in the core
D: core diameter
(1000 MWe class reactor):

BWR: 2.2 cm, 5.6×10^{-3}
PWR: 1.8 cm, 5.3×10^{-3}
LMFBR: 5.0 cm, 2.8×10^{-2}

7. In addition, attention should be paid to the following hardware features:
    1. Since the melting point of sodium is 98 °C, piping and components in which sodium circulates must be equipped with thermal insulation and preheating equipment to prevent sodium freezing.
    2. Since sodium is opaque and chemically active, fuel must be exchanged by remote control.
    3. Sodium has good compatibility with structural material, and corrosion and stress corrosion cracking are unlikely to occur in purity-controlled sodium.
    4. Since sodium has high heat conductivity and small heat capacity, measures must be taken against thermal stress in both steady and transient conditions. Specifically, thermal shield plates should be attached to the walls that come into contact with sodium coolant.

For the light water reactor, neutrons on the periphery of the core are sufficiently moderated to the thermal neutron energy, and the leak of high-energy neutrons from core is insignificant. On the other hand, for the fast reactor, the leak of high-energy neutrons from core is significant because neutrons are not greatly moderated in the core. Because high-energy neutrons cause severer neutron damage to the in-vessel structural materials in comparison with thermal neutrons, neutron shielding is important for the fast reactor. In addition, the need for shielding against activated sodium, such as ^{24}Na, is a characteristic of the fast reactor [1].

Reference

1. Japan Atomic Energy Agency, *Feasibility study on commercialized fast reactor cycle systems*, Technical Study Report of Phase II-(1) Fast Reactor Plant Systems, JAEA-Research 2006-042, April 2006

Chapter 4
Policy of Safety Assurance (Design Constraints and Additional Functional Requirements)

Yoshio Kani

This chapter describes the policy of safety assurance, on which the highest priority should be placed in the system design of fast reactor (FR) plants. It is essential that safety approaches and safety measures for nuclear reactor facilities should be taken based on fundamental philosophy common to overall reactor facilities as well as full consideration of technological features and safety characteristics of the relevant reactor facility. Further, in consideration of lessons learned from the Fukushima Dai-ichi nuclear power plant (NPP) accident, it is necessary to significantly enhance safety, including the provision of severe accident measures; and it is required that ceaseless efforts be made on a continuous basis to pursue the highest level of safety in the world. This chapter starts with the fundamental philosophy of safety assurance for general reactor facilities, discusses technological features of FRs in view of safety in comparison with light-water reactors (LWRs) followed by a description of the basic policy for FR safety assurance, and finally clarifies the safety measures specific to FRs in concrete terms.

The relation between safety philosophy and system design is shown in Fig. 4.1. Prevention of abnormality, i.e., the first level of defense in depth discussed below, is of essential importance in assuring reactor safety. Hence requirements of integrity and reliability (constraint conditions) should be imposed on various structures, systems, and components (SSCs), which are provided to achieve the original functions of FRs, such as power generation, fuel breeding, and waste reduction, in order to prevent the occurrence of abnormal states in the reactor facility. These SSCs are mostly in the category of abnormality prevention systems (hereinafter referred to as "PSs") and normally operated/used. Major SSCs performing PS functions are the reactor core and fuel, structures, and boundaries, including nuclear steam supply system (NSSS) and power conversion system (PCS). They are

Y. Kani (✉)
Tokai University, Kanagawa, Japan
e-mail: kani.yoshio@tokai-u.jp

N. Kasahara (ed.), *Fast Reactor System Design*, An Advanced Course in Nuclear Engineering 8, DOI 10.1007/978-981-10-2821-2_4

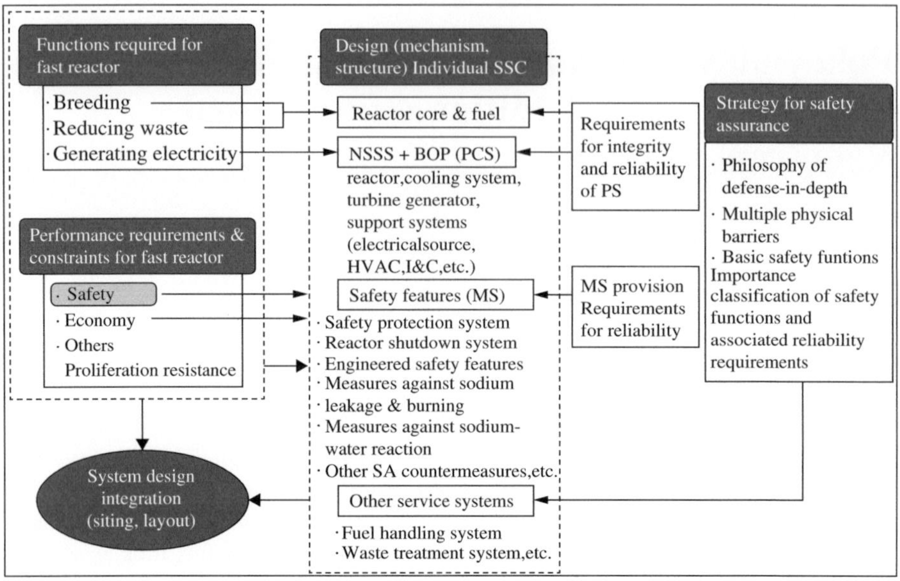

Fig. 4.1 Relation of safety and system design

classified into appropriate component categories according to the importance of their safety functions, and certain design considerations have to be made so that a properly deserved level of integrity (allowable standards or judgment criteria) is ensured. These considerations give rise to constraints in system design. In the case of advanced new type reactors, system design is, in general, forced to provide large safety margins for prevention of abnormality due to poor knowledge and information, which results in severe constraint conditions. Further, abnormality mitigation systems (hereinafter referred to as "MSs") are required in system design assuming the occurrence of abnormal conditions. On this basis so-called safety features performing MS functions have to be incorporated into system design. They include the safety protection system (instrumentation and control system), the reactor shutdown system, the engineered safety features, etc. They are basically standby systems for exclusive use as MS functions; however, in some cases, they perform the MS function along with normally operating systems. The design of such PS and MS occupies an important position in the system design in view of safety assurance.

4.1 Fundamental Philosophy of Safety Assurance for Nuclear Reactor Facilities

In short, the safety of a reactor facility is described as no exteriorization of the hazardous nature of the contained radioactive materials. In "Fundamental Safety Principles" (SF-1) [1] of the IAEA, the fundamental safety objective is to protect people and the environment from the harmful effects of ionizing radiation. Further, the following statement is given: To ensure that facilities are operated and activities conducted so as to achieve the highest standards of safety that can reasonably be achieved, measures have to be taken:

(a) To control the radiation exposure of people and the release of radioactive material to the environment
(b) To restrict the likelihood of events that might lead to a loss of control over a nuclear reactor core, nuclear chain reaction, radioactive source, or any other source of radiation
(c) To mitigate the consequences of such events if they were to occur

The fundamental philosophy for assurance of such safety for reactor facilities is discussed below.

4.1.1 Safety Goals and Risk Management

Generally there exists no absolute safety in any engineered artifact, and safety is regarded as a condition under which the potential of harm to people or damage to materials, namely, the risk, is suppressed to a permissible level (JIS Z 8115-2000). Also an international safety standard (ISO/IEC Guide 51, 1999) regards safety in a similar way as freedom from unacceptable risk. Most commonly the risk is defined as the combination of the probability of occurrence of harm and the severity of that harm.

How far the risk should be reduced to ensure safety can be a perpetual issue, yet there is some manner of definition of the permissible risk. It is defined as the risk that can be accepted under given conditions based on the social values of a particular time. In the nuclear engineering field, the question of "How safe is safe enough?" has been discussed, and safety goals have been investigated for some time. The permissible risk, i.e., safety goal, will be achieved by iterative application of the risk reduction process with risk assessment, namely, by risk management. Risk management consists of "risk assessment" and "risk reduction policy (safety protection policy)."

The safety goals proposed in Japan's safety regulation system are described as follows:

1. Qualitative goal

The possibility of damage to public health by the emission of radiation or the release of radioactive materials accompanying activities for utilization of nuclear energy should not meaningfully increase the risk of damage to the public's health in daily life.

2. Quantitative goals

The mean value of acute fatality risk by radiation exposure resulting from an accident at a nuclear installation to individuals living in the vicinity of the boundary of the installation site should not exceed the probability of approximately 1×10^{-6} per year; and the mean value of fatality risk by latent cancer caused by radiation exposure resulting from an accident at a nuclear installation to individuals living in the area but some distance from the nuclear installation should not exceed the probability of approximately 1×10^{-6} per year.

The quantitative indices for the performance goal have been established to represent the level of safety assurance and to evaluate conformance to the above safety goals. These indices (per site of NPP) include (1) core damage frequency (CDF) approximately 10^{-4}/year and (2) containment failure frequency (CFF) approximately 10^{-5}/year, and both indices have to be met for the performance goal of nuclear power reactors. Furthermore, following the Fukushima Dai-ichi NPP accident, a condition was added, namely, that the frequency of accident with an amount of ^{137}Cs release of more than 100 TBq should be suppressed below approximately 10^{-6}/reactor· year [2].

The risk management of nuclear reactor facilities is accomplished by taking risk reduction (safety protection) measures and examining the adequacy of those measures based on risk assessment (safety evaluation), and iterating the processes for continuous improvement, while bearing in mind the achievement of safety goals or performance goals. Major elements of risk reduction (safety protection) policy, as discussed in the next section, include the application of defense in depth as a strategy, the provision of multiple physical barriers against radioactive materials, and the fulfillment of fundamental safety functions. It is noted that safety evaluation includes deterministic and probabilistic assessments (PSA, Probabilistic Safety Assessment/PRA, Probabilistic Risk Assessment). The relationship among those elements is shown in Fig. 4.2.

4.1.2 Policy of Safety Assurance (Defense in Depth and Fundamental Safety Function)

The concept of defense in depth has been adopted internationally as a fundamental philosophy for nuclear reactor safety assurance strategy [3]. This concept provides multiple layers of defense lines against potential events posing threats to reactor safety. Should one layer (level) of defense be broken through, the subsequent layer comes into play and interdicts event progression. Thus, this concept aims to ultimately ensure the containment of radioactive materials and the protection of the surrounding public and environment. Since a defense line is generally provided based on a certain supposition or postulation, there exist uncertainties in the effectiveness of the defense line itself due to omissions in the supposition or knowledge uncertainties. For this reason, the concept of defense in depth is applied in order to enhance the effectiveness (success probability) of overall protection measures, by provision of the subsequent measures of successive defense lines. What is essential to defense in depth is that different defense lines perform their

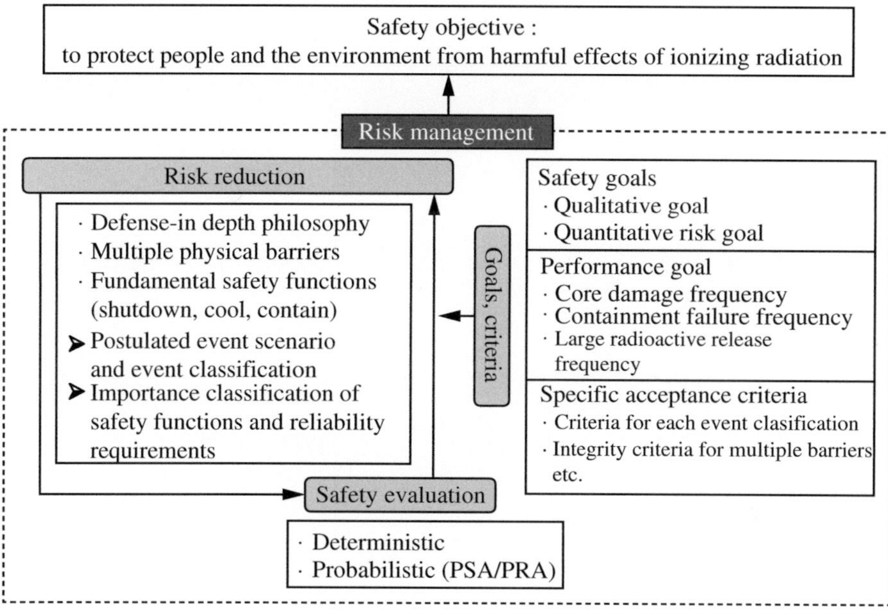

Fig. 4.2 System for safety assurance of nuclear reactor facilities

functions independently and effectively. The number of layers and the thickness of each layer in defense in depth should be determined based on the safety goal (risk reduction goal) to be achieved taking into account the extent of the hazards in the reactor facility and the current understanding of associated uncertainties.

Recently, the following five layers (levels) are considered (see Fig. 4.3):

1. Prevention of abnormality: Prevent abnormalities due to malfunction of components, operator error, etc. Design, manufacture, and install the systems and components with good quality and high reliability.
2. Prevention of escalation and progression to accidents: Should abnormalities occur, detect them early and provide measures, such as interlock for stopping control rod withdrawal, automatic reactor shutdown, etc.
3. Prevention of abnormal release of radioactive materials during accident: Even if abnormalities lead to accident conditions, cool the reactor by the independent and redundant decay heat removal systems, confine the radioactive materials in the containment vessel, and eventually accommodate the accident within the design basis condition (prevention of core damage, no significant radiation exposure risk, etc.).
4. Control and management of severe accident: Provide measures for the prevention and mitigation of severe accident (SA) consequences such as damage or meltdown of reactor core, and ultimately prevent the large release of radioactive materials.
5. Emergency preparedness: Mitigate the radiological consequences of radioactive releases that could potentially result from severe accident conditions.

Strategy	Accident prevention			Accident mitigation	
Operational state of the plant	Normal operation	Anticipated operational occurrences	Design basis and complex operating states	Severe accidents beyond the design basis	Post-severe accident situation
Level of defence in depth	Level 1	Level 2	Level 3	Level 4	Level 5
Objective	Prevention of abnormal operation and failure	Control of abnormal operation and detection of failures	Control of accidents below the severity level postulated in the design basis	Control of severe plant conditions, including prevention of accident progression, and mitigation of the consequences of severe accidents, including confinement protection	Mitigation of radiological consequences of significant releases of radioactive materials
Essential features	Conservative design and quality in construction and operation	Control, limiting and protection systems and other surveillance features	Engineered safety features and accident procedures	Complementary measures and accident management, including confinement prodection	Off-site emergency response
Control	Normal operating activities		Control of accidents in design basis	Accident management	
Procedures	Normal operating procedures		Emergency operating procedures	Ultimate part of emergency operating procedures	
Response	Normal operating systems		Engineered safety features	Special design features	Off-site emergency preparations
Condition of barriers	Area of specified acceptable fuel design limit		Fuel failure; Severe fuel damage	Fuel melt; Uncontrolled fuel melt	Loss of confinement

Fig. 4.3 Overview of defense in depth (IAEA INSAG-12)

In designing a nuclear reactor facility, as in the case of any engineered artifact, assumed design conditions have to be established. In safety design, events to be considered (i.e., design basis events (DBEs)) have to be postulated, and the appropriate design measures shall be provided for the prevention of their occurrence and the mitigation of their consequences should they occur. The following criteria have to be met; (1) the core shall not be damaged considerably, and adequate coolable geometry of the core shall be maintained, and (2) the radiological risk to the off-site public shall be acceptably low (5 mSv as guideline dose value). These design measures correspond to the defense lines of up to the third level, and they have been strict regulatory requirements heretofore. It is noted that the design basis events include, obviously, not only abnormal events occurring within the reactor facility (internal events), such as malfunctions and failures of the pump, piping, etc., but also natural hazards, such as earthquake, tsunami, volcano eruption, typhoon, tornado, flood, snowfall, etc., and external human-made events, such as airplane accident, etc., (external events). Furthermore, measures for accident management (AM) shall be provided to prevent escalation to severe core damage or meltdown far beyond design basis events, i.e., severe accidents (SAs), or to mitigate the consequence of the SAs. AM measures correspond to the defense line of the fourth level of defense in depth. These measures in the fourth level have just been incorporated into regulatory requirements as severe accident measures based on the lessons learned from the Fukushima Dai-ichi NPP accident.

A discussion of the fundamental nuclear reactor safety functions follows. A nuclear reactor utilizes fission to produce a large amount of energy with a small amount of fuel, while it inevitably generates fission products (FPs) (see Fig. 4.4). Since the FPs are unstable and decay while emitting energy (i.e., radiation), they remain a source of heat (decay heat) even after termination of nuclear fission and, along with fuel materials, continue to be the source of radiation. During the rated power operation of a nuclear reactor, decay heat accounts for 7% of the total thermal energy generation. Following reactor shutdown decay heat decreases according to the half-life of the radioactive nuclides in the FP (decay heat in a reactor decreases to approximately 1/5 in 1 h, approximately 1/15 in 1 day, approximately 1/40 in 1 month, and approximately 1/70 in a half year).

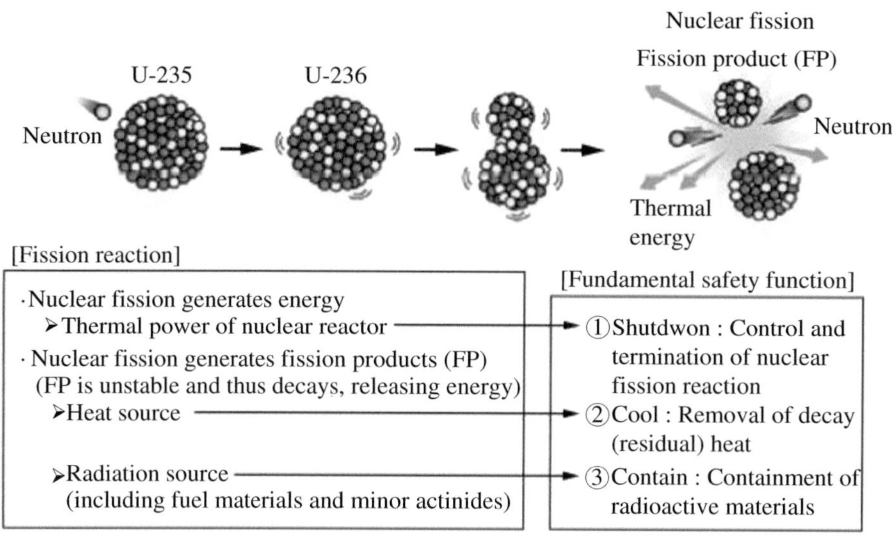

Fig. 4.4 Nuclear fission reaction and reactor safety

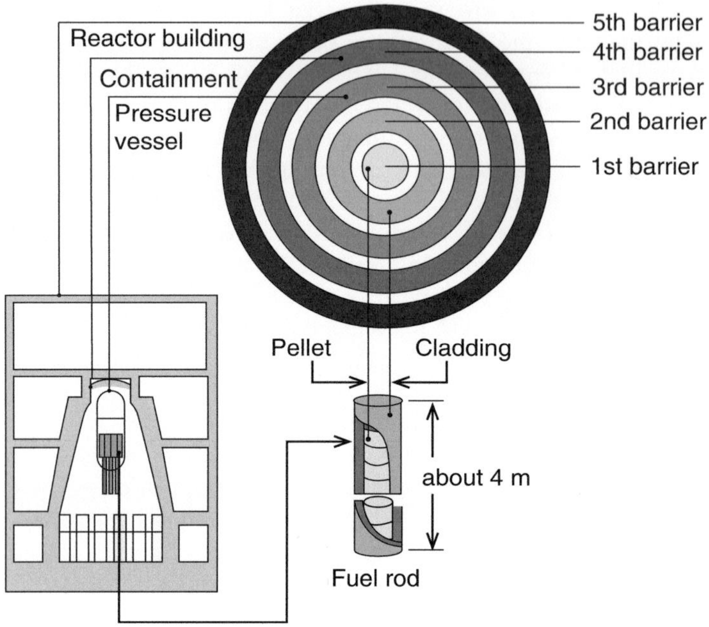

Fig. 4.5 Barriers against release of radioactive materials

The following three functions can be defined as fundamental safety functions. Should abnormalities occur, at first (1) the reactor has to be shut down (termination of nuclear fission reaction), secondary (2) the reactor has to be cooled (decay heat removal), and finally (3) the radioactivity has to be contained (containment of radioactive materials). If the first two fundamental safety functions ((1), (2)) are lost, damage to reactor core or core meltdown could occur, and it could affect the containment structure, resulting in a significant loss of containment function, and eventually the third safety function, (3), is unlikely to be achieved.

As shown in Fig. 4.5, multiple barriers (fivefold barriers) are provided in a nuclear reactor as a measure to contain radioactive materials. They are physical barriers against radioactive materials, and the strategy to assure reactor safety (i.e., to protect people and the environment from the harmful effects of ionizing radiation) by maintaining the integrity of the physical barriers is the concept of defense in depth with five layers of defense lines.

The first barrier is a fuel pellet, which is sintered with uranium oxide or mixed oxide of uranium and plutonium. It is very hard and stable even at high temperatures. The second barrier is fuel cladding, which is a hermetically closed metallic cylindrical vessel that holds fuel pellets. As the third barrier, a steel-made reactor (pressure) vessel is provided to prevent the release of radioactive materials to the environment even in the case of fuel cladding failure. Moreover, as the fourth barrier, a reactor containment vessel is installed to store the reactor (pressure) vessel. The fifth barrier is a reactor building outside the reactor containment vessel.

The independence among the five physical barriers is of essential importance. In case of a failure of some inner barriers, if the fundamental safety functions of "shutdown" and "cool" are not accomplished, then the risk of dependent failures of the remaining outer barriers comes into being. Therefore, it is indispensable to ensure the fundamental safety functions.

The function of "shutdown" is most commonly achieved by detecting abnormality and automatically inserting control rods into the reactor core (reactor scram). Moreover, as backup measures, dedicated systems are provided. These are based on the characteristics of the relevant reactor facility, e.g., other diverse control rod systems, injection of borated water absorbing neutrons, etc.

The function of "cool" is discussed taking a light-water reactor (LWR) as an example. As shown in Fig. 4.6, since the overheating and failure of fuels can be prevented as long as fuels are submerged in water, at first (1) *injection of water into the reactor core* is essential. Therefore, safety features such as the emergency core cooling system (ECCS) are provided. In the case of a boiling-water reactor (BWR), water in the condensate storage tank is used as a water source for the ECCS. After the water in the tank is exhausted, the ECCS injects water from a pool in the suppression chamber at the bottom of the reactor containment vessel. The cooling mechanism in this process is the latent heat of the vaporization of water. The injected water is vaporized by heat from fuels and then the reactor pressure vessel pressure increases with the accumulation of steam. Safety relief valves are provided in the reactor pressure vessel to prevent vessel failure due to overpressure, and through the valves, steam is led into the water pool in the suppression chamber to be condensed for pressure suppression. (If piping failures near the reactor are postulated, steam can be ejected directly into the containment vessel.) In due course, the temperature of water pool increases, and the containment vessel is filled with steam

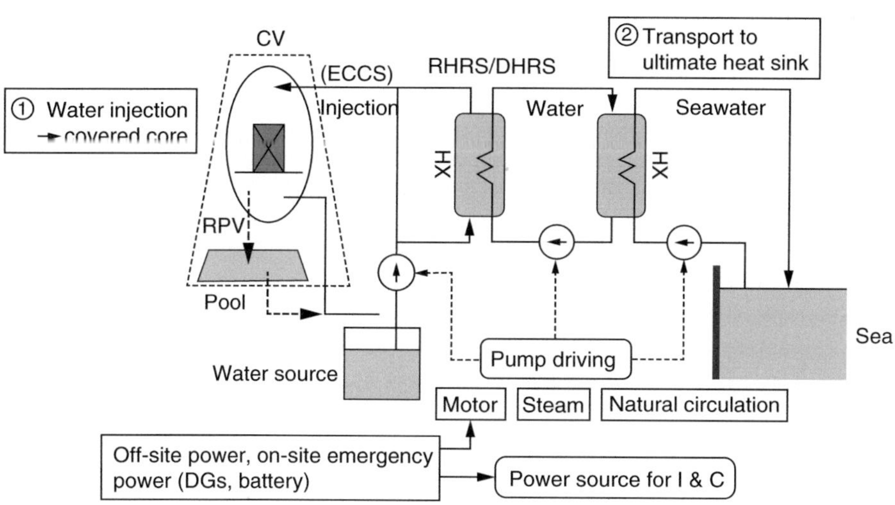

Fig. 4.6 Concept of decay heat removal (LWR)

and is eventually pressurized. Unless the thermal energy in the containment vessel is released to the outside, the water in the reactor pressure vessel is gradually lost, and the reactor core should be exposed, resulting in core damage or meltdown, and the containment vessel could fail due to overheating and overpressure. Therefore, (2) it is essential to *transport heat to an ultimate heat sink*. A heat transport system, consisting of the residual heat removal system (RHRS) and the associated support cooling systems, is provided in order to transport heat from the water pool in the containment vessel to sea, the ultimate heat sink, through heat exchangers, while the ECCS injects water into the reactor core and recirculates the water to keep the core submerged in water.

The pumps have to be driven to operate the systems associated with cooling. (Some pumps would not be needed if natural circulation is possible.) Although most of pumps are driven by electric motors with an alternating-current (AC) power source, the steam turbine-driven pumps, which can be operated even without an AC power source, are provided for the injection of water into the reactor core. Moreover, a direct-current (DC) power source, i.e., batteries, would be essential for the operation of valves and the instrumentation and control (I&C) system. Should the off-site external power source (AC power source) be lost, emergency diesel generators would be automatically activated to supply AC power.

In order to perform the above safety functions, it is required that multiple systems having the same function should be provided (i.e., redundancy) and, if practicably possible, other systems with different attributes (mechanism, manner of operation, etc.) should be arranged (i.e., diversity). Furthermore, it is required that such multiple systems should not fail together, namely, that they be free from simultaneous functional impediment caused by common or dependent factors (i.e., independency).

4.1.3 Measures against Severe Accident

Severe accident is defined as an event far beyond the design basis with the condition under which core cooling or reactivity cannot be adequately controlled by means postulated in evaluation of safety design, and eventually leading to severe core damage. Measures for the prevention of progression to severe accidents or mitigation of consequence of severe accidents (i.e., accident management (AM) measures) are provided as the fourth layer of defense in depth. In Japan AM measures were developed in the 1990s and early 2000s. These measures were to be implemented flexibly on the fly in an emergency situation and were positioned as voluntary measures for safety of operators (licensees) rather than as official regulatory requirements. In addition, only internal events had been considered as potential initiators of severe accident in the development of the AM measures.

In the Fukushima Dai-ichi NPP accident, each layer of defense in depth was successively broken through. More specifically, (1) the loss of off-site power, abnormal event, occurred due to damage to the receiving circuit breakers, etc. at the switchyard during the earthquake. (2) Although the reactor was automatically shut down after detecting the high acceleration due to the earthquake, tsunami-induced water immersion caused loss of functions of onsite emergency power sources leading to station blackout, and thus the abnormal event was escalated and progressed to the accident. (3) Since the total loss of electric power, including DC power, continued for a long period of time, regular safety features with the function of injecting water into the reactor core (e.g., Reactor Core Isolation Cooling (RCIC) system, High-Pressure Core Injection (HPCI) system, etc.) had come to a successive stop, and eventually the cooling function was lost as well as the ultimate heat sink was lost due to the damage of seawater pumps induced by tsunami. Subsequently the severe damage to and the meltdown of the reactor core occurred, and it can be described that the third layer of defense lines in defense in depth was broken through. (4) Although the operators of the NPP made great efforts in the field to mitigate the accident consequences by taking the measures such as "alternative water injection using fire extinguishing system" and "containment vessel venting" developed before as AM measures, it was very difficult to cope with the situation in a timely and appropriate manner under severe environmental conditions such as the effects of damage induced by earthquake or tsunami and the failure of reactor buildings due to hydrogen detonation. Thus, the control and management of severe accident, i.e., the fourth layer of defense line, was also executed unsuccessfully.

Concerning the multiple barriers against radioactive materials, the independence of the barriers could not be maintained because the cooling function was lost and the accident led to core damage and meltdown, as described above, and as a result, the containment function of reactor pressure vessel and containment vessel had deteriorated significantly, resulting in the large release of radioactive materials to the environment.

Based on the lessons learned from this accident, it is required [4] that the postulation of the design basis events associated with natural hazards, such as earthquake, tsunami, etc., should be rectified and that the fourth layer of defense line be strengthened for the SA measures (prevention of occurrence and mitigation of consequence). In the past the SA measures had been provided on a voluntary basis by the AM almost exclusively as operational management, but at present the SA measures (i.e., AM measures) are positioned as regulatory requirements; and they must be thoroughly performed, improved, and strengthened, including the modification and addition of equipment. Further, in the international context [5, 6], there exists the thought that a certain range of severe accidents, including significant core damage, should be postulated as a "design extension condition" (DEC) and that some design measures should be provided against these. In particular it is important to reduce threats to the containment system and to strengthen containment capability. Overall reinforcement of defense in depth is important as well, enhancing independence between all levels of defense in depth through diversification in addition to the strengthening of each of these levels.

In the new Japanese regulatory criteria following the Fukushima Dai-ichi NPP accident, the following are required as basic policies concerning severe accidents [4].

1. Multitier protective measures, such as "prevention of core damage," "conservation of containment function," "controlled release by venting," and "suppression of diffusion of radioactive materials."
2. Basically the measures should be provided with portable equipment, and reliability should be enhanced in combination with permanent equipment.
3. Reinforcement of protective measures for spent fuel pool.
4. Strengthening of tolerance of emergency stations, enhancement of reliability and durability of means of communication, and enhancement of reliability and durability of instrumentation systems including those for the spent fuel pool. (Reinforcement of command/communication and instrumentation systems.)
5. Development of procedures, securement of personnel, implementation of training, etc., should be required, with a focus on achieving functions in an integrated manner using both hardware (equipment) and software (field operations).
6. As for measures against an intentional aircraft collision, dispersed storage and connection of portable equipment are required, and also a "specified facility for severe accident measures" should be introduced as a backup measure for enhancement of reliability.

Furthermore, in terms of severe accident prevention, reinforced diversity and independence of safety systems as well as the utilization of passive safety features are strongly desired. As for the safety system, in particular, the cooling system to remove decay heat or residual heat, it is extremely important to exclude the loss of specified functions due to any common-mode failure by enhancing the diversity of the relevant system elements, such as heat transport system, pump driving system, electric power source, ultimate heat sink, etc. For example, the requirement could be met through diversification of power source such as deployment of different emergency power sources (air-cooled diesel generators, gas turbine generators, etc.) and power supply cars, the assurance of an alternative heat sink cooled by air, and so on. Moreover, it is most essential to maintain independence among multiple systems or components by physical separation with scattered configuration, etc., so that those systems or components do not lose their functions simultaneously due to common-mode failures and dependent failures.

The passive safety feature uses as its driving source natural phenomena such as gravity, compressible gas (nitrogen, air), natural convection, evaporation, condensation, etc., and does not rely on active systems, such as pumps, which need external driving power. Utilization of passive safety features enables simplified design, since the system configuration (e.g., cooling path) can be established only by simple operation, such as valve handling, and it is not necessary to operate active components, such as AC power supplies, heating and ventilating air conditioners (HVACs), etc.

4.2 Characteristics of Fast Reactors and Basic Policy of Safety Assurance

First this section describes technological characteristics of fast reactors in comparison with LWRs in view of safety and then discusses the basic policy of safety assurance for fast reactors based on the fast reactor characteristics.

4.2.1 Safety Characteristics of Fast Reactors [7]

A sodium-cooled fast reactor (SFR) uses fast neutrons without providing neutron moderators and thereby produces an abundance of surplus neutrons by sufficiently reducing the neutron capture reactions compared to the fission reactions and thus effectively generates plutonium as new fuels and burns minor actinides. Sodium used as coolant has high heat transport capability and enables the design of low-pressure cooling systems with high performance. The safety characteristics of SFR are shown in Table 4.1 in contrast with those of LWRs.

As for the reactivity characteristics of reactor core, the LWR core is provided with an optimal (maximum) reactivity geometry by optimizing the ratio of moderator (light water) to fuel. While in the SFR core, neutron moderator is not used, fuel rods are separately placed in an appropriate manner for ensuring heat removal by sodium, and eventually, the core is not arranged in the most reactive configuration. In addition, positive reactivity can be inserted if sodium voids occur in the central part of the reactor core. Should sodium boil there, the reactor power would increase to cause fuel damage and melting which may lead to core compaction due to the coherent movement of molten fuel, and as a result, re-criticality may occur to release significant mechanical energy. Therefore the re-criticality in the core disruptive accident (CDA) has been the key safety issue of SFR from the beginning of its development history. As a result of safety research of fast reactors over the years, it has been understood with high confidence that the reactor core behavior

Table 4.1 Safety characteristics of SFR

Item	LWR	SFR
Reactivity of core	Optimal (maximum) reactive configuration	Not most reactive configuration
Coolant pressure	High	Low (ordinary pressure)
Margin for coolant boiling	Low (or zero)	About 350 °C
Natural circulation capability	Low	High
Ultimate heat sink	Sea (river, lake)	Atmosphere (air)
Chemical reactivity of coolant	Low	High

following fuel melting could result in fuel dispersal rather than fuel compaction and, accordingly, could lead to reactivity reduction. The reliability of the reactor shutdown system is important for enhanced capability of preventing core damage by preventing sodium boiling and appropriately coping with abnormal reactivity behavior.

Characteristics of sodium used as coolant include low pressure, quite large margin to sodium boiling, excellent compatibility with metallic materials and high tolerance against corrosion and erosion (little concern about stress corrosion cracking (SSC) and flow-assisted corrosion), and so on. Since the sodium coolant is single phase and low pressure, the possibility of piping break is low compared with the LWR, and even in case of coolant leakage, there is no flashing. Therefore it is possible to ensure the coolant inventory required for reactor core cooling by maintaining the reactor coolant level with provisions of passive components, such as the guard vessel. (The SFR does not require coolant injection-type ECCSs, which are required for the loss of coolant accident (LOCA) of the LWRs, as shown in Fig. 4.6, and the SFR does not need any equipment or operation to reduce the system pressure.)

Furthermore, the ultimate heat sink of decay heat for the SFR is atmosphere (air) in most cases, and air coolers are installed. The heat flux from the higher temperature fluid (liquid) to the lower temperature fluid (air) at a heat transfer tube of air cooler is given by the product of overall heat transmission coefficient and the temperature difference between both fluids. The overall heat transmission coefficient is determined by the heat transfer coefficients between each fluid and the heat transfer tube wall and the ratio of thickness to heat conductivity of the wall. Since the heat transfer between air and metal is overwhelmingly inferior compared with that between liquid and metal, the overall heat transmission coefficient is largely dominated by the heat transfer coefficient of the air side, regardless of the liquid coolant (sodium or water) of the higher temperature fluid. On the other hand, in the case of SFRs, the temperature difference between the liquid sodium and air can be raised sufficiently, and thus the decrease in the overall heat transmission coefficient due to the use of air with low heat transfer capability can be significantly compensated by the large temperature difference. Therefore, it is possible to suppress the necessary heat transfer area to a reasonable level while ensuring an appropriate heat flux. (In the case of LWRs, the temperature difference between the water and air is at most several tens of degrees centigrade since the water used for decay heat removal would not be pressurized. Consequently, air-cooled type equipment like those for SFRs would have to be unrealistically large. It is likely that an air cooling method using the latent heat of evaporation of water, which is similar to the cooling tower type equipment used in the USA or European plants, would be necessary.) Decay heat in the reactor core is cooled by the primary sodium and transported to the secondary sodium through intermediate heat exchangers (IHXs), and then the heat in the secondary sodium is cooled and released to the atmosphere by air coolers. This heat transport system is relatively simple as compared with that of LWRs transporting decay heat to the ultimate heat sink of sea (e.g., RHR, reactor auxiliary cooling system, reactor auxiliary seawater cooling system). Moreover, since the sodium coolant is liquid in a single phase and the temperature difference between the inlet and outlet of reactor vessel or heat exchangers can be raised significantly, it is possible to transport decay heat to the atmosphere by driving all cooling fluids (sodium and air) with natural circulation, on the condition of appropriately providing the difference between the heat transfer central points (i.e., difference in height).

On the other hand, the chemical reactivity of sodium coolant is quite high. Should sodium make contact with air and water, chemical reaction would occur. Thus, the design consideration has to prevent the occurrence of such reactions and mitigate their consequences, so that the chemical reaction could not affect the fundamental safety functions. It is important to provide measures against the burning of leaked sodium and the sodium-water reaction during heat transfer tube failure in the steam generator (SG).

As for the characteristics of the containment system of the SFR, while the LWR uses high-pressure water and the containment boundary of the LWR can be directly pressurized by the ejected steam in case of the failure of reactor pressure boundary, the containment boundary of the SFR is hardly pressurized by sodium burning following sodium leakage because the atmosphere around the reactor coolant boundary is filled with inert gas (nitrogen gas). Concerning the containment system against severe accidents, while the containment of LWRs may be affected by the potential of hydrogen generation, deflagration and detonation, steam explosion, etc., the containment of the SFR may be affected by the potential of mechanical energy release due to re-criticality during the CDA.

4.2.2 *Safety Approach of Fast Reactors*

(1) Safety assurance of sodium-cooled fast reactor (SFR) and its system design

The relationships among various elements related to safety assurance of SFRs are shown in Fig. 4.7.

The basic strategy for safety assurance of fast reactors is based on the concept of defense in depth (DiD), and the functions required to realize safety are identified as the safety functions. In the classification of importance of safety functions, structures, systems, and components (SSCs) bearing the safety functions are classified into two categories according to the nature of their safety functions: abnormality prevention systems (PSs) and abnormality mitigation systems (MSs). SSCs falling into PS and MS categories are further classified into classes 1, 2, and 3 according to the importance of their safety functions. The classification of importance associated with seismic safety is also developed to be basically consistent with the classification of importance of safety functions. The PSs are SSCs whose failure could cause abnormal conditions at nuclear reactor facilities, thereby leading to undue radiation exposure to the public or site personnel. They include the reactor core and fuel; the reactor coolant boundary, such as reactor vessel; the nuclear steam supply system

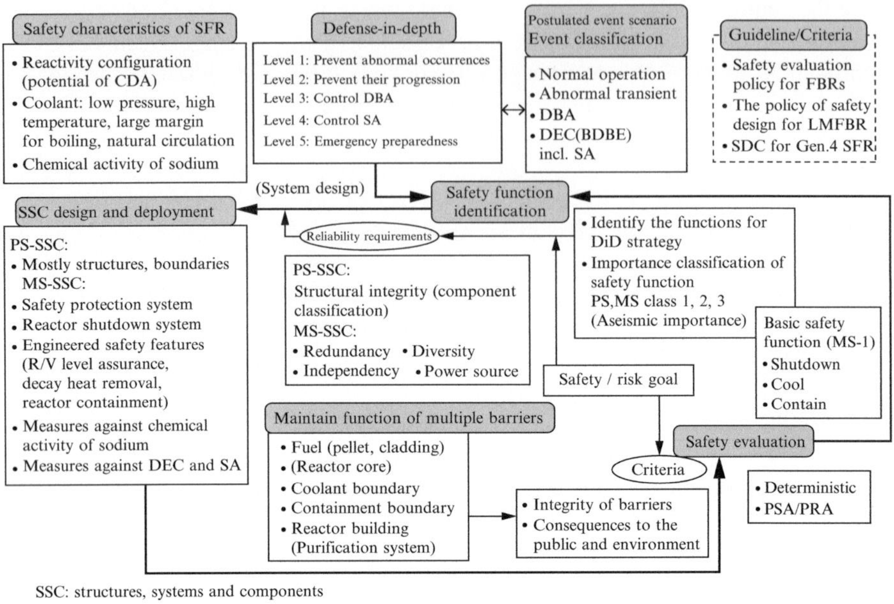

Fig. 4.7 Basic policy for safety assurance of SFRs

(NSSS); and the balance of plant (BOP) or power conversion system (PCS). The MSs are SSCs whose function is to prevent, in case of abnormal conditions at nuclear reactor facilities, an escalation of such conditions or put such conditions under control immediately and thereby prevent or mitigate possible undue radiation exposure to the public or site personnel. They perform functions which are added for safety assurance, in addition to the original functions required for fast reactors. They fall into the category of so-called safety features and include the safety protection system, the reactor shutdown system (RSS), the engineered safety features (ESF) for decay heat removal and the containment functions, measures against sodium leakage or sodium-water reaction, etc. The SSCs bearing the fundamental safety functions (shutdown, cool, contain) are categorized into the most important class of MS (i.e., MS-1).

Reliability requirements are imposed, according to the classification of importance of safety functions, on the SSCs bearing the function. Major SSCs bearing PS functions are structure or boundaries, and they are appropriately categorized into the component classification. Thus, the duly commensurate level of structural integrity can be maintained. Structural integrity is evaluated and confirmed based on the postulation of thermal loads and the number of cycles, relying on the condition equivalent to the classification of postulated events for safety evaluation (i.e., thermal transient evaluation). In addition to the cyclic thermal loads, all the loads postulated during reactor operation such as seismic forces and self-weights are taken into consideration in the evaluation of the structural integrity of SSCs. The design and evaluation are made for core fuels, in a similar way, to meet the acceptable design limits which should be established so that the fuel claddings do not fail and the fuels can be reused during the "normal operation" and "abnormal transient" specified in the event classification. On the other hand, as for SSCs bearing the MS functions, duly high reliability is required according to the importance of functions, and hence the redundancy, diversity, and independence of systems and components as well as the reliability of associated supporting power source should be prescribed as requirements. Of course the part of passive boundary constituting the MS is required to keep the structural integrity similarly to the SSCs bearing the PS functions. (Application of the event classification and component classification is different from the case of PS because the SSCs of MS are activated and operated following the occurrence of abnormal conditions.) Such policy for assurance of structural integrity is described in Chap. 5.

It is essentially important to maintain the performance and integrity of multiple barriers against radioactive materials. Fuel claddings and the reactor coolant boundary are classified into the category of SSCs bearing PS functions, while the reactor containment boundary and the reactor building, including the annulus recirculation and exhaust equipment of the emergency gas treatment system, are classified into the category of SSCs bearing MS function. It should be kept in mind that the fundamental safety functions of "shutdown" and "cool" play definitely vital roles in keeping the integrity of these barriers, and the loss of these safety functions would cause the failure of multiple barriers, resulting in a large release of radioactive materials. (The integrity of fuels and coolant boundary is assured on the premise of shutting down the reactor as needed by activation of RSS during the associated abnormal transients.)

(2) Basic policy of safety design for fast reactors

This description is made based on the philosophy of regulatory guides developed by the former Nuclear Safety Commission (NSC) and actual licensing performance or experiences concerning the fast reactors. First, safety approach and basic safety design of SFRs are discussed, mainly referring to the prototype fast breeder reactor (FBR) Monju.

The safety philosophy of Monju has gradually matured from the one adopted for the experimental fast reactor Joyo. During this process, together with the consideration of evolvement of the regulatory guides for LWRs to achieve at least the same level of safety as that of LWRs, a variety of safety philosophies from overseas fast reactors (FRs) have been assimilated with an aim of enhancing the safety of Monju. The fundamental philosophy concerning the safety design and evaluation of Japanese FRs is given in the "Safety Evaluation Policy for Fast Breeder Reactors (FBRs)" [8], which was issued in November 1980 by the former NSC of Japan to give a basic guidance for the licensing of Monju. The main body of the policy describes the basic philosophy of examining the safety of FR plants and the way of applying the existing regulatory guides but did not establish specific and concrete standards.

Attachment of the Safety Evaluation Policy discusses both the safety design and safety evaluation. Safety design requires the licensing authority to pay particular attention and take into careful consideration 11 items specific to FRs, which include reactor core, fuels, chemically active sodium coolant used under low-pressure and high-temperature condition, positive void reactivity (it is important to prevent sodium boiling and cover-gas entrainment), highly reliable reactor shutdown system, reactor coolant boundary and cover-gas boundary, intermediate heat transport system (secondary cooling system), decay heat removal, containment, elevated temperature design, and aseismic design in consistency with thin-wall structure. Regarding the safety evaluation, six articles are provided, which include representative events to be evaluated (i.e., design basis events (DBEs), such as "abnormal transient during reactor operation" and "accident," examples of those events, judgment criteria, factors to be taken into account in the analysis (neutronic, thermal-hydraulic, mechanical, and chemical factors), etc.). One of the most distinguishing characteristics in this policy is Article 5, which stipulates the evaluation of beyond design basis events (BDBEs) as cited below. (The event is referred to as an "Article 5 event" and is described later in detail.)

Since the operational experience of liquid metal-cooled FRs is limited, safety evaluation should be conducted for postulated events with lower probabilities and higher consequences than those described as accidents in Article 2 above. Careful evaluation should be conducted in view of initiating events and preventive measures against accident progression to ascertain that the release of radioactive materials to the environment is appropriately limited.

Moreover, Article 6 requires siting evaluation, by postulating two kinds of accident: major accident and hypothetical accident, taking into consideration the BDBEs.

Besides, the applicant of Monju developed on its own initiative "The Policy of Safety Design for liquid metal-cooled FRs," referring to "Guides for the safety design of LWR plants" and fully considering the characteristics specific to SFR. In the application document for reactor establishment license of Monju, the applicant describes this Policy and discusses the conformance of Monju design to the requirements in the Policy [9]. The Policy consists of about 60 articles and includes the following contents:

- Design consideration on the usage of sodium as coolant
- Integrity of sodium boundary and prevention and mitigation of sodium leak
- Assurance of reactor sodium coolant
- Independence and capability of reactor shutdown system
- Intermediate heat transport system
- Decay heat removal capability
- Structural design at elevated temperature and low pressure

In the next place, the safety approach of future SFRs is discussed, mainly referring to the JSFR which is under research and development by Japan for commercialization [10].

Goals of safety and reliability for the Generation IV nuclear energy systems are as follows: (1) Gen IV nuclear energy systems will excel in operational safety and reliability, (2) Gen IV nuclear energy systems will have a very low likelihood and degree of reactor core damage, and (3) Gen IV nuclear energy systems will eliminate the need for off-site emergency response. The following are pointed out: an integrated strategy is essential, the defense-in-depth concept should be kept in system design, the design process should rely on a risk-informed approach, and modeling and simulation should play major roles in addition to mock-up testing and demonstration. Further, system improvements for safety enhancement should be based on the safety design philosophy of "built-in" during the basic design stage rather than "added on" to the finalized plant system.

As part of the IAEA safety standards, a specific safety requirement, "Safety of Nuclear Power Plant: Design" (SSR-2/1) [5], was published in 2012. This mainly requires that each of the different levels of defense line should independently achieve the effectiveness and severe accidents should be considered in design by providing the design extension conditions (DECs) including severe accidents based on the plant state classification (event category) as shown in Table 4.2. In the Generation IV International Forum (GIF), the safety design criteria (SDC) of SFR have been developed [12], taking into account the characteristics of SFRs as well as referring to the framework of IAEA standard SSR-2/1 and further incorporating lessons learned from the Fukushima Dai-ichi NPP accident.

Thus, based on the abovementioned situations, the design requirements concerning the safety of JSFR have been settled as follows:

1. Observance of fundamental principles for safety design

No significant radiation exposure risk should be imposed on the off-site public due to a range of DBEs, by providing safety measures based on the defense-in-depth philosophy.

2. Prevention of situations which require activation of off-site emergency response

Core damage should be avoided for any anticipated transient without scram (ATWS) by the provision of passive safe shutdown mechanism. Should core damage occur, the consequence of the core damage should be accommodated within a containment facility.

3. Achievement of quantitative risk goal

The frequency of occurrence of core damage should be less than 10^{-6}/reactor·year. The frequency of containment failure following the core damage should be less than 10^{-7}/reactor·year. (Note: These goals are set as two orders of magnitude

Table 4.2 Plant state classification (considered in design) – IAEA SSR-2/1 [5]

DiD level 1	DiD level 2	DiD level 3	DiD level 4
Operational states		Accident conditions	
Normal operation	Anticipated operational occurrences (AOOs)	Design basis accidents (DBAs)	Design extension conditions (DECs)
			(Including severe accidents)

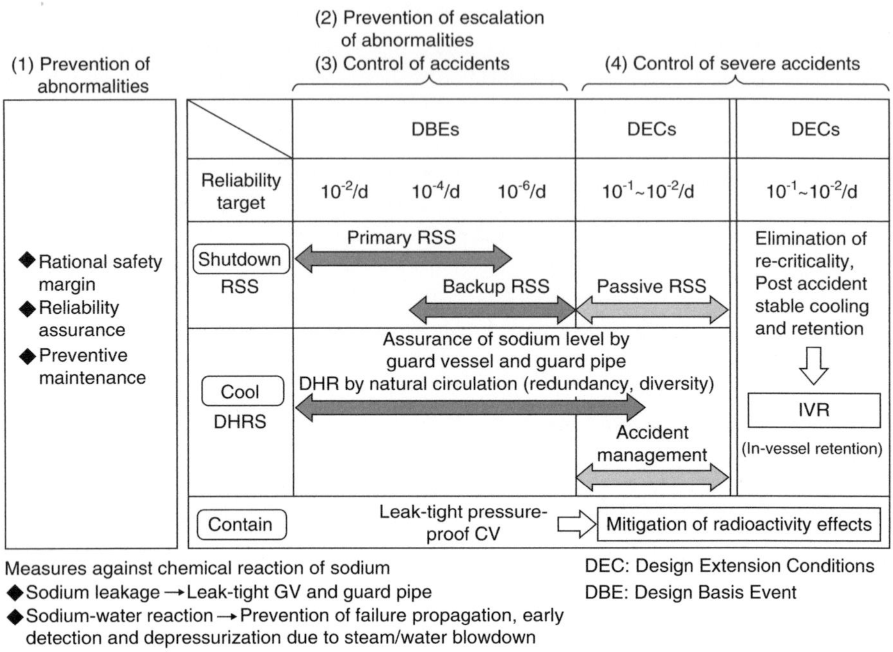

Fig. 4.8 Example of safety design concept for JSFR

higher (better) than the performance goals indicated in Sect. 4.1.1. This is because the values of these goals are defined as frequency per reactor unit rather than per site, and these goals have been developed for a future reactor.)

The basic framework of safety design approach is illustrated in Fig. 4.8 [10]. The design extension conditions (DECs, i.e., beyond DBEs) have been defined so that the prevention and mitigation measures for severe accidents may be "built in" the basic design.

The reactor shutdown system (RSS) is provided with two independent systems against DBEs: primary shutdown system and backup shutdown system, moreover with self-actuated shutdown system (SASS) against DECs. The SASS has a passive feature making use of the Curie point of an electromagnet. When the core outlet sodium temperature rises excessively and the temperature of the electromagnet holding a control rod exceeds the Curie point (corresponding to a predetermined activating temperature), then the magnet force should be naturally lost, and the control rod should be dropped into the reactor core. Thus, the SASS is able to shut down the reactor even if the safety protection system loses the function of sending the shutdown signal. As for decay heat removal during accidents, the reactor sodium level is passively maintained without sodium coolant injection by providing a double structure, i.e., covering the reactor coolant boundary with the guard vessel or outer piping, and the redundant and diversified systems are provided for decay heat removal by the passive feature of natural circulation of sodium and air. A safety function can be achieved even without any external operation or any input of signals by an inherent mechanism in the system, which is realized by the intrinsic physical and chemical properties of materials constituting the system or action of natural laws (e.g., gravity, natural convection) in the system, where such features are referred to as "passive safety." The abovementioned SASS, passive assurance of reactor sodium level, and decay heat removal by natural circulation are the passive safety features. It is noted that the "inherent safety" means the inherent characteristic that is inclined to return to the original state when the system moves on from the normal state to an abnormal state and usually refers to the "self-controllability of reactor" to suppress a reactor power change.

4.2.3 Safety Requirements and Safety Evaluation of Fast Reactors

As indicated in Fig. 4.7, safety design means an iterative process such that safety evaluation for the designed SSCs is made to validate the SSC design in light of safety goals or risk targets and safety requirements (judgment criteria/acceptance criteria) associated with event classification, and the design should be modified as necessary. In this section, safety evaluation performed in the licensing process is discussed, taking Monju as an example.

The events shown in Table 4.3 were evaluated as DBEs including the "abnormal transient during reactor operation" and the "accident." In addition, the abovementioned Article 5 events as "beyond DBE" and the accidents for siting evaluation were evaluated.

Table 4.3 Design basis events (DBEs)

(1) Abnormal transient during reactor operation	(2) Accident
Abnormal change in reactivity and power distribution in the core (e.g., control assembly withdrawal during power operation)	Reactivity increase in the reactor core (e.g., rapid withdrawal of control assembly, gas bubble passage through core)
Abnormal heat generation or removal from the reactor core (e.g., loss of off-site power, decrease in main feed-water flow)	Degradation of cooling capability of the reactor core (e.g., pump seizure, coolant leakage)
	Accidents associated with fuel handling
	Accidents associated with waste management facility
Chemical reaction of sodium (small leakage from heat transfer tubes of steam generator)	Chemical reaction of sodium (e.g., sodium leakage in primary system, heat transfer tube failure in steam generator)
	Accidents associated with reactor cover-gas system

The judgment criteria (acceptance criteria) for safety evaluation are basically the same as those for LWRs. As for the "abnormal transient during reactor operation," a postulated event shall be accommodated without causing damage to the core, and the event shall be placed under control in a condition where normal operation can be resumed. (Neither mechanical failure of cladding nor melting of any fuel pellet should occur.) As for the "accident," the core shall not be damaged considerably, and adequate coolable geometry of the core shall be maintained, and also no significant radiation exposure risk (guideline dose of 5 mSv) should be imposed on the off-site public.

The following approach was taken for the evaluation of Article 5 events (beyond DBE):

- Select the potential initiators leading to core damage (CDA).
- Consider the events as beyond the design basis, not aiming for the verification of a particular equipment design.
- Ascertain the tolerance and safety margin of overall reactor facility even against the severe events beyond DBE.
- Refer to the guideline dose for "Hypothetical Accident" in the Guide for Reactor Siting, as the criteria of judging that the release of radioactive materials to the environment is appropriately limited.
- Evaluate the realistic event sequence on a nominal basis without conservative analytical conditions (i.e., superposition of single failure assumptions or loss of off-site power and worst combination of conditions or parameters).

The following three events were evaluated as Article 5 events.

1. Local fuel failure

This event postulates the misloading of fuel pellets whose linear heat rate is higher than the normal state and local flow blockage (2/3 planar blockage) in the fuel assembly. The results of evaluation showed that the fuel pin failure is restricted locally with no possibility of the occurrence of significant core damage and the release of radioactive materials to the environment.

2. Large piping break in the primary heat transport system (PHTS) (or primary cooling system)

This event is referred to as "LOPI" (loss of piping integrity). Break size is postulated to be a size exceeding the design basis (Dt/4; D, pipe diameter; t, pipe thickness) up to double-ended guillotine. Realistic event sequence of the reactor core and cooling systems was analyzed giving appropriate credit for the function of non-safety class components. In addition, sodium burning analysis was conducted considering the spray and pool fires in order to evaluate the integrity of the primary cells filled with inert (nitrogen) atmosphere and the containment vessel. Furthermore, the amount of radioactive material released and the exposure dose were evaluated. The results of evaluation showed that the reactor core is cooled without resulting in any serious core damage or significant radiation risk to the surrounding public, while a low percent of fuels in the whole core fails.

3. Loss of reactivity control function (i.e., anticipated transient without scram (ATWS))

The postulated events include the loss of reactivity control function during decrease in primary coolant flow (ULOF: Unprotected Loss of Flow) and the loss of reactivity control function during abnormal withdrawal of control rods (UTOP: Unprotected Transient Overpower), where "unprotected" means "without scram." The results of evaluation showed that even if the events lead to CDA generating the mechanical energy (energetics), the integrity of reactor coolant boundary and containment boundary is maintained and the debris of molten fuel is accommodated to be cooled and retained in the reactor vessel, and eventually the amount of radioactive material released is appropriately limited.

The event progression scenario of the ULOF is as follows:

- In the nominal case, the event enters into the transition phase (TP*[1]) without releasing much mechanical energy.
- In the conservative case, where the energy released in the initiating phase (IP*[1]) was conservatively assumed, a maximum of approximately 330 MJ of mechanical energy was released, while the integrity of the reactor vessel can be maintained against an energy of 500 MJ.
- The energy produced by re-criticality in the TP was estimated to be less than the conservative value of 330 MJ in the abovementioned IP.
- Post-accident heat removal (PAHR) is achieved, and core materials are retained in the reactor vessel.
- The containment vessel withstands the pressure increase due to a spray fire of sodium ejected through the gaps of reactor closure head plugs, and the integrity of the containment vessel is maintained.

Further, the "Hypothetical Accident" in the siting evaluation was defined only as the source term (radiation source) enveloping all events up to the Article 5 events, not designating a particular accident. Assuming that 100% of noble gas, 10% of iodine, and 1% of plutonium of the radioactive inventory in the reactor core are placed on the containment vessel floor, the exposure dose was evaluated by taking the function of containment boundary, the purification function of annulus recirculation, and the exhaust equipment and exhaust stack into consideration.

[1]"IP" is the early phase of a core damage and meltdown process, where the wrapper tube of each fuel assembly maintains the essential integrity and molten core materials move only in the axial direction within the wrapper tubes. When the "IP" does not lead to significant power rise due to prompt criticality, gradual melting of wrapper tubes subsequently causes the mutual movement of molten core materials among fuel assemblies, eventually resulting in a condition where the reactivity changes due to the whole core motion. The subsequent process is referred to as "TP."

The result confirmed that the exposure dose falls below the guideline dose (e.g., 0.25 Sv for whole body).

In safety evaluation associated with the reactor core, it is important to estimate with high accuracy the parameters for core safety analysis, such as reactivity coefficients and control rod worth. In particular, when fuels bearing the minor actinides (MAs) have to be dealt with or variations in plutonium composition have to be taken into consideration as in the case of a reactor core designed for reduction in the toxicity of radioactive waste or actinide management, it is vital to appropriately understand how the neutron spectrum varies and how the variation affects the parameters for core safety analysis by sufficient investigation associated with the reactor physics.

While the abovementioned deterministic safety evaluation was used mainly in the licensing process, the probabilistic safety evaluation (PSA/PRA) has been applied as part of research and development of Monju in the stage of detailed design, construction, and commissioning. Thus, PSA/PRA confirmed that Monju would achieve the international level of safety goals, such as IAEAs. In addition, the risk reduction measures for the events beyond DBEs have been investigated by analyzing the dominant factors contributing to the risks, and the results have been appropriately reflected in the detailed design and operational procedures (corresponding to the development of AM measures).

For the safety evaluation of Generation IV fast reactors, in a way similarly to the requirements indicated in the "Safety of Nuclear Power Plant: Design" (SSR-2/1) [5] of IAEA, comprehensive deterministic and probabilistic safety assessments are to be implemented in the whole process of design in order to ensure that all safety requirements associated with plant design will be satisfied during all stages over the entire life of a plant.

4.3 Measures to Satisfy the Requirements of Safety Functions Specific to Fast Reactors

Measures to satisfy the requirements of safety functions specific to fast reactors are described more concretely, mainly taking Monju as an example.

4.3.1 Characteristics of Core Fuels and Associated Safety Measures

The fast reactor utilizes fast neutrons in terms of fuel breeding and uses plutonium-uranium mixed oxide (MOX) as fuel, and attention needs to be paid to the close-packed reactor core with high neutron flux, power density, and burnup conditions. Fuel claddings have been developed to be provided with strong tolerability against fast neutron irradiation, and the design of fuels has been made in consideration of creep and swelling effects due to the internal pressure of fuel cladding. In addition, neutron shield should be installed surrounding the core.

The design should consider possible deformation of the fuel assemblies to prevent the blockage of the coolant channels. It is important to prevent fuel failure due to local flow blockage caused by entrainment of foreign substances, fuel pin swelling, etc., since the reactor core is closed packed and the specific power is high. In order to prevent local fuel failures in the core from propagating and progressing to whole-core damage and meltdown, it is essential to provide fuel failure detection devices (cover-gas monitoring, delayed neutron detection, etc.) and to perform emergency reactor shutdown as appropriate.

A fast reactor core is provided with self-controllability (role of delayed neutron, inherent power suppression characteristics) as with an LWR core. Due to sodium properties (high heat conductivity, excellent heat transfer characteristics, single-phase flow) and negative reactivity coefficients caused by the Doppler and fuel expansion effects, the fast reactor core has an inherent feature of negative reactivity feedback working quickly during power transient over all operational ranges. Although the excess reactivity of the core and the reactivity variation due to burnup are small, consideration is to be given to a possibly positive sodium void reactivity in the central part of the reactor core, and hence the prevention of sodium boiling and gas bubble inflow (cover-gas entrainment) is highly important.

As already described, an important feature is that the reactor core is not in the most reactive configuration. There is a possibility that re-criticality occurs, leading to the release of mechanical energy due to reactivity increase caused by fuel damage, melting, and aggregation followed by core compaction. Such a core disruptive accident (CDA) is touched upon in the severe accident measures described later.

It is vitally important to ensure the reliability of reactor shutdown system including safety protection system to enhance the core damage prevention capability by preventing sodium boiling and appropriately coping with a reactivity anomaly.

(1) Anomaly detection

Anomalies are clearly detected by the instrumentation and control (I&C) system in an early phase. The detection of fuel failure, sodium leakage, and water leak in the steam generator (SG) is primarily important. The safety protection system activating the reactor shutdown system (RSS) and the engineered safety features (ESFs) has redundancy or diversity, independency, and a fail-safe feature.

(2) Reactor shutdown system (RSS)

Monju has two independent reactor shutdown systems: the primary shutdown system and the backup shutdown system, either of which is able to shut down the reactor and keep the subcriticality even in a low-temperature condition. The primary shutdown system performs both functions of reactivity control and emergency shutdown and consists of three fine control rods and ten coarse control rods. The backup shutdown system only performs the emergency shutdown function and

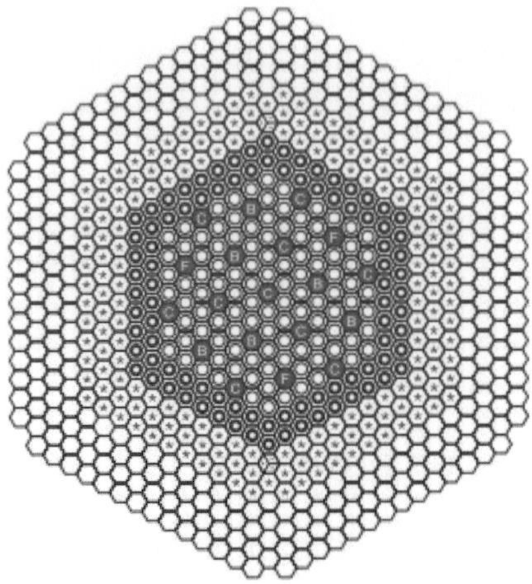

Core assemblies		Mark	Number
Core FA	Inner		108
	Outer		90
Blanket FA		*	172
CR	Fine CR	F	3
	Coarse CR	C	10
	Back-up CR	B	6
Neutron source			2
Neutron shield			316
Surveillance assemblies			8

Fig. 4.9 Reactor core configuration of Monju

comprises six backup shutdown rods (see Fig. 4.9). Different structure and acceleration methods are used for the control rod driving mechanisms of the two reactor shutdown systems, and thus diversity is provided as practicably as possible.

4.3.2 Safety Measures Associated with Usage of Sodium

This section describes various safety provisions based on adequate consideration of the use of sodium as coolant. Essential points of sodium include its excellent heat transfer capability, the availability in low-pressure conditions even at high temperatures because of its high boiling point, and the need to pay attention to its chemical reactivity. As for the last feature, it is noted that the inert gas (argon gas) is used as cover gas covering the free surface of sodium coolant in the components to avoid the chemical reaction of sodium during normal operation.

(1) Decay heat removal system (DHRS)

The SFR, as already described, dissipates decay heat to the atmosphere through the secondary sodium using air coolers. The DHRSs are categorized into some types according to the location of components designed for removing decay heat (i.e., location of the secondary sodium piping or the air cooler of the DHRS). The type of system removing decay heat directly from the reactor vessel is referred to as Direct Reactor Auxiliary Cooling System (DRACS). The type of system removing decay heat from the primary heat transport system (PHTS, i.e., primary cooling system), for example, primary-side plenum of IHX, is referred to as Primary Reactor Auxiliary Cooling System (PRACS). The type of system removing decay heat from the intermediate heat transport system (IHTS, i.e., secondary cooling

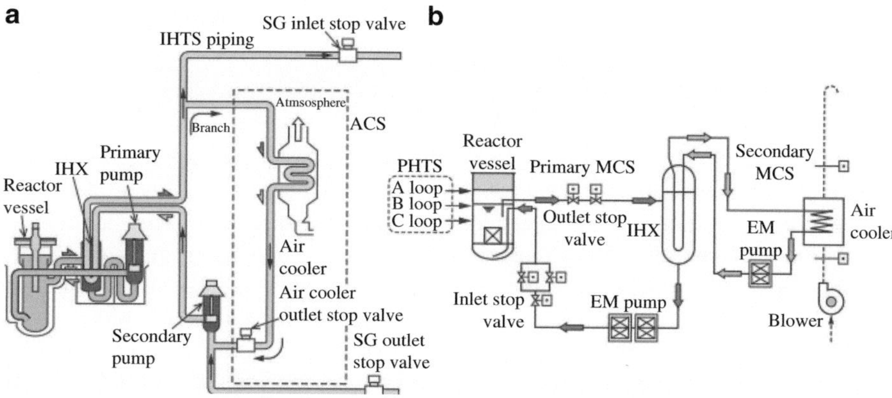

Fig. 4.10 Decay heat removal system of Monju (**a**) Auxiliary cooling system (ACS), (**b**) Maintenance cooling system (MCS)

system) is referred to as Intermediate Reactor Auxiliary Cooling System (IRACS). Moreover, there exists the type of system removing decay heat from the water/steam side of SG using air cooler, which is referred to as Steam Generator Auxiliary Heat Removal System (SGAHRS).

Monju adopted the IRACS-type DHRS, where decay heat removal after reactor shutdown from both normal operating and abnormal or accident conditions can be accomplished by the air cooler of the auxiliary cooling system (ACS) branching off from the IHTS. As shown in Fig. 4.10a, the transport path of decay heat generated at the reactor core includes the PHTS, the IHX, part of the IHTS, and the air cooler releasing the decay heat to the atmosphere, ultimate heat sink. The heat removal system (ACS) comprises three independent systems, each corresponding to the three loops of the main cooling system (PHTS and IHTS).

Even in case of the loss of off-site power, the pony motors driving the PHTS and IHTS pumps and the blowers for air coolers can be powered by the emergency power source with diesel generators and continue to circulate the sodium coolant and air for decay heat removal.

Furthermore, even in case of a further loss of emergency power leading to the total blackout, decay heat can be removed by natural circulation of sodium coolant and air. This is due to the fact that sodium coolant flows in a single-phase (liquid), the temperature difference between inlet and outlet of the reactor vessel is large, a different feature from the LWRs, and the difference in heat transfer central points (difference in height) among the reactor core, IHXs, and air coolers is adequately considered in design.

Since the ultimate heat sink for the SFR is not sea unlike in the case of the LWR, the SFR does not need to depend on the reactor auxiliary seawater cooling system and the reactor auxiliary cooling system, and therefore, the configuration of the heat transport system for decay heat removal can be relatively simple. It is noted that Monju is provided with the maintenance cooling system (MCS) for decay heat removal during maintenance which is the DRACS-type system with a small heat removal capacity (see Fig. 4.10b).

(2) Assurance of reactor coolant inventory

In the event of the leakage of primary sodium coolant, it is absolutely necessary to assure the reactor coolant inventory and maintain the emergency level of sodium in the reactor vessel for decay heat removal. The size of design basis leak due to piping failure (area of failure hole) was postulated as $Dt/4$ (D piping diameter, t piping thickness), i.e., 22 cm^2, not the double-ended guillotine like the LWR. This postulation is based on the concept in which the feature of "leak before break (LBB)" can be assured mainly because the fast reactor piping is a low-pressure system (moderate-energy fluid system as described in the Standard Review Plan of USNRC). (The justification of LBB is given later in Sect. 5.3.) A leakage of sodium coolant never leads to a rapid loss of coolant by boiling due to pressure reduction, since the system pressure during operation is low and the boiling point of sodium coolant is high. Thus, the coolant injection-type ECCS like LWRs is not required for SFRs. In order to keep the sodium level in the reactor vessel higher than both the top of reactor core (for submerging the core) and the predetermined emergency level (above the outlet nozzle of reactor vessel) to ensure sodium circulation for reactor cooling, the guard vessels are installed to hold the leaking sodium as shown in Fig. 4.11. In addition, for assurance of the reactor vessel sodium level, it is essential to trip (stop) the main motors that drive pumps which affect the balance of sodium level due to pressure loss, and the main motors are automatically tripped by the signal of a low reactor vessel sodium level in the safety protection system.

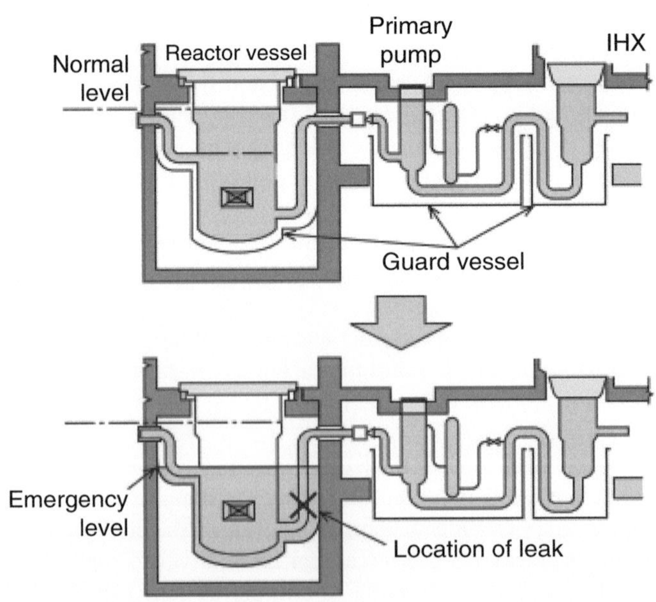

Fig. 4.11 Assurance of coolant level in reactor vessel

(3) Safety measures against sodium leakage
1. Primary cooling system

The primary main cooling system (the PHTS) is installed in the cells filled with inert gas (nitrogen) within the reactor containment vessel to suppress the sodium burning during sodium leakage. (Basically, the atmosphere around the components containing radioactive sodium is inert gas, so that sodium fire may not occur even in the event of sodium leakage.) The cells filled with nitrogen are provided with steel liners on the whole surface of floor, wall, and ceiling, and thereby the contact of leaking sodium with concrete (leading to hydrogen generation) is prevented and the leak tightness of nitrogen-filled cells is maintained. The measures for assurance of reactor sodium level and decay heat removal in the event of sodium leakage are as already described in previous items (1) and (2).

2. Secondary cooling system

The safety requirement against sodium leakage in the secondary cooling system (including the IHTS) is the assurance of isolation and independence among the three decay heat removal systems (ACS) so that an intact system will not be adversely affected and will not lead to the loss of function as a result of sodium leakage in any of the remaining systems. Since the secondary sodium is not radioactive, the cells containing the sodium components are in an air atmosphere so that inspection and maintenance can be performed during reactor operation. In order to mitigate the consequence of the burning of leaked sodium, Monju provides sodium burning suppression measures such as the transport and storage of leaked sodium, etc., and suppression measures against the diffusion of white smoke (aerosols). Further, a floor steel liner is provided to prevent the contact of leaked sodium with concrete. The measures against secondary sodium leakage have been modified based on the lessons learned from the sodium leakage accident of Monju, as shown in Fig. 4.12. Monju has implemented a series of measures for early leak detection (cell monitors) and associated operator support, improvement of HVACS, reinforcement of sodium drain function, strengthening of burning suppression function (nitrogen gas injection), etc.

(4) Measures against sodium-water reaction

The IHTS (secondary main cooling system) is provided for the SFR to prevent the consequence of sodium-water reaction due to water leak in the SG from affecting the reactor core directly. Monju is provided with various water leak detection systems that monitor the hydrogen concentration in sodium and the cover-gas pressure and detect the activation of pressure relief (rupture) disc. Further, the storage tank of reaction products, such as hydrogen and sodium hydroxide, is installed, where hydrogen is ignited to be released to the atmosphere as water. In addition, equipment for blowing down water/steam in the SG to outside the system is provided (see Fig. 4.13).

The design basis water leak was established taking into account the failure propagation of heat transfer tubes in the SG. The magnitude of initiating leak which determines the initial spike pressure is equivalent to a double-ended rupture

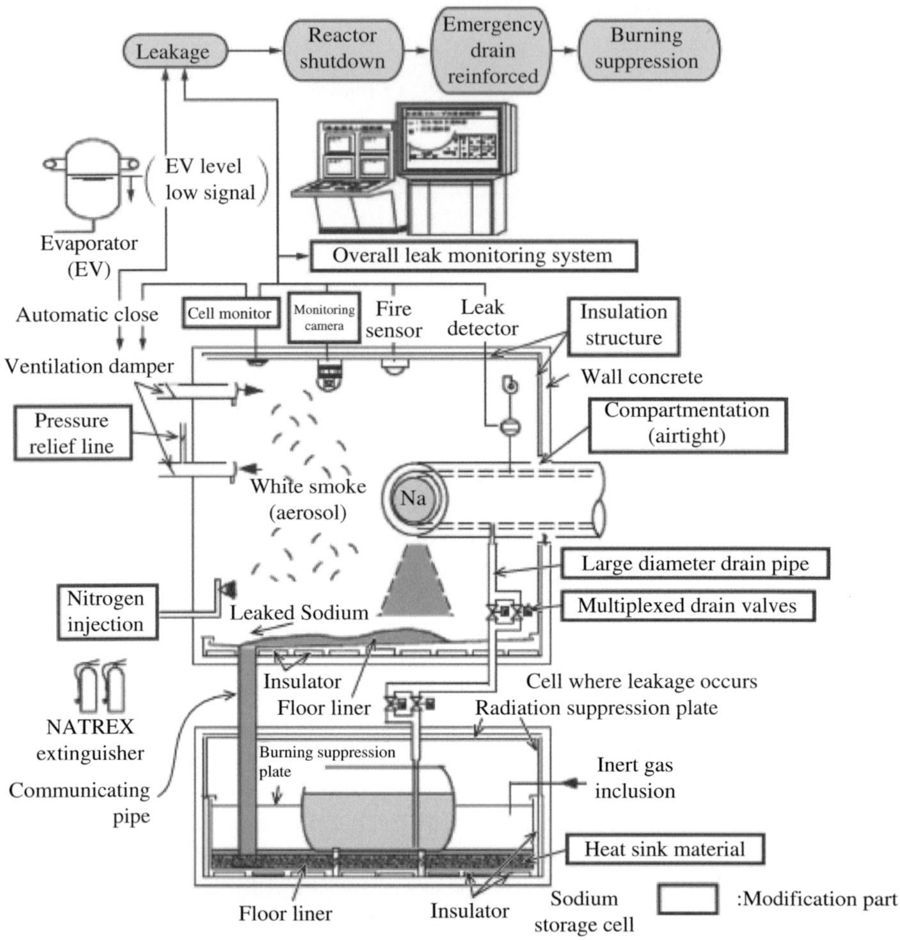

Fig. 4.12 Modification of measures against secondary sodium leakage

of one heat transfer tube. This was postulated to be followed by additional three-tube double-ended rupture. The residual pressure was determined from the magnitude of water leak equivalent to that due to four-tube double-ended rupture.

(5) Containment system

As already described, multiple barriers against the release of radioactive materials are provided in a way similar to the LWR (see Fig. 4.14).

The points clearly different from LWRs are as follows. First, the third closed barrier is formed by the combination of the reactor coolant and cover-gas boundaries, and the primary cells are provided outside the third barrier (reactor coolant boundary) as the 3+ barrier, since sodium is used as primary coolant and the liquid surface of sodium is covered by cover gas (argon gas). This latter barrier, as described previously, consists of reinforced concrete and cell steel liner preventing

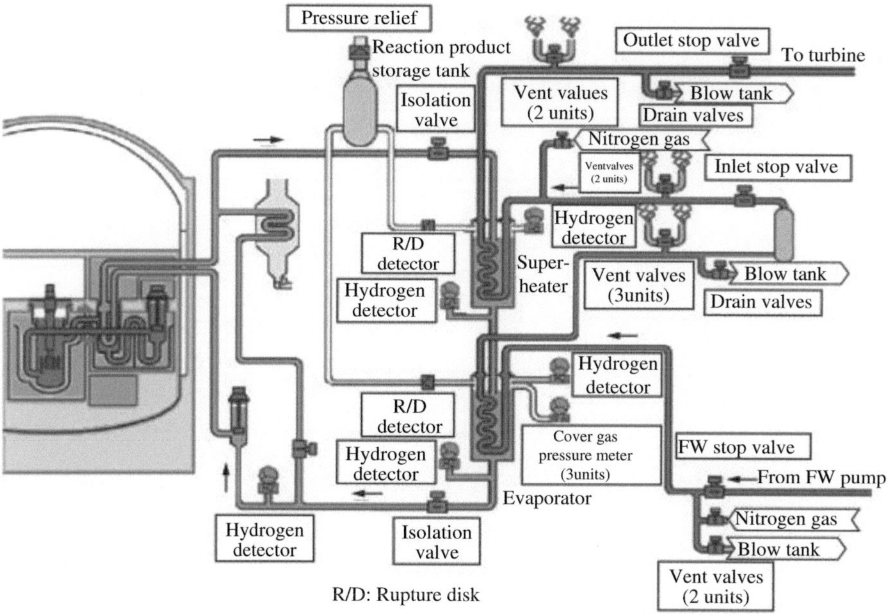

Fig. 4.13 Measures against sodium-water reaction

Fig. 4.14 Multiple barriers against release of radioactive materials

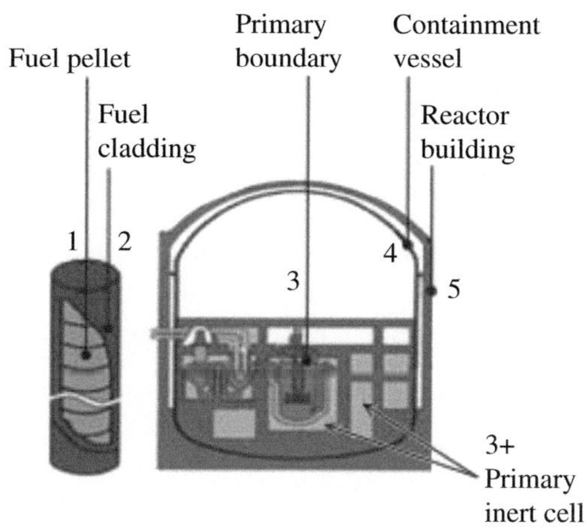

the contact of sodium with the concrete and encloses the nitrogen atmosphere in the primary cells to cope with leak and burning of the primary sodium with radioactivity. Further, the containment vessel of Monju is relatively large, i.e., 49.5 m in diameter and 80 m in height. The design basis events for the containment vessel include primary coolant leakage and earthquake, and additionally, the containment vessel is designed to be as robust as reasonably possible. Consequently, the

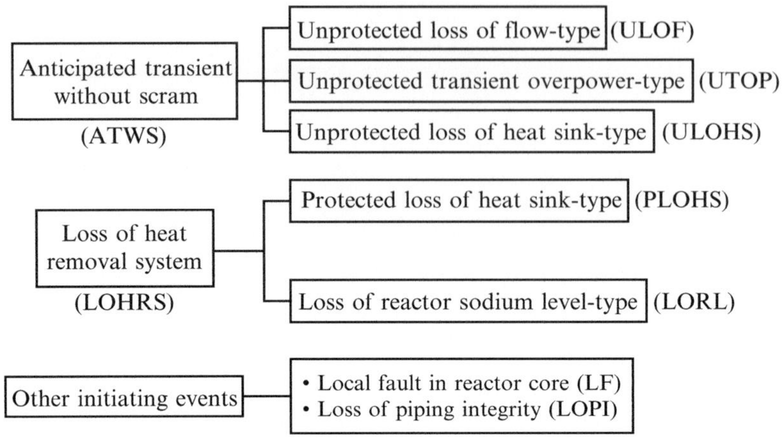

Fig. 4.15 Event category of potential initiator leading to core damage

containment vessel can withstand the loading of beyond DBEs leading to the core disruptive accident (CDA) as described later.

4.3.3 Severe Accident Measures of Fast Reactors

The potential initiating events in the fast reactors leading to core damage accidents are broadly divided into two categories: anticipated transient without scram (ATWS)-type events and loss of heat removal system (LOHRS)-type events. These are moreover subdivided into smaller categories, as shown in Fig. 4.15. In addition, other potential initiators are local faults in the reactor core (LF) and large piping break in the primary heat transport system (LOPI).

(1) Severe accident measures of Monju

As already described, taking the characteristics of fast reactor core into consideration, the mechanical energy (energetics) release due to re-criticality during CDA has been one of the most important safety issues of the fast reactor from the beginning of its development history. Energetics means that reactivity increases with core melt and compaction, thereby exceeding prompt criticality, and then the thermal energy generated by excursion of reactor power is converted into the mechanical energy which could be released and potentially cause damage to structures. It has been crucially essential to resolve this issue of energetics, further in terms of achieving at least the safety level equivalent to the LWR. In the case of Monju, the initiators (ULOF/UTOP) leading to CDA and the event progressions have been evaluated and then the design considerations have been substantially taken against CDA measures. Eventually it was confirmed that the integrity of reactor coolant boundary could be maintained against the release of mechanical energy and that the molten fuel debris can be cooled and retained within the reactor vessel (as described in the safety evaluation of Sect. 4.2.3). In the SFR, the event progression to core melt due to the inability of cooling, i.e., LOHRS, including the loss of function of decay heat removal system (the protected loss of heat sink: PLOHS) and the loss of reactor level (LORL), is significantly slow, and hence the time margin is large enough to enable recovery by operators. Therefore, the ULOF/UTOP has been treated as representative events, since the ULOF/UTOP could progress rapidly and exceed the prompt criticality in a short time.

The role of evaluating the so-called Article 5 events can be regarded as the confirmation of the following two matters in the licensing review while treating them as beyond design basis events. One matter is that the release of radioactive materials is appropriately limited without unacceptably large consequence of the CDA, and another matter is that the other potential initiators, LF and LOPI, do not lead to CDA (i.e., prevention of severe accident). As examples of design considerations against such beyond design basis events, CDA measures include strong fixing of reactor vessel plug, structures suppressing sodium ejection, lower structures of reactor core for cooling and retention of molten fuel debris, etc. Also the LOPI measures include the isolation structure between the space of guard vessel (or guard pipe) around the reactor vessel inlet piping and the space of main body of guard vessel on the side of reactor vessel, recovery path for sodium spillover, addition of rotary inertia to the MG (motor and generator) set of main pumps, isolation of cover gas in the reactor vessel, etc.

In this way, a certain range of severe accidents had been coped with by providing various measures against them much earlier than the LWRs, so Monju can be regarded as ahead of its time for reinforcement of the fourth layer defense line in the defense-in-depth concept.

Furthermore, the application study on the PSA had been carried out over the stage of construction and commissioning of Monju, and the severe accident management (AM) measures have been developed based on the knowledge obtained through the PSA study. The AM measures have been provided [11], as shown in Table 4.4, considering only internal events (excluding earthquake and tsunami) with an emphasis on phase 1 AM (i.e., prevention of severe accidents).

Furthermore, based on the lessons learned from the Fukushima Dai-ichi NPP accident, Monju has provided emergency safety measures [11], as shown in Table 4.5, in order to prevent core damage and spent fuel failure and to suppress the release of radioactive materials by assuring an electrical power source even in the loss of safety functions due to tsunami.

Monju is quite unlikely to be affected by tsunami, because the ultimate heat sink is atmosphere (air) and the heat transport performance can be accomplished by natural circulation of sodium and air thanks to the characteristics of the SFR, and the sodium components important to safety are installed at a relatively high position (at 21 m above sea level). Moreover, extremely long time margins are expected to be ensured for the implementation of safety measures. However, from the perspective of long-term plant safety, electric power sources for instrumentation to monitor accident conditions and heaters to prevent sodium from freezing would be necessary.

Table 4.4 Accident management (AM) measures of Monju

Safety function		Accident management (AM) measures
"Shutdown" reactor shutdown function		Manual trip
		Shutoff of power source for holding control rods, etc.
"Cool" reactor cooling function	Reactor sodium level assurance	Primary pump trip
		Isolation of primary argon gas system
		Primary coolant siphon break for PHTS
		Primary coolant siphon break for maintenance cooling system, etc.
	Decay heat removal	Transition to natural circulation for auxiliary cooling system (remote or on the spot)
		Emergency startup of maintenance cooling system
		Decay heat removal by using steam generator
"Contain" function of containing radioactive materials		Operation of containment isolation valves (remote or manual control)
Supporting safety function		Restoration of power source
		Assurance of power source by accommodating HVAC chillers

Table 4.5 Emergency safety measures of Monju

Safety function	Emergency safety measures
"Cool" cooling function for spent fuels	Assurance of EVST cooling (operation for natural circulation cooling and forced circulation cooling following restoration of power source)
	Assurance of cooling water for fuel pond
Supporting safety function	Assurance of power source (strengthening of monitoring function at central control room, supply of power source for EVST cooling system, etc.)
	Reinforcement of personnel and systems for emergency response
	Assurance of means of communication during an emergency
	Securement of working environment at central control room

(2) Severe accident measures of future generation IV fast reactors (SFRs)

The following approach is taken in the safety design criteria (SDC) of the Generation IV fast reactors [12].

In order to achieve the goal of eliminating the need for off-site emergency response, the Generation IV reactor should pursue safety design ensuring practical elimination of a large release of radioactive materials and strengthening of the prevention and mitigation of design extension conditions (DECs) in the fourth layer of the defense-in-depth concept. As for the design policy for DECs, provision should be made sufficiently considering the common-mode failures in common with the safety-related components for design basis accidents (DBAs), and then utilization of passive safety mechanisms should be required based on SFR characteristics. It would be essential to provide a proper balance of design provision with an appropriate combination of active and passive safety systems in order to enhance safety for the whole spectrum of events, including DBAs and DECs. In parallel, the AM measures with developed procedures should be incorporated as necessary if there is a sufficient time margin in the progress of an event. The provision for the AM measures should be built into the design to enable the implementation of the AM measures when needed.

A fast reactor, whose core is not in the most reactive configuration, has the potential of leading to CDA due to the insertion of excessive positive reactivity when sodium coolant boiling, relocation of fuel cladding, and fuel compaction are assumed to occur in the DECs. The measures for prevention and mitigation against possible conditions during such CDAs have to be incorporated into (built in) safety design. The requirement for the DEC is to provide measures to prevent core damage (hereinafter referred to as Category 1) and to assure the containment function (hereinafter referred to as Category 2). The DEC is regarded as a condition beyond design basis, for example, multiple failure of safety provisions, and several representative conditions are postulated as DECs taking into account the similarity and representativeness of those conditions. Such postulation is made based on deterministic evaluation, with reference to the information obtained by the PSA/PRA concerning the comprehensiveness of the conditions and the possibility of their occurrence in addition to operational experience with the SFR.

The DECs need to be determined for each event category of the abovementioned potential core damage initiators taking the characteristics of respective events into consideration. As for the ATWS-type event, design measures of both Category 1 and Category 2 should be provided because there is an essential difference between design measures of the two categories (i.e., reactor shutdown by passive or inherent safety feature for Category 1, reduction in mechanical energy released and cooling/retention of damaged core for Category 2, respectively) and because there is only a small time margin before the occurrence of core damage. On the other hand, as for the LOHRS-type event, the assurance of reactor sodium level and decay heat removal should be absolutely necessary to maintain core cooling regardless of the presence or absence of core damage, and there is no difference in principle between design measures of the two categories. Moreover, there is a considerably large time margin before the occurrence of core damage. Therefore, diversified measures can be taken for the LOHRS-type event according to the conditions of events. The degree of core damage depends on the time margin to core damage from the initiation of the loss of the decay heat removal function. Such approach to the LOHRS-type event can be applied in a similar way to the installation of equipment for storing spent fuel in sodium outside the reactor vessel.

An approach to the safety design against severe accident is illustrated in Fig. 4.16 and explained below.

In the DECs associated with ATWS-type events, since they could lead to core damage in a short time, the design measures for cooling/retention of damaged core materials within reactor vessel have been introduced as the Category 2 measures. By maintaining the reactor coolant boundary, involatile fission products (FPs) and

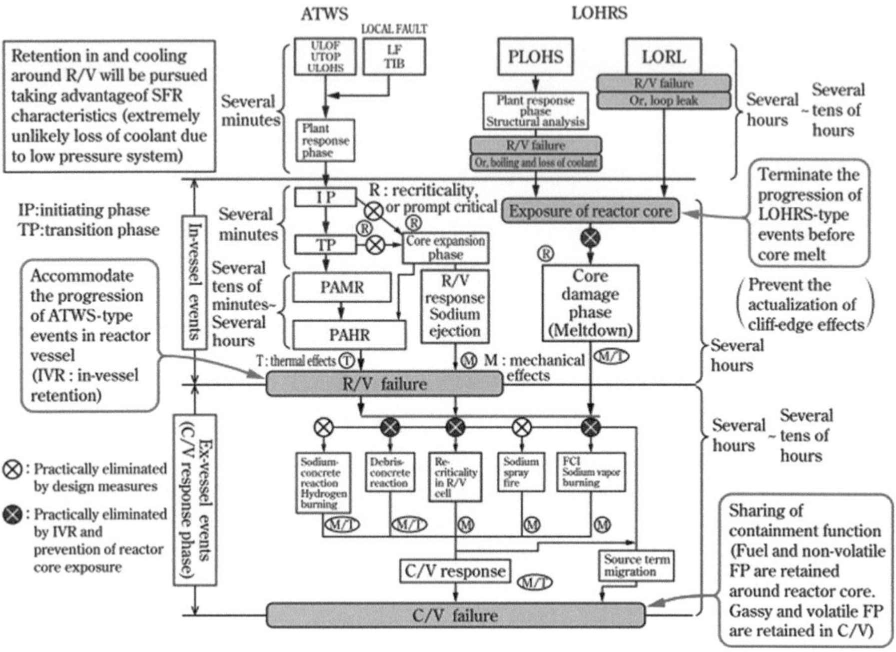

Fig. 4.16 Approach to safety design against severe accident

fuel materials can be retained within the reactor vessel, while gaseous and volatile FPs should be accommodated within the containment vessel. In order to practically eliminate the generation of mechanical energy (energetics) caused by re-criticality during CDA, which acts as an impediment to in-vessel retention, prevention measures should be built in against prompt criticality due to coolant boiling or re-criticality due to molten fuel compaction. In case of the loss of reactor level (LORL)-type event included in the LOHRS-type events, when the reactor core becomes exposed in a short time and leads to meltdown, an extremely large amount of radioactive materials could be released to the atmosphere in the containment vessel, and accordingly, it is difficult to suppress the release of radioactive materials to below a level requiring evacuation even if the containment function can be kept. As for the PLOHS-type event, the coolant boundary could fail by creep damage due to temperature rise of the whole coolant boundary to cause bypass leakage in the piping penetrating through containment vessel prior to fuel melting, and reactor vessel failure and core melt could occur while involving sodium boiling and imposing significant loads on the containment vessel. As a result, if remaining untouched, a cliff edge would be reached, leading to an overall loss of the containment functions. In order to avoid such situations, therefore, significant core damage should be practically eliminated in the LOHRS-type event to maintain the containment function.

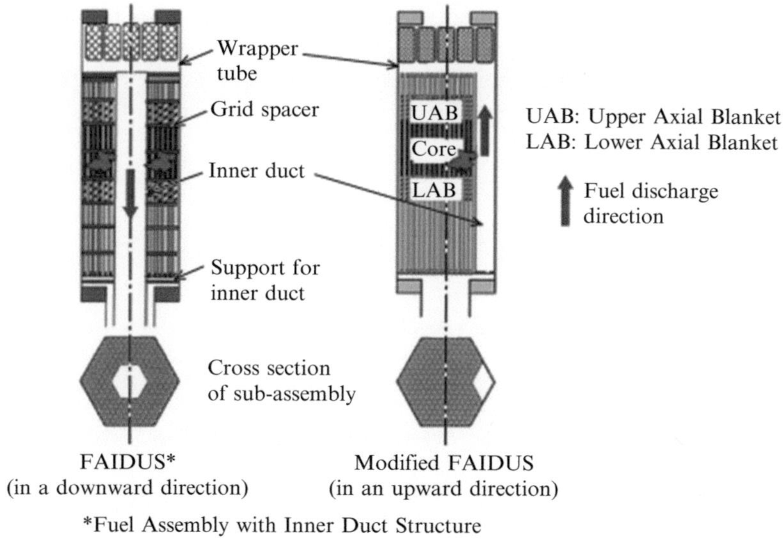

Fig. 4.17 Concept of FAIDUS

JSFR is designed to be provided with the self-actuated shutdown system (SASS) utilizing a Curie point electromagnet with a passive feature against DECs. Further, concerning the core damage caused by the ATWS-type events, JSFR pursues the elimination of re-criticality (to prevent the generation of significant mechanical energy) and the cooling/retention of molten fuel debris in the reactor vessel by maintaining the integrity of the reactor vessel against CDA (IVR: in-vessel retention). As for a large fast reactor with an increased fuel inventory, the release/dispersal of molten fuel from the reactor core has to be enhanced for the elimination of re-criticality. For this purpose, introduction of the fuel assembly concepts indicated in Fig. 4.17 has been investigated [13]. This concept discharges molten fuel early by providing a duct within the fuel assembly and is referred to as FAIDUS (Fuel Assembly with Inner Duct Structure). Design study is in progress also for the installation of plates (e.g., multistage pans) to hold, cool, and retain fuel debris.

The event spectrum of severe accidents is considerably broad, and hence a key issue is to closely examine and determine how far the severe accidents should be regarded as DECs and be incorporated in design based on SFR characteristics, past experiences associated with operation and safety evaluation (licensing), and lessons learned from the Fukushima Dai-ichi NPP accident. Basically, emphasis should be placed on prevention measures for severe accident, and mitigation measures for severe accidents should be studied as a backup. It is necessary to deliberately investigate how far the mitigation measures should be provided. It is important to judge whether it is adequate or not to practically eliminate certain events (conditions) from a range of events to be considered in design, i.e., to put the certain events (conditions) into the area of residual risk. It is described [5] that "the possibility of certain conditions occurring is considered to have been practically eliminated if it is physically impossible for the conditions to occur or if the conditions can be considered with a high level of confidence to be extremely unlikely to arise." To justify this, it is necessary to demonstrate that the relevant events could not occur by carefully providing independent defense lines with diversity and robustness.

References

1. *Fundamental Safety Principles*. IAEA Safety Standard (2006)
2. Major matters discussed on the safety goal before the previous Commission Meeting. (April 3, 2013) (in Japanese), Document prepared for the 2nd NRA (Nuclear Regulation Authority) Commission meeting (April 10, 2013)
3. *Basic Safety Principles for Nuclear Power Plants*. 75-INSAG-3 Rev.1, INSAG-12, IAEA (1999)
4. Rules for the criteria of location, structure and equipment of commercialized power reactors and associated facilities (in Japanese), Nuclear Regulation Authority (2013)
5. *Safety of Nuclear Power Plants: Design*. IAEA Requirements No. SSR2/1 (2012)
6. *Safety Objectives for New Power Reactors*. WENRA (Western European Nuclear Regulator's Association) (2009)
7. R. Nakai, *Design and Assessment Approach on Advanced SFR Safety with Emphasis on Core Disruptive Accident Issue*, Proceedings of international conference on fast reactors and related fuel cycles: Challenges and opportunities FR09, Kyoto, Japan, December 7–11 (2009)
8. *Safety Evaluation Policy for Fast Breeder Reactors (FBRs)*. Regulatory Guide of the former Nuclear Safety Commission (1980)
9. *Application Document for Reactor Establishment License of Prototype Fast Breeder Reactor Monju* (in Japanese) (1980)
10. H. Yamano, S. Kubo, Y. Shimakawa, K. Fujita, T. Suzuki, K. Kurisaka, *Conceptual Design for a Large-Scale Japan Sodium-Cooled Fast Reactor (2) Safety Design and Evaluation in JSFR*. Proceedings of 2011 International Congress on Advances in Nuclear Power Plants (ICAPP 2011), Nice, France (May 2–5, 2011) No.11219
11. Document prepared for the 2nd investigation committee meeting on Monju comprehensive safety evaluation (in Japanese), Japan Atomic Energy Agency (JAEA) (2011)
12. *Investigative Study on Safety Design Criteria of Generation-IV Sodium-cooled Fast Reactor* (fiscal year 2012) (in Japanese), Special Expert Committee of Atomic Energy Society of Japan (AESJ) (2013)
13. I. Sato, et al., *Elimination of severe recriticality events in the Core Disruptive Accident of JSFR aiming at In-Vessel Retention of the core materials*, Proceedings of international conference on fast reactors and related fuel cycles: Challenges and opportunities FR09, Kyoto, Japan, December 7–11 (2009)

Chapter 5
Mind-Set Required to Ensure Structural Integrity (Design Constraints)

Naoto Kasahara and Hiroyuki Ohshima

The assurance of structural integrity is a prerequisite for the first to third layer of safety based on the defense-in-depth concept. Design should preclude abnormality in reactor structure, systems, and components and thereby prevent reactor core damage and abnormal release of radioactive material to the environment for any events assumed within the design basis. In addition, the assurance of structural integrity is an essential condition for the maintenance of plant reliability and economic competitiveness. For this reason, plant design should prevent significant damage. There may be differences in the types of components that are regarded as important and in how the margin of safety is set between the above two design perspectives; however, it is fundamental for both to provide a conservative design with an appropriate margin to the ultimate strength.

These requirements provide the constraints on operating temperature, material type, geometry, and various other parameters in designing plant components. This chapter introduces the mind-set required to ensure the structural integrity of nuclear reactor facilities and then describes the features of FR loading conditions. Such conditions include high temperature and low pressure and the failure modes that are assumed based on these features.

Following these, measures to ensure structural integrity peculiar to FRs which include the high-temperature strength against thermal stress and sufficient seismic resistance of thin-walled structure, as well as the resulting design constraints, are described.

N. Kasahara (✉)
The University of Tokyo, Tokyo, Japan
e-mail: kasahara@n.t.u-tokyo.ac.jp

H. Ohshima
Japan Atomic Energy Agency, Ibaraki, Japan
e-mail: ohshima.hiroyuki@jaea.go.jp

N. Kasahara (ed.), *Fast Reactor System Design*, An Advanced Course in Nuclear Engineering 8, DOI 10.1007/978-981-10-2821-2_5

Some of these measures require new functions in design. Such measures are thermal stress relaxation and seismic isolation, and examples of these functions are described.

5.1 Mind-Set Required to Ensure the Structural Integrity of Reactor Facilities

5.1.1 Integrated Management in Terms of Material, Design, Manufacturing, Inspection and Maintenance

The assurance of structural integrity of a reactor facility requires integrated management in consideration of material, design, manufacturing, inspection, and maintenance (in-service inspection and repair based on the assessment), as shown in Fig. 5.1 [1]. This is due to the fact that there are various types and features of component failure modes, and effective measures depend on the main cause of failure, i.e., material, design, manufacturing, inspection, or maintenance.

Since structural design can be translated into a variety of efforts to prevent failures under assumed conditions, we consider the effectiveness of efforts, below. The first step of a design process is to assume failure scenarios and modes, that is, what types of loads are applied during operation and how components may fail due to those loads. Although attention needs to be paid to making appropriate assumptions in design based on operating experience and research and development results, separate consideration is required to address the lack of assumption and

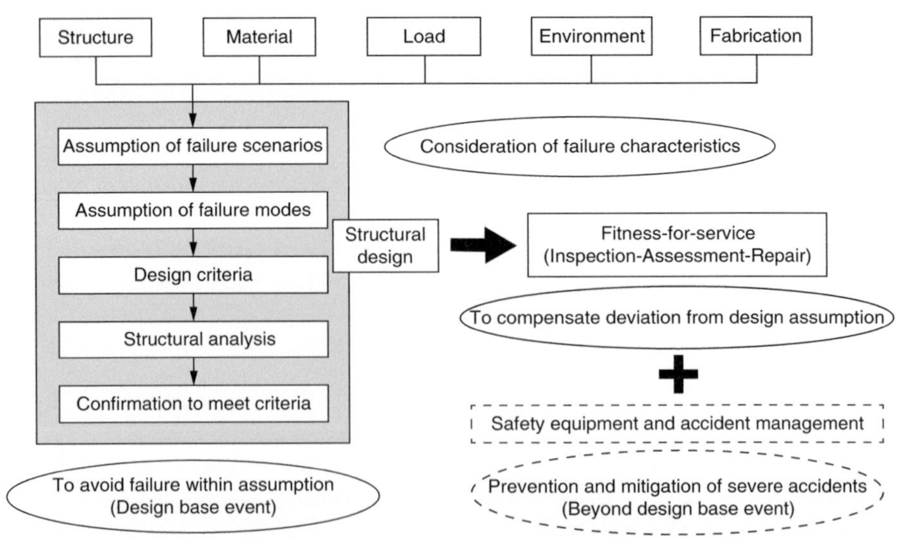

Fig. 5.1 Approach to ensuring structural integrity of reactor facilities [1]

deviation from assumed conditions. Next, design criteria for assumed failure modes are determined by limiting loads within an allowable value that has a margin for the ultimate strength. Here, for some failure modes that do not significantly depend on the magnitude of loading, measures in design are ineffective.

Maintenance activities are carried out to cope with deviation from and/or change in design assumption conditions. Safety is ensured by finding occurrence of load and changes in conditions not assumed in design, identifying signs of damage through inspection and addressing them, including through repair, if they do not lead to rapid destruction. Reliability and economic competitiveness are also maintained through early restoration.

In addition, because some failure modes are sensitive to material type and manufacturing method, care must be taken to ensure that these are appropriately selected to cope with such failure modes.

The above description clearly suggests the need for integrated management in material, design, manufacturing, inspection, and maintenance, and the combination of these factors must be optimized to ensure integrity through appropriate resource allocation. The System Based Code (SBC) concept, which determines the most reasonable combination of factors, material, design, manufacturing, inspection, and maintenance under the condition of a certain failure probability, was proposed as an optimization method to achieve the required reliability [2].

Recently, it has become necessary to change design consciousness, from "the prevention of a major (severe) accident" to "the taking of measures assuming the occurrence of a major accident" as a lesson learned from the Fukushima nuclear power plant accident [3]. Moreover, "Safety of Nuclear Power Plants" [4], an IAEA standards published in 2012, defines the design extension conditions (DECs) in the accidents beyond design basis accidents (beyond DBAs or BDBAs) and requires a conservative assessment of DBA while requiring a best estimate assessment of DEC. To arrive at the best estimation, approaches that differ from conventional design assessment using conservative factors, such as integrity assessment based on the ultimate strength, are required in the near future.

The major failure modes assumed for the reactor structure and their features are described below.

1. *Brittle fracture*: The main cause is local stress due to pressure, thermal stress, earthquake force, and other loads. Brittle fracture is sensitive to material factors, such as crystal structure and defects. Material strength is reduced at low temperatures and under neutron irradiation.
2. *Ductile fracture*: Destruction occurs due to unstable loading, such as constriction, after significant plastic deformation (however, it can be limited by plastic collapse). This type of fracture is caused by load-controlled loads, such as pressure and dead weight, and is not caused by displacement-controlled loads, such as thermal stress. Material strength is reduced due to cracking and thinning.
3. *Buckling*: An unstable phenomenon that occurs when a slender or thin-walled structure is subject to compressive stress or shear stress. The main causes are

pressure (especially external pressure), earthquake, tsunami, etc. Buckling is sensitive to defective shaping during manufacturing.

4. *Excessive deformation*: Deformation grows in a particular direction due to cyclic loading. The typical example is thermal ratchet, which causes deformation in a direction in which pressure is applied when a constant pressure and cyclic thermal stress are simultaneously applied.
5. *Fatigue*: Cracking occurs and progresses gradually due to repeated stress. (Since fatigue rarely causes rapid destruction, inspection is effective.) The main cause is local stress (sensitive to stress concentration) due to pressure, thermal stress, earthquake force, and other loads. It is difficult to predict loads due to vibration and fluid temperature change. Fatigue is sensitive to weld and surface conditions. (Construction management is required.)
6. *Corrosion*: Cracking and thinning occur by the effect of material, environment, and stress, as is represented by stress corrosion cracking. Since corrosion is greatly affected by local stress that is difficult to control, such as welding residual stress, priority is given to measures from the perspective of material and the environment.

As explained above, each failure mode has its own characteristics, and effective measures differ among the modes. Consequently, different sets of measures are taken from the perspective of material, design, manufacturing, inspection, and maintenance to prevent the occurrence of the respective failure modes, as shown in Table 5.1.

Table 5.1 Relationship between assumed failure modes and limiting elements

Failure modes	Material	Design	Fabrication	Inspection	Fitness for service
Brittle fractures	◎	○Requirement for ductility	△Delayed fracture Heat treatment	○Defect size	○Defect assessment Monitoring of null ductility transition temperature (NDTT)
Ductile fractures	△	◎		△Plate thickness	△Thinning assessment
Buckling		◎	△Tolerance		△
Excessive deformation		◎			△
Fatigue		◎	○Finishing	○Defect size	○Defect assessment monitoring of operating transients

◎ Main limiting element, ○ sub-limiting element, △ indirect limiting element

5.1.2 Structural Design Methods for Nuclear Components

This paragraph explains the main structural design effort to ensure structural integrity. There are two approaches to structural design: "design by rule" and "design by analysis" [5].

"Design by rule" is a traditional approach to design a pressure vessel that "confirms that the circumferential stress in membrane does not exceed the allowable stress using a simple standard calculation" and is expressed by the following equation:

$$S = \frac{BPD}{2t} < \frac{R}{DF} \tag{5.1}$$

The membrane stress of a cylindrical shell S calculated from the stress coefficient B, maximum operating pressure P, outer diameter D, and wall thickness t shall be limited to no greater than the allowable value determined from the material strength R and the design coefficient DF.

This approach does not provide strict requirements for the material/shape/manufacturing process. The effects of these factors are incorporated into the design coefficient in reference to knowledge accumulated on boilers and other components over years of experience in design, manufacturing, and operating to realize a simple assessment method. This is a proven approach and is applied to the design of general and nuclear plant components with the exception of main components (class 1 components).

The "design by analysis" approach is peculiar to nuclear components. It was proposed by the American Society of Mechanical Engineers (ASME) in its structural design codes for nuclear components in the 1960s [5, 6]. A brief description of this approach is "to assume all possible failure modes and prepare corresponding design standards to allow for a detailed assessment of structural integrity by analysis." Its framework consists of the following four steps:

1. Assumption of failure mode

Define the failure modes that should be taken into account in design. In the "Standards for Nuclear Power Generation Equipment – Design and Construction Standards" issued by the Japan Society of Mechanical Engineers (JSME) for light-water reactors, the ductile fracture, buckling, excessive deformation, and fatigue that are shown in Table 5.1 are assumed. Here, brittle fracture is eliminated by selecting ductile material. In addition to the above, possible failure modes at high temperatures where creep phenomena will occur are assumed in the standards for FRs [7, 8].

2. Formulation of failure mechanisms for each failure mode

The factors (stress, strain, etc.) that govern the relevant failure mechanism for each failure mode are selected to formulate the failure limit based on strength theory or other methods.

3. Calculation of stress, strain, and other parameters by analysis

The methods of calculating the factors that govern the relevant failure mechanism by stress (strain) analysis are specified. Basically, elastic analysis is used for the stress (strain) analysis.

4. Determination of limiting values of stress, strain, and other parameters

The limiting values of stress, strain, etc., are determined by combining the material strength standards based on material strength data and design coefficients that take into account the significance (component classification) and the frequency of occurrence of loading (service conditions). In this regard, attention should be paid to the fact that there are different perspectives on the significance between safety design in which the relative significance of various safety functions is determined by safety concerns and structural design (Table 5.2) in which significance is determined to ensure the structural integrity of the relevant components under assumed service conditions. The frequency of occurrence of loading is classified by service conditions, as listed in Table 5.3.

When applying the "design by rule" approach, walls become thicker with the increase in pressure and diameter. Thicker structure is disadvantageous when thermal stress is applied because it constrains thermal expansion deformation; however, this disadvantage cannot be assessed by Eq. (5.1). This is an unreasonable aspect of the "design by rule" approach. Since the "design by analysis" approach allows for a more reasonable and detailed assessment, it is applied to the main nuclear components (class 1 components). On the other hand, since there is a risk of

Table 5.2 Component classification based on safety significance

Class 1 components	Components forming the reactor coolant boundary (e.g., reactor vessel, IHX, primary main pipe)
Class 2 components	Components forming the reactor cover gas boundary, safety systems, etc. (e.g., secondary main pipe, guard vessel, primary cover gas system)
Components of less safety significance	Class 3 components, support structure, core support structure (e.g., steam generator, secondary cover gas system)
Class MC vessels	Reactor containment vessel

Table 5.3 Condition classification based on the frequency of occurrence of loading

Service condition I	Normal operating conditions of reactor facility
Service condition II	Conditions deviated from normal operating conditions due to a single failure of component, misoperation, or other causes
Service condition III	Conditions requiring emergency shutdown due to a failure or abnormality of reactor facility
Service condition IV	Conditions where an abnormal situation assumed in reactor safety design occurs
Testing condition	Conditions where pressure exceeding the maximum service pressure is applied during a pressure test

failure when the assumptions of service conditions or loading patterns are not conservative, consideration is given to ensuring structural integrity in combination with severe requirements for the selection and maintenance of material and manufacturing method.

5.2 Features of Structural Design Conditions of FRs

5.2.1 Loading Conditions Specific to the Use of Sodium

Since the boiling point of the liquid metal sodium used as FR coolant is approximately 880 °C, system pressurization is not required at temperatures less than this. Therefore, the operating temperature tends to be determined at high temperatures that allow a better thermal efficiency as long as core safety and structural integrity are ensured (normally the reactor outlet temperature of 500–550 °C) [9] (Fig. 5.2).

For this reason, the structural design specifications of FRs differ greatly from those of light-water reactors, as shown in Table 5.4 [10]. In short, an FR plant can be described as a high-temperature low-pressure system with the following features: First, since the operating temperature is high, it is necessary to take into consideration the reduction in material strength due to creep and other phenomena. Second, the main load is thermal stress due to the fact that the large change in coolant temperatures occurring during the change in plant operating conditions is easily transmitted to structural material because of the large temperature difference in a plant [9] (Fig. 5.3) and the specific thermal characteristics of sodium (much larger thermal conductivity and smaller heat capacity (the product of density and specific heat) compared with water). Since the temperature gradient in a structure (e.g., in a through-thickness direction) will cause thermal stress, a thin-walled structure is more advantageous in reducing thermal stress. Third, the stress caused by internal pressure is not significant because the internal pressure exerted on vessels and piping is only 1 MPa even at high-pressure portions due to the fact that sodium (the boiling point is approximately 880 °C) does not require

Fig. 5.2 Reactor outlet coolant temperature

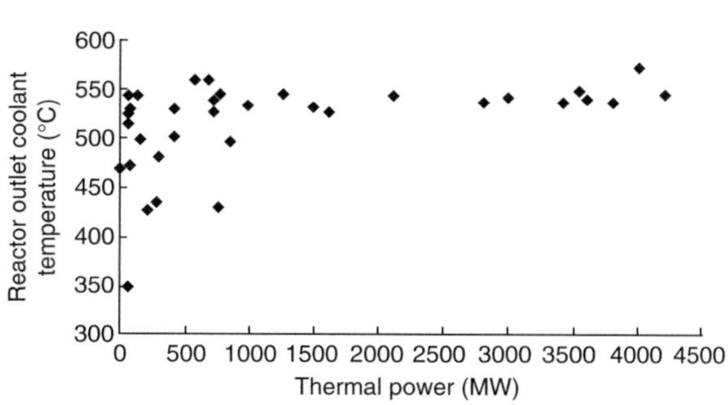

Table 5.4 Comparison of design specifications between LWR and FR

	LWR	FR (sodium-cooled reactor)
Reactor outlet temperature	PWR: approx. 325 °C,	Monju: 529 °C,
	BWR: approx. 286 °C	Demonstration reactor: approx. 550 °C
Inlet/outlet temperature difference	PWR: approx. 35 °C ,	Monju: 132 °C,
	BWR: approx. 70 °C	Demonstration reactor: approx. 150 °C
Operating pressure	PWR: approx. 16 MPa,	Not exceeding 1 MPa
	BWR: approx. 7 MPa	
Reactor vessel shell wall thickness (radius/wall thickness ratio: R/t)	PWR: approx. 220 mm, $R/t \fallingdotseq 10$	Monju: 50 mm, $R/t \fallingdotseq 70$
	BWR: approx. 160 mm, $R/t \fallingdotseq 20$	Demonstration reactor: 50 mm, $R/t \fallingdotseq 100$

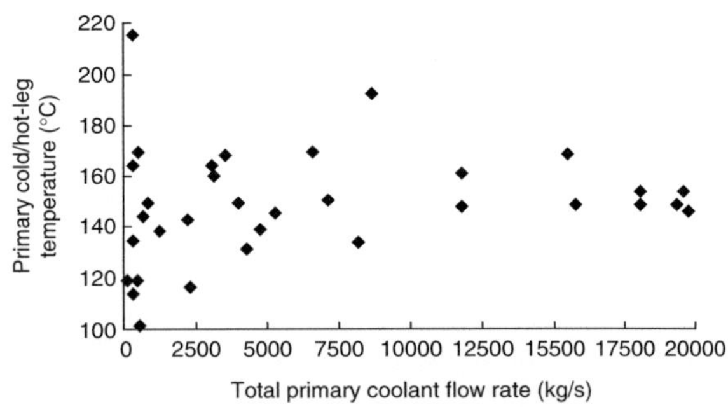

Fig. 5.3 Primary cold-/hot-leg temperature difference

pressurization. Fourth, since the system pressure is low, a thin-walled structure, which is advantageous for thermal stress, can be adopted [10].

These features that differ from those of light-water reactors lead to differences in loading conditions, and the characteristic FR design specifications are shown in Fig. 5.4. The dominant load for light-water reactors is the internal pressure, a load-controlled stress, while that for FRs is the thermal load, a displacement-controlled stress.

When load-controlled (primary) stress or displacement-controlled (secondary) stress is applied to a structure, it undergoes an elastic plastic response. As an example of load-controlled stress, let's consider a bar with an increasing force applied at its end, as shown in the left side of Fig. 5.4. Axial normal stress occurs in the bar, which is determined by an equilibrium equation with the external force. Strain increases proportionally to stress according to the theory of elasticity and rapidly increases due to plastic deformation when the stress reaches the yield stress.

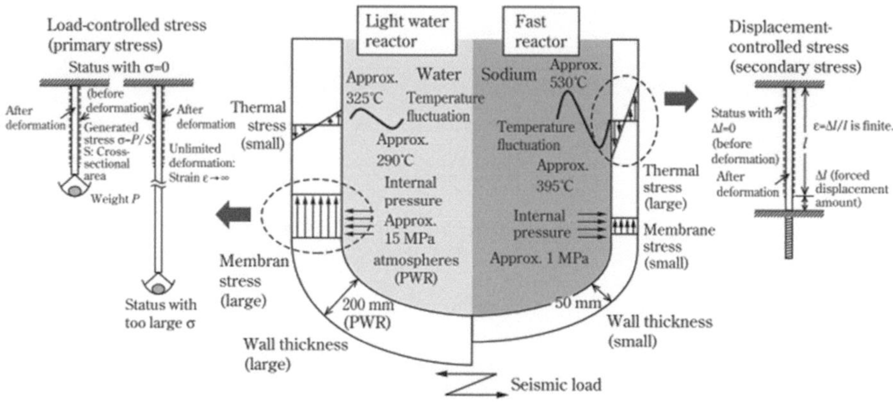

Fig. 5.4 Comparison of loading conditions between LWR and FR

As an example of displacement-controlled stress, let's consider a bar with a forced displacement applied at its end, as depicted in the right side of Fig. 5.4. The strain in the bar is determined by geometric compatibility according to the amount of forced displacement and increases at a constant rate. The stress also increases in proportion to the elastic strain. Until the stress reaches the yield stress, the strain remains elastic and is therefore proportional to the stress. After yielding, the growth rate of stress starts decreasing due to the presence of plastic strain. Therefore, displacement-controlled stress hardly leads to rapid failure.

Since the dominant loads differ from each other from the perspective of a structural mechanics, the failure modes caused by the respective stresses are also different [11]. Therefore, both the structural design assessment method and the resulting constraints for FRs are unique to the reactor type. More details about this are given in the later sections.

5.2.2 Types and Prediction of Thermal Loads

1. Types of thermal loads

The types of thermal loads that are dominant in FR environments and prediction methods are explained. Thermal loads occurring in an FR plant are classified into two types: those occurring during normal operating conditions (during power operation) and those occurring during transients, such as reactor scram. Typical examples are shown in Fig. 5.5 [11]. In addition to the stress caused by thermal expansion of piping, the typical thermal loads that occur during normal operation (power operation) include (a) thermal striping (temperature fluctuation) phenomenon and (b) the thermal load caused by axial temperature distribution near the free surface of coolant in the reactor vessel. The typical thermal loads that occur during transients include (c) thermal stratification phenomenon, (d) hoop force due to the

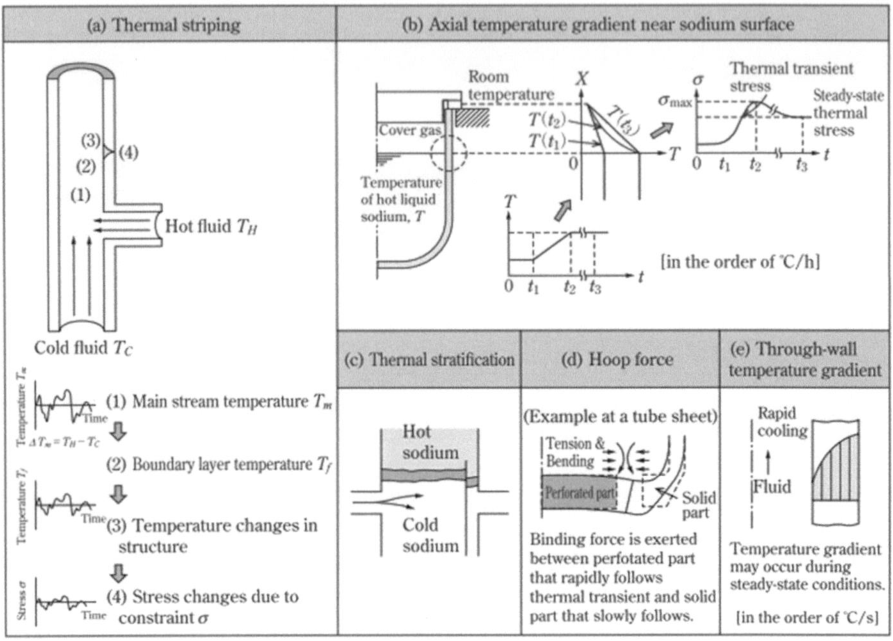

Fig. 5.5 Representative thermal loads in fast-reactor plants

difference in temperature response of structural members, and (e) thermal shock associated with through-thickness temperature gradient generated during a rapid change in fluid temperature. Generally speaking, the effects of thermal transients on structural material in FRs are more severe than those in LWRs due to the fact that the temperature differences that generate thermal loads are relatively larger (Table 5.4), the sodium coolant has approximately a 100 times the heat transfer capacity of the water coolant used in LWRs, and that the structural material, stainless steel, has relatively lower thermal conductivity while having a large thermal expansion rate. For this reason, structural and thermal hydraulic design seeks to mitigate thermal loads to the fullest extent possible, and full consideration of this point is given to determining the operation procedures. In design, it is necessary to accurately predict the changes in fluid temperature during transient operations that determine the thermal loading conditions for structures.

2. Prediction of thermal loads

In order to confirm the structural integrity of parts where thermal loads occur, thermal transient strength tests using specimens cut from the concerned parts and thermal stress analysis are conducted. The thermal loads (stresses) shown in Fig. 5.5 are caused by the temperature change in the structure that is transferred from the fluid temperature fluctuation via structure surface. To predict thermal loads, therefore, it is important to accurately predict not only the structural stresses in concerned parts but also the temperature transfer process from fluid to structure,

Table 5.5 Main dimensionless quantities used for determining test conditions

Symbol	Name	Remarks
Bi	Biot number	Same form as Nu (note: thermal conductivity of solid is used)
Fr	Froude number	Effects of inertial force and buoyancy
Gr	Grashof number	Effects of buoyancy and viscous force
Nu	Nusselt number	Effect of heat transfer rate
Pe	Peclet number	Effects of convective heat transport and heat conduction ($=Re \cdot Pr$)
Pr	Prandtl number	Effect of working fluid (thermal diffusivity)
Re	Reynolds number	Effects of inertial force and viscous force
Ri	Richardson number	Effects of buoyancy and inertial force ($= Gr/Re^2$)
St	Strouhal number	Fluctuation (oscillation) frequency

as well as to understand the mechanism of the occurrence of temperature fluctuation in fluid. The most certain method is to confirm the concerned phenomenon by testing using (full-scale equivalent) equipment that accurately simulates the actual system. In general, however, since it is often difficult to conduct such large-scale testing, scaled-down testing focusing on a region where severe thermal loads occur is usually performed to understand the status of fluid temperature change and thermal loads in structure under actual system conditions to confirm the effects of these phenomena. Therefore, full attention needs to be paid to the design of test equipment and the determination of test conditions to reproduce physical phenomena of interest. To achieve this, it is necessary to select the physical quantities that control the concerned phenomena, identify the relevant dimensionless quantities using dimensional analysis, and to determine the test conditions based on the similarity rule to the actual system conditions. Table 5.5 shows the dimensionless quantities that are used relatively often to determine testing conditions or reduce measured data [12]. However, since thermal hydraulic phenomena in a nuclear reactor are complex, such as the flow rate (flow velocity) condition, for example, more than one condition, such as the consistent Reynolds (*Re*) number condition, the consistent flow velocity condition and the consistent Richardson (*Ri*) number condition can be determined. It is important to determine appropriate conditions by identifying dominant factors with regard to the similarity between the test equipment and the actual system. When it is not possible to determine a representative quantity, it is necessary to conduct experiments under more than one condition to carefully analyze the observed physical phenomena.

Recently, moreover, progress in parallel computing technology, advances in numerical simulation technology, and remarkable progress in computer processing capabilities have made it relatively easy to estimate not only time-averaged quantities by steady-state analysis but also instantaneous states of a thermal hydraulic field even in a large-scale system. Previously, as the first step, a thermal hydraulic simulation was performed exclusively for a fluid field under the adiabatic condition on a wall surface to obtain fluid temperature information near the wall; and then, based on the fluid temperature information (numerical simulation results), the temperature profile in structure (assuming a temporally and spatially constant profile although this is not so in reality) was calculated taking the heat transfer coefficient into account. Recently, however, it has become possible to comprehensively analyze both fluid and structure regions in consideration of heat transfer between fluid and structure (conjugate heat transfer) to estimate temperature fields within fluid and structures that change continuously. It has also become possible to estimate thermal stress by coupling a structural analysis code. By applying numerical simulation methods, which are validated through the abovementioned scaled-down testing, to actual system analyses, it has become possible to predict physical phenomena that occur in the actual system. Recently, more than a few troubles caused by the lack of knowledge and experience have been found in an improved environment in which it is relatively easy to use a numerical simulation. For this reason, the importance of rulemaking to ensure the reliability of numerical simulation codes has been recognized worldwide, and the guidelines that provide the procedures to be followed in verification (i.e., a process to confirm that an analysis code is mathematically correct) and validation (i.e., a process to confirm that an analyzed result is physically reasonable), respectively, were issued by the American Society of Mechanical Engineers (ASME) [13, 14]. For the future, analytical assessment of an actual system requires a verification and validation (V&V) process that follows guidelines such as the ones above in advance.

As an example of predictive assessment of the thermal load obtained through combination with experiments and numerical simulations, an assessment result of the thermal striping phenomenon (high-cycle thermal fatigue phenomenon occurring in the lower part of the core instrumentation plate (CIP)) in the large-scale sodium-cooled FR that is currently under development in Japan (JFSR) [15] is shown.

Figure 5.6 shows a schematic of thermal striping phenomenon occurring in the lower part of upper in-core structure (UIS) of the JSFR [16]. Thermal striping occurs when hot sodium flowing out of fuel subassembly outlets and cold sodium flowing out of blanket subassembly outlets or control-rod channels are mixed. The concerned region for thermal stress assessment is a perforated plate called CIP located at the bottom of the UIS and is within a circle with a radius of approximately 40–50 cm with the center at a control-rod channel or one of the outermost fuel subassemblies in the core (note that the diameter of a fuel subassembly outlet is approximately 10 cm).

Accurate assessment of physical phenomena in the concerned region requires that the entire flow condition in the upper plenum be taken into consideration. Since it is not efficient to require detailed space accuracy similar to that required in the concerned region for the phenomena in the upper plenum, it is reasonable to determine appropriate space accuracy corresponding to the space-time scale of the flow. For the purpose, firstly, as shown in Fig. 5.7, a thermal hydraulic simulation of the entire upper plenum region is performed with relatively coarse mesh to obtain information on the flow structure in a large scale. Secondly, a fluid-structure thermal interaction analysis code (MUGTHES) [17] is applied to the local

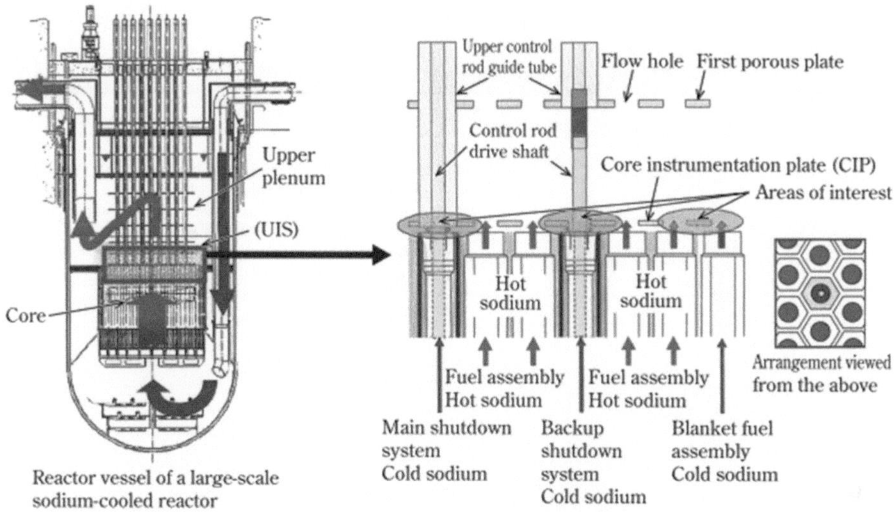

Fig. 5.6 Schematic diagram of thermal striping phenomenon in the lower part of core internals of the JSFR (Japanese sodium-cooled fast reactor)

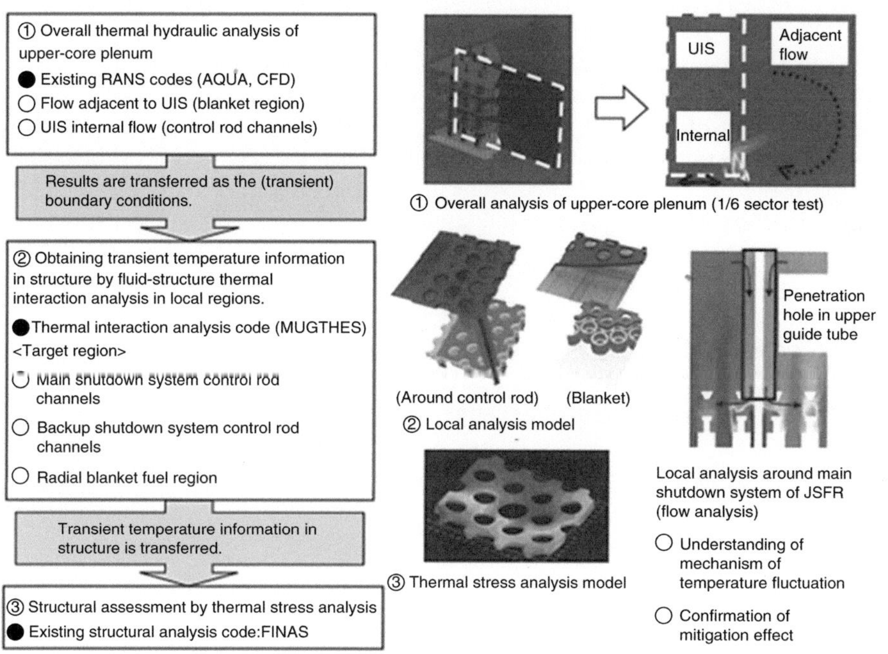

Fig. 5.7 Procedure for analytical estimation of actual reactor

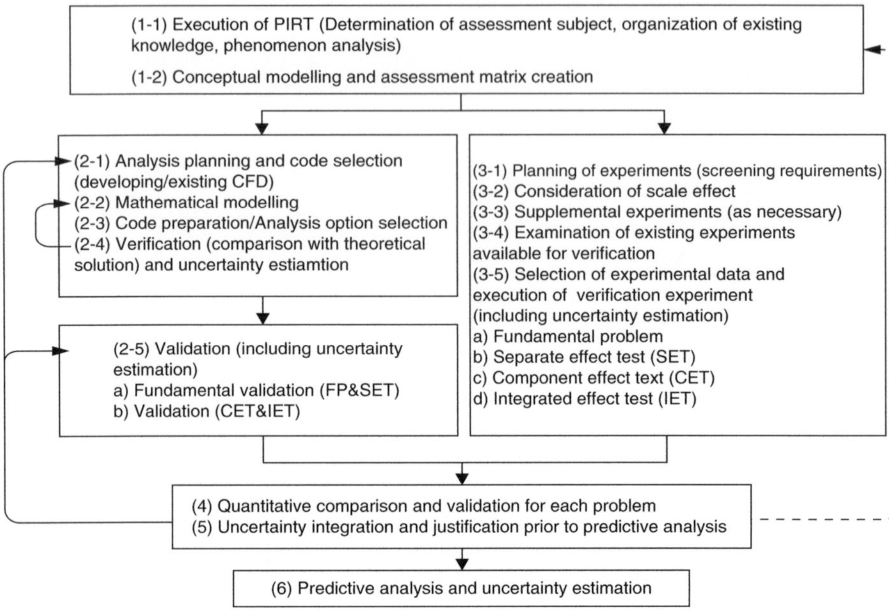

Fig. 5.8 Procedure for verification and validation (V&V) of numerical analysis code

region that includes the concerned region using the flow information in the upper plenum as the boundary condition to obtain transient temperature information in structure (a part of the CIP) by simulating thermal hydraulics and unsteady heat conduction in structure taking the conjugate heat transfer into consideration. Finally, using the transient temperature information in structure (a part of CIP) analyzed by the MUGTHES, thermal stress analysis in structure is performed to assess the structural integrity. To complete a series of analyses, determine V&V procedures [13, 14] that follow the abovementioned ASME guidelines; one conducts verification step using a basic problem, fundamental validation step using a relatively simple existing experiment as an example problem, and validation step using an experiment that simulates a scaled-down mockup and then performs the assessment of the actual system, as shown in Fig. 5.8.

5.2.3 *Failure Modes Assumed in FRs*

FRs operate at high temperatures where creep deformation of material can occur. Therefore, in FRs, failure modes taking into account the creep effect are added to those assumed in LWRs. The failure modes assumed in the structural design standards of FRs are listed in Table 5.6 [7, 8].

The feature of thermal stress, that is, the main load in FRs, is such that a rapid rupture is unlikely to occur by a single load because the thermal stress is a

Table 5.6 Failure modes assumed in the fast-reactor structural design standards

Common to those for LWRs (at low temperatures: non-creep range)	Ductile fracture, plastic collapse
	Excessive plastic deformation
	Fatigue failure
	Elasto-plastic buckling
Specific to FRs (at high temperatures: creep range)	Creep rupture
	Excessive creep deformation
	Creep-fatigue failure
	Creep buckling

displacement-controlled stress. Accordingly, the main failure modes assumed as those that are more likely to occur are the creep-fatigue cracking and excessive deformation due to cyclic loading.

Next, since the components and piping in FRs have a thin-walled structure that is advantageous in terms of thermal stress because of low internal pressure, seismic resistance decreases. For the reactor vessel, in particular, in which the ratio of wall thickness to diameter becomes quite small, care needs to be taken to prevent buckling due to horizontal acceleration. In addition, it is necessary to restrict the horizontal and vertical relative displacements between control rod and core that are both supported by the reactor vessel because it affects the proper insertion of control rods [10].

Since piping also has a thin-wall structure and a large diameter, attention has to be paid to seismic resistance. Based on experimental research, it is believed that the most probable failure mode for piping is the occurrence and development of cracks due to low-cycle fatigue.

Basic information required to understand the failure modes specific to FRs is outlined below:

1. Creep deformation and creep rupture

When metallic materials are subject to high temperatures, yield stress decreases; plastic deformation is more likely to occur; and creep deformation can occur. The increase in such deformation may lead to significant displacement and creep rupture. Therefore, it is necessary to have sufficient resistance to such deformation and rupture. Creep deformation increases with time even under constant stress (load). The relationship between creep strain obtained under a constant tensile stress σ at a constant temperature, ε_c, and time t is called the creep curve (Fig. 5.9).

The creep strain increases with stress at the same temperature and increases with temperature under a same stress condition. A typical creep strain equation that describes the above relationships is Norton's law:

$$\dot{\varepsilon}^c = A\sigma^n \tag{5.2}$$

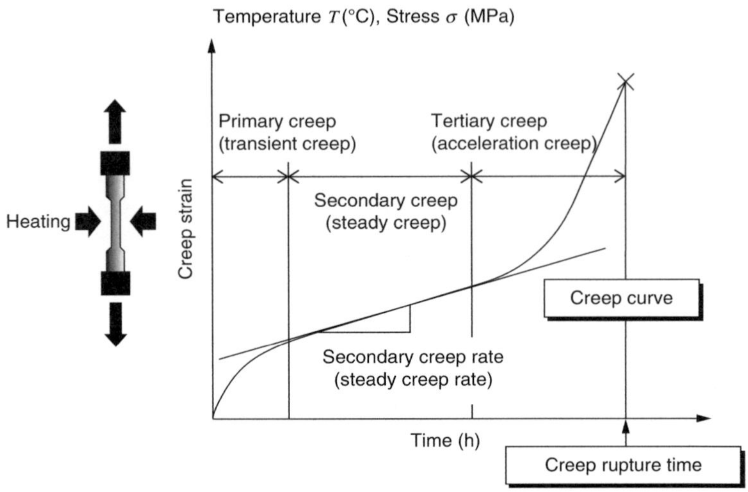

Fig. 5.9 Creep curve

Norton's law is the simplest equation that describes the creep strain velocity $\dot{\varepsilon}^c$ with parameters A and n depending, respectively, on material and temperature. Since n exceeds 5 for materials generally used for nuclear structure, the creep strain has strong nonlinearity with stress. For this reason, the creep strain in a structure with a stress distribution tends to occur selectively in regions where a high stress is applied.

Creep deformation increases even under constant stress and ultimately leads to rupture of material if it is not stopped. For this reason, the creep strength of material is defined by the creep rupture time, the time required until material ruptures under constant stress. The creep rupture time is affected by stress and temperature in a manner similar to creep strain. The plot of the data of creep rupture time t_R of SUS304 steel measured at various temperatures under various stress levels in stress-t_R diagram is shown in Fig. 5.10 [11]. As is shown in the figure, at a constant temperature, the creep rupture time tends to decrease monotonously with the increase in stress. In addition, the figure shows that as the temperature increases, the creep rupture time generally tends to decrease and the stress-t_R curve tends to shift downward.

Since the feature of thermal stress is self-controllability, when a thermal stress is applied once, material does not lead to rupture; however, when thermal stress is repeatedly applied, (creep) fatigue cracks occur and develop.

The typical failure mode of structure in the case where cyclic loading is applied at high temperatures is creep fatigue. Creep effect will significantly affect fatigue life. When a strain-controlled fatigue test is performed by providing the retention time on the tensile side (i.e., trapezoid wave), the hysteresis loop of the stress-strain curve becomes as shown in Fig. 5.11. During the period of retention, the stress is relaxed by creep; however, creep damage reduces fatigue life.

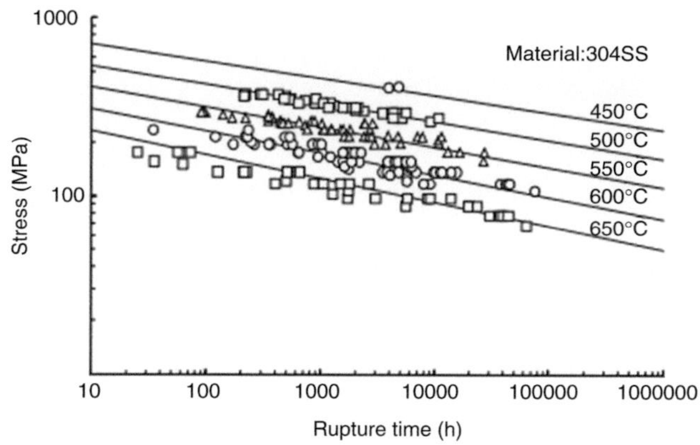

Fig. 5.10 Relationship between creep rupture time and temperature/stress [11]

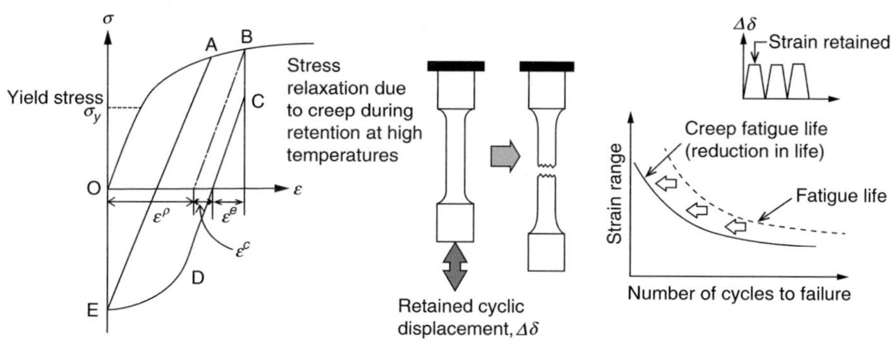

Fig. 5.11 Creep-fatigue damage [11]

2. Sodium environmental effect

In sodium for which the oxygen concentration is controlled at a low level, the effect of sodium corrosion is insignificant, and a tendency similar to that in vacuum is found unless the temperature exceeds 600 °C. In addition, decarburization or carburization, reverse reaction, of steel may occur in sodium. Decarburization reduces creep strength in addition to generated stress associated with softening. Carburization reduces ductility and suppresses plastic deformation due to hardening. Since these phenomena have conflicting effects on strength and occur in a region extremely close to the wall surface, it was verified by experiment that neither of these significantly reduces lifetime under actual plant conditions [10].

3. Neutron irradiation environmental effect

In a condition under which neutron energy exceeds 0.1 MeV, atoms on the metal crystal lattice are knocked out by collision with neutron to produce a disturbance of the crystal lattice (i.e., lattice defect). These defects hinder the dislocation movement and lead to a reduction in ductility due to hardening. For this reason, the "Rules for Fast Reactors" specifies the upper limit of neutron fluence ($E >$ 0.1 MeV) depending on temperature and creep effects (e.g., 5×10^{25} n/m^2 for SUS304 base metal at 400 °C) as a condition to allow for application of the material strength standards, etc., without any specific correction. Besides, thermal neutrons convert minute impurities in steel into helium (He). Since the generated He atoms enhance the aggregation of vacancy to the grain boundary, etc., a reduction in ductility and creep strength results. Accordingly, when the thermal neutron fluence ($E <$ 0.4 eV) exceeds the specified value (10^{19} n/m^2), the correction factor depending on the fluence is applied to the limiting values related to creep rupture time and creep strain rate [10].

4. Thermal aging

When metallic material is maintained at a high temperature, structural change consisting of the formation and coarsening of deposits (carbide, etc.) and modification phase (Laves phase, etc.) occurs. As a result, changes in mechanical properties, such as an increase in static tensile strength and yield stress, a softening of cyclic stress-strain characteristics, and a reduction in ductility and toughness, are likely to occur. It is necessary to take these changes into consideration when a plant's lifetime is extended and these changes in mechanical properties significantly affect the assumed failure modes [10].

5.3 Measures for Ensuring Structural Integrity of FRs

5.3.1 Approaches from Material, Design, Manufacturing, Inspection, and Maintenance Perspectives and the Resulting Constraints

1. Material

For structural material of FRs, steel types that have excellent high-temperature strength, compatibility with sodium, and sufficient ductility when used at high temperatures under neutron environment for a long period of time are selected. Therefore, austenitic stainless steel is often selected. However, although stainless steel has high intensity, its thermal conductivity is low and thermal expansion coefficient is high. These properties will work against thermal stress design. For this reason, ferritic high-chromium steels that have higher thermal conductivity and lower thermal expansion coefficient compared with austenitic stainless steel are used in some cases, although their ductility and high-temperature strength are somewhat less attractive. Steel materials that were selected for the reactor vessel are shown as examples in Fig. 5.12 [9]. Since the heat transfer tubes of a steam generator are exposed to water environments, it is necessary to select materials that maintain a certain degree of high-temperature strength and have excellent

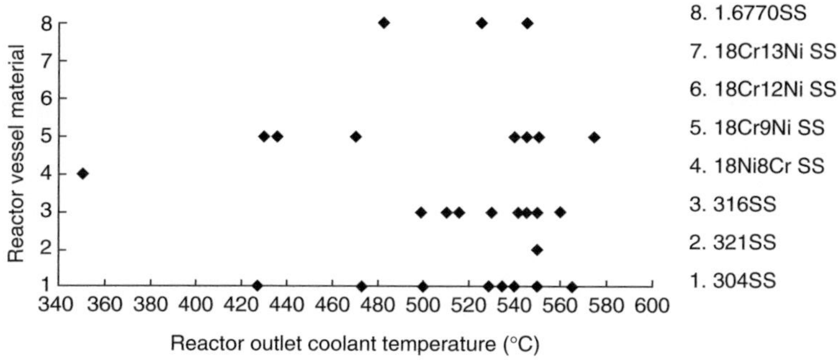

Fig. 5.12 Reactor vessel materials

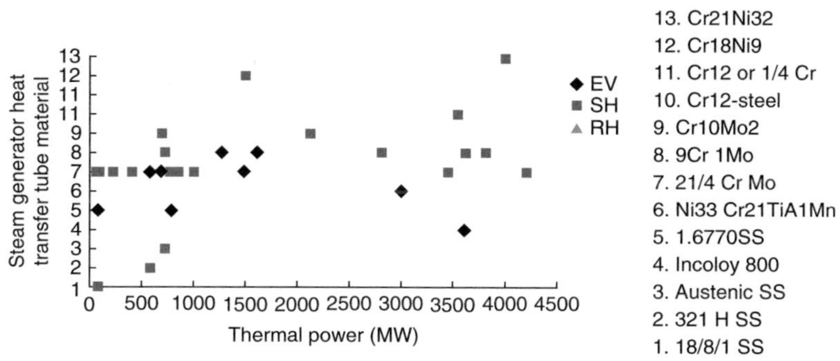

Fig. 5.13 Steam generator materials

resistance to SCC, such as chromium-molybdenum steels. Steel materials that were selected for the steam generator are shown as examples in Fig. 5.13 [9]. As described above, since requirements for high-temperature strength, compatibility with sodium, and ductility must be simultaneously met, there are constraints on the selection of materials.

2. Design and manufacturing

In a design process, the assessment of each of the failure modes listed in Table 5.6 is based on the procedures for "design by analysis." Since the prevention of excessive deformation and creep fatigue due to cyclic thermal stress and buckling due to seismic load often become design constraints, thermal stress design and aseismic design are described in detail in later sections.

In a manufacturing process, since the creep-fatigue strength to cyclic thermal stress is sensitive to weld shape and metallurgical discontinuity, when welding important components, it is necessary to limit applicable welding processes to those with a small heat input, such as TIG, specify weld surface finish, and limit the location of weld line away from regions where high stress is applied.

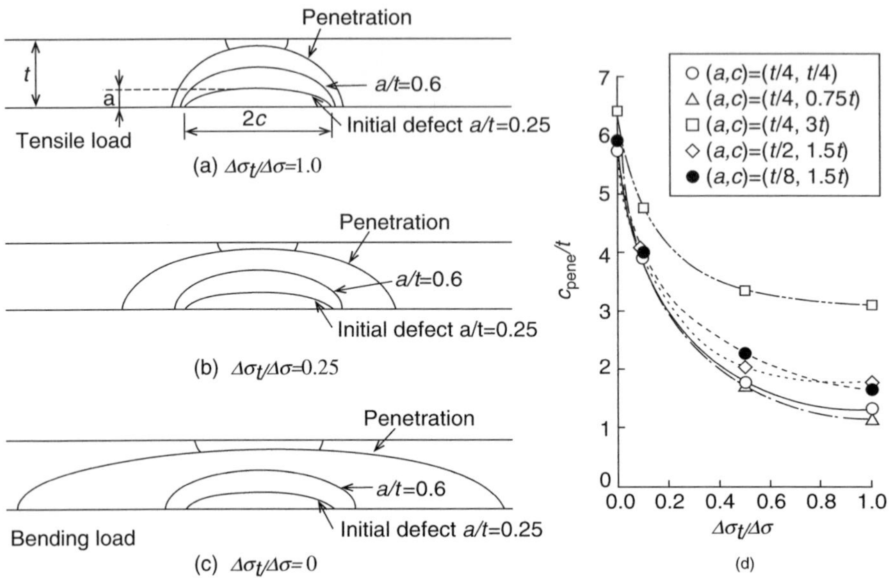

Fig. 5.14 Effects of initial defect size and loading condition on crack length at penetration [19]

3. Inspection and maintenance

Since inspection and maintenance are explained in detail in Chap. 9, Maintenance, only the distinguishing features are described here. The use of ductile material in the manufacture of FR components and piping along with operation at high temperatures under low pressures makes the occurrence of instantaneous break (rupture) unlikely. The boundary failure that should be assumed is a case in which cracks occur and progress due to cyclic thermal stress, etc., resulting in penetration. Even in this case, because sodium does not boil at a reduced pressure due to its high boiling point, it is possible to avoid massive break if operation is stopped when leak is detected. That is, the concept of leak before break (LBB) is applicable [18]. To confirm the applicability of the LBB concept, we assess by fracture mechanics that pipe break does not occur due to external forces, such as earthquake, when a crack penetrates through the pipe wall. Figure 5.14a–c shows predicted results obtained by fracture mechanics for the crack growth pattern when pure tensile load, pure bending load, or the superposition of the two loads is applied to the initial semielliptical defect (a, depth; c, half length) that was assumed on the inner surface of a plate with a thickness of t and a/t of 0.25. This figure suggests that the crack length at penetration becomes the longest when pure bending load is applied. Next, Fig. 5.14d shows predicted results obtained by fracture mechanics for the crack length at penetration depending on the depth a and half-length c of the initial semielliptical defect. The horizontal axis of the figure shows the rate of tensile stress to total stress (i.e., tensile stress plus bending stress). The plots at the left end of the figure suggest that the crack length becomes the longest, approximately 12 times the plate thickness t, regardless of the initial defect size when pure bending stress is applied [19].

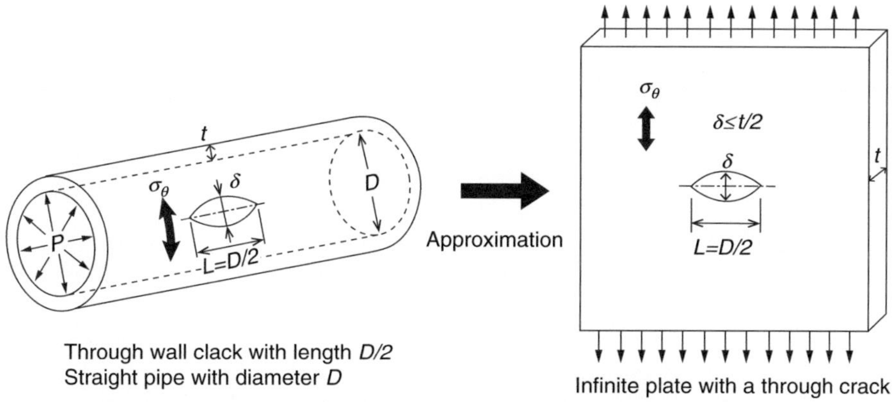

Fig. 5.15 Estimation of crack opening area under the design condition of Monju

Low-pressure pipes used for FRs are thin-walled and large diameter, and the diameter-to-thickness ratio, D/t, is 24 or greater in Monju. Therefore, the crack length L relative to the pipe diameter would be no greater than $D/2$, suggesting that the break of ductile material is unlikely to occur at low pressures.

Next, we assess the crack opening area that is used as a basis for the estimation of the amount of leak that affects safety design. We assume that the opening size is δ when there is a through crack with L equal to $D/2$ in a cylinder with a diameter of D and a thickness of t, as shown in Fig. 5.15. δ can be conservatively predicted by the theoretical solution of a model in which a tension load equivalent to the θ-direction stress that is the maximum stress component of the cylinder is applied to an infinite plane with a rectangular through crack with a length L of $D/2$. Calculation under the conditions of Monju's piping systems shows that $\delta \leq t/2$. Multiplication of this size δ and the crack length L shows that the maximum opening area $L\delta \leq Dt/4$ [19].

As described above, the LBB concept is applicable. In addition, the crack opening area is not significant, and rapid crack growth is unlikely to occur in case of leakage. Consequently, there is no considerable constraint on in-service inspection (ISI). For this reason, continuous monitoring of coolant leakage is effective for ISI [18].

5.3.2 *Structural Design for High Temperatures and Thermal Stress*

The specific phenomena occurring in high-temperature structures include deformation and reduction in strength due to plasticity and creep at high temperatures and thermal stresses caused by the difference in temperature in structures. Full attention needs to be paid to these phenomena in high-temperature structural design. A

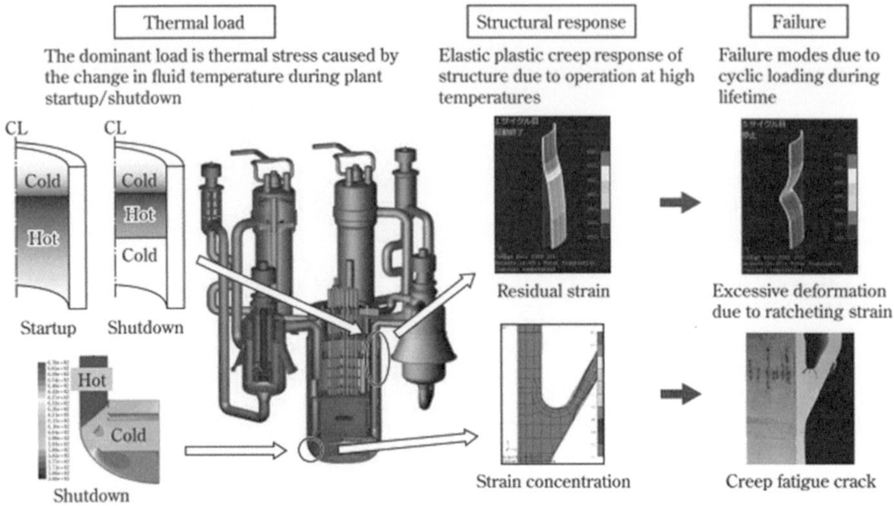

Fig. 5.16 Schematic flow of reactor vessel design in consideration of thermal loading (JSFR)

design flow of the reactor vessel of JSFR is given as an example (Fig. 5.16) [20]. The overall design process consists of three stages: (1) determination of thermal loads, (2) assessment of structural response to the loads, and (3) structural assessment when the loads are repeatedly applied in the lifetime. In stage (1) aiming for determination of thermal loads, sodium temperature distributions in reactor during plant transient operation and their time history responses are estimated. The main factors for the thermal loads in reactor vessel are the vertical temperature distribution that occurs near the sodium surface during start-up and the difference in temperature across the core support plate that occurs during shutdown. Accordingly, the vertical temperature distribution and temperature difference are determined by detailed system design. Next, in stage (2) the temperature responses in structure and the resultant responses of thermal stress and strain are predicted. Finally, in step (3) structural strength against the assumed failure modes is assessed. When thermal loads are repeatedly applied during operation at high temperatures, excessive deformation due to the accumulation of strain and the occurrence of cracks due to creep fatigue are assumed as failure modes. An explanation of each method is given below.

1. Prediction of thermal loads

FR systems are subject to steady-state thermal stresses and thermal transient stresses due to the change in system temperature depending on operational conditions.

To predict thermal transient loads, plant dynamic analysis is performed for events determined by system design as those to be taken into consideration during operation. Next, thermal hydraulic analysis in components is performed using the predicted conditions at the inlet/outlet of the components as the boundary

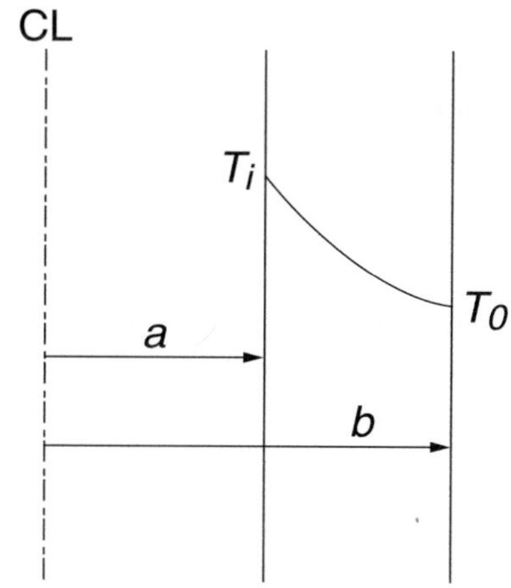

Fig. 5.17 Cylinder with a through-thickness temperature gradient

conditions to obtain detailed temperature distribution, flow velocity distribution, heat transfer coefficient, etc. Specific procedures for these processes are as described in the previous section on thermal hydraulic design.

2. Prediction of thermal stress and inelastic strain

After the prediction of structural temperature response based on the surface thermal boundary conditions, thermal stresses are predicted based on the obtained temperature distribution. Here, it should be noted that the uncertainties of the predicted temperature responses affect all results to be obtained in the subsequent prediction processes.

In a simple structure, thermal stress can be estimated using the theoretical solution. When there is a through-thickness temperature distribution T in a long cylinder with an inner and outer radii of a and b that is fixed at both ends (Fig. 5.17), the circumferential thermal stress is obtained from the thick-walled cylinder theory as follows [10]:

$$\sigma_\theta = \frac{\alpha E}{1-\nu}\left(\frac{1}{r^2}\frac{r^2+a^2}{b^2-a^2}\int_a^b Trdr + \frac{1}{r^2}\int_a^r Trdr - T\right) \tag{5.3}$$

Here, E Young's modulus, α coefficient of linear expansion, ν Poisson's ratio, T temperature, and r radius. When $T_i < T_o$, σ_θ becomes tensile at the inner surface and compressive at the outer surface. When $T_i > T_o$, vice versa. The magnitude of the stress reaches maximum for σ_θ at the inner surface. When assuming that the through-thickness temperature distribution is linear, the following expression holds:

$$\sigma_\theta = \sigma_z = \pm\frac{\alpha E}{2(1-\nu)}(T_i - T_0) \tag{5.4}$$

Here, + is for the stress at the outer surface and – for the stress at the inner surface. Equation (5.4) is often sufficient for an approximate estimate of thermal stress due to a through-thickness temperature gradient.

Thermal loads that are problematic for FRs often cause nonsteady, transient thermal stresses. Estimation of these stresses requires, in most cases, thermal stress analysis using the finite element method.

Next, since creep and elastoplastic deformation are likely to occur in high-temperature structures, these phenomena must be taken into consideration in estimating stress and strain. To estimate the elastoplastic creep behavior of material, two methods are available: a method to estimate inelastic behavior using a simplified method based on elastic analysis results and a method to directly obtain the solution for elastoplastic creep using the inelastic finite element method. Of these methods, the former method based on elastic analysis is more often used in design. Neuber's law is explained below as a representative simplified inelastic analysis method that is adopted by "Rules for Fast Reactors." This law is used to simply evaluate the strain concentration at a local structural discontinuity, as shown in Fig. 5.18. Originally, this law was proposed to evaluate the strain concentration at the notched end of a notched rod that is subject to shear strain; however, it has been widely used in design since it is capable of evaluating the elastoplastic creep strain concentration in quite a simple manner, as shown in the following equation:

$$\left(\frac{\sigma}{\sigma_n}\right)\left(\frac{\varepsilon}{\varepsilon_n}\right) = K^2 \tag{5.5}$$

where, σ and ε are stress and strain at the notched end, σ_n and ε_n are stress and strain in the smooth part, and K is elastic stress concentration factor. When the elastic solution is obtained as shown in Fig. 5.18, it is possible to simply predict the elastoplastic solution from the intersection between the hyperbolic stress-strain

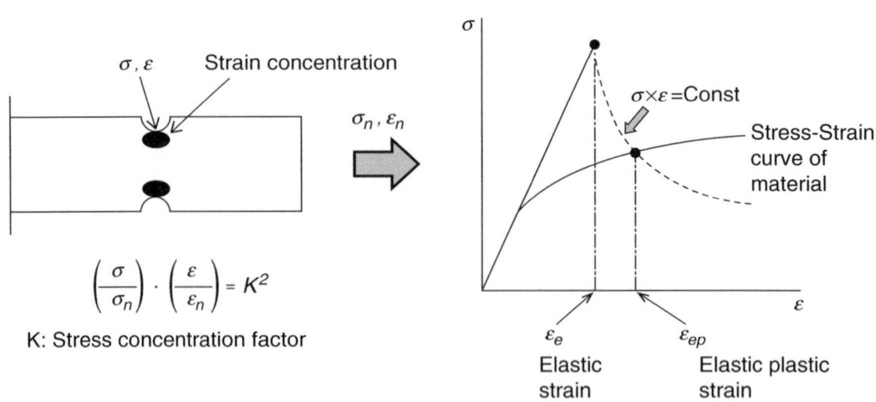

Fig. 5.18 Neuber's law [11]

redistribution curve according to Eq. (5.5) and the stress-strain relationship curve of the material.

Research and development is underway for direct application of inelastic analysis using the finite element method in consideration of elastoplastic creep behavior of material to design. This is because an excessively large evaluation margin associated with the use of a simplified evaluation method based on elastic analysis impairs the economy of plant design. Previously, simplified evaluation methods have mainly been used for design evaluation, and inelastic analysis using the finite element method is often used for confirmation of the validity and the margin of the simplified evaluation methods. The reasons for this include the fact that the inelastic analysis method is more complex, requires a higher level of knowledge and experience, takes more calculation time, and requires greater cost than the elastic analysis method. In addition, the inelastic analysis method has a technical problem in that the solution depends on the selection of constitutive equations and load history that is unknown in the design stage.

3. High-temperature strength assessment

The "design by analysis" approach that was proposed in Section III of the ASME (American Society of Mechanical Engineers) Boiler and Pressure Vessel Design Code and which has become the basic concept of structural design of the nuclear components requires "detailed design in consideration of the failure modes." For this reason, criteria for strength evaluation against the failure modes listed in Table 5.6 are prepared in the design code for high-temperature structures [7, 8]. A creep-fatigue evaluation method is shown as a typical example. Creep-fatigue damage is prevented by the linear cumulative damage rule that limits the sum of the following two calculated values: (i) usage factor D_f and (ii) creep damage factor D_c. The D_f and D_c values are calculated separately based on the time period while the strain changes (corresponding to the plant start-up and shutdown) and the time period while the strain remains constant (corresponding to the steady-state plant operation), which comprise the trapezoidal strain waveforms, as shown in Fig. 5.19:

$$D_f + D_c \leq D \quad D_f = \sum_i \frac{n_i}{N_{fi}} \quad D_c = \sum_j \frac{t_j}{t_{Rj}} \tag{5.6}$$

Here, the limiting value D equals 1 when pure fatigue damage or creep damage solely occurs; however, consideration is given to the fact that D becomes less than 1 when both types of damage occur simultaneously due to their interaction. The usage factor D_f is obtained according to what is called Miner's rule, which linearly sums $n_i/N_{f\,i}$. Here, the damage caused when the strain range $\Delta\varepsilon_i$ is repeatedly applied n_i times is designated φ_{fi}. Then, φ_{fi} is represented by n_i/N_{fi}, the ratio of n_i to the fatigue life N_{fi} with regard to $\Delta\varepsilon_i$.

Creep damage factor D_c is obtained by a time consumption method that linearly sums φ_{ci} $(=t_j/t_{Rj})$. Here, the damage caused when stress σ_j is applied for the time

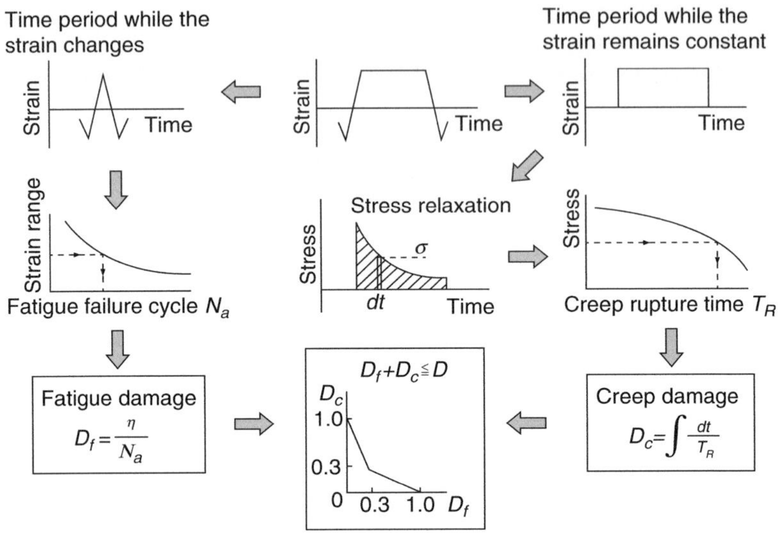

Fig. 5.19 Creep-fatigue damage estimation method [10]

period t_j is designated φ_{ci}. Then, φ_{ci} is represented by t_j/t_{Rj}, the ratio of t_j to the creep rupture time t_{Rj} with regard to σ_j [10].

By design assessment consisting of the abovementioned procedures, the shape of the concerned structure, operating temperature, time required for start-up and shutdown, and the combination of the number of cyclic loads are constrained.

5.3.3 Aseismic Design of Thin-Walled Structures

In FRs that operate at high temperatures under low pressure, thin-walled structures, in which thermal stresses caused by the change in coolant temperature are easily mitigated, are more advantageous; however, as walls become thinner, the earthquake resistance deteriorates due to the reduction in rigidity. Accordingly, it is important in the structural design of FRs to harmonize thermal load design and aseismic design, as shown in Fig. 5.20 [10].

1. Earthquake-proof structure

The representative component of FRs that has a great diameter-to-thickness ratio is the reactor vessel. Regardless of whether the loop or pool type is employed, the reactor vessel is often suspended from the pedestal of the reactor building at the top flange (Fig. 5.21). In this configuration, when the reactor vessel containing the core that concentrates weight in the lower part is subject to horizontal load during earthquake, flexural buckling and shear buckling, the buckling modes specific to a thin-walled cylinder, are likely to occur, as shown in Fig. 5.22. Rolling also causes relative displacement between control rods that are suspended from the upper part

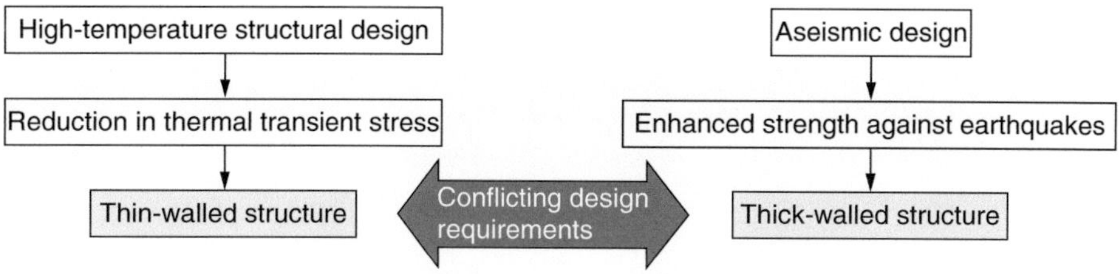

Fig. 5.20 Harmonization of thermal load design and aseismic design

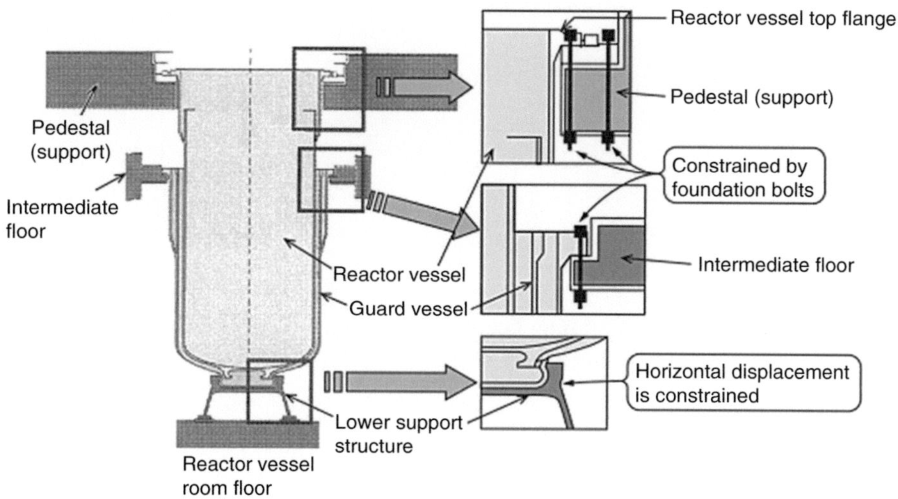

Fig. 5.21 Method of supporting reactor vessel

Fig. 5.22 Buckling mode of thin-walled cylinder

of the reactor vessel and the core that is supported at the lower part of the reactor vessel, affecting the insertability of control rods.

For this reason, to prevent rolling, the bottom of the vessel is supported by the lower support structure or the body of the vessel is supported with shearing die structure (Fig. 5.21).

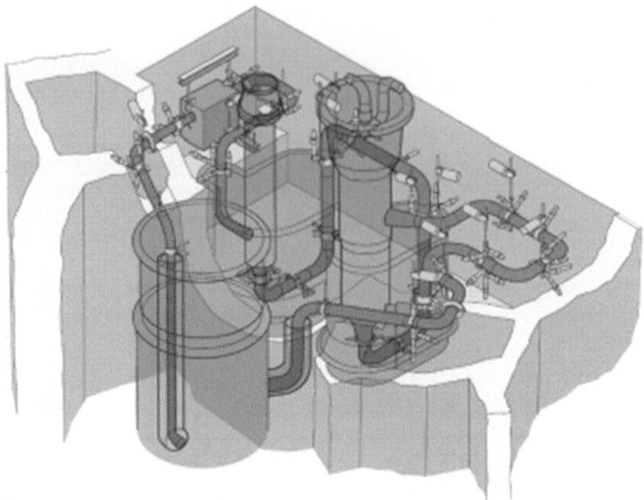

Fig. 5.23 Fast reactor piping system and support

Buckling of a vessel and piping become likely with a larger diameter and thinner wall thickness, and accordingly, the shape (diameter-thickness ratio) is limited to prevent buckling.

FR piping tends to be long and accompanied by numerous elbows to allow for a flexible release of thermal expansion deformation at high temperatures. For this reason, FR piping tends to have a smaller natural frequency than that of an LWR and needs more support equipment to ensure earthquake resistance (Fig. 5.23). To this end, anti-vibration equipment that freely follows low-velocity displacement, such as thermal expansion, and restrains high-velocity movement, such as earthquake, is used (Fig. 5.24). Oil dampers are typically used; however, in a region where material deterioration due to radiation is a factor, mechanical snubbers need to be applied [10]. As described above, aseismic design of piping constrains the layout and the method of supporting piping systems.

2. Collision vibration of core

For FR cores, supporting methods that do not restrict radial displacement are often used to release deformation due to irradiation and creep. In this condition, fuel subassemblies vibrate while colliding during an earthquake. To ensure the insertion of control rods into the assemblies even in an earthquake, it is necessary to predict the collision vibration behavior and limit the relative displacement between the control rods and assemblies to within the allowable range.

3. Impact evaluation of fluid

Since a reactor vessel contains a significant amount of liquid metal sodium and has a liquid surface, attention needs to be paid to fluid behavior during an

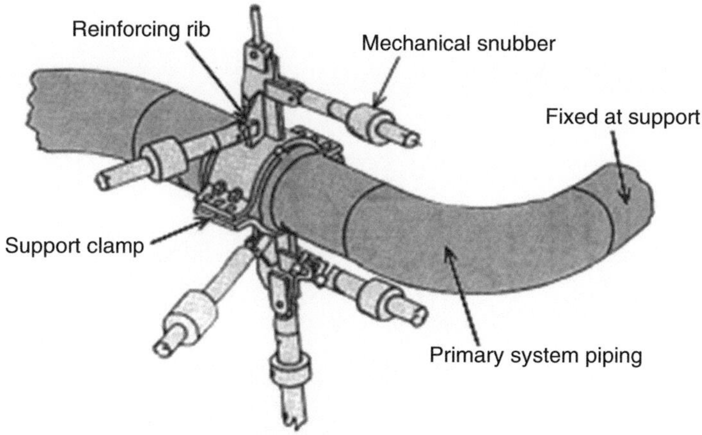

Fig. 5.24 Pipe supports to prevent vibration while allowing thermal expansion

earthquake. As a countermeasure for sloshing in the vessel, a dip plate may be installed near the free surface. In addition, consideration needs to be given to the position of nozzles, etc., to prevent cover gas entrainment even in an earthquake during emergency cooling with a lowered liquid level. Furthermore, it is necessary to take the effect of fluid force into consideration using the added mass method, etc., in assessing the abovementioned collision vibration of a core.

5.4 Measures for the Mitigation of Loads

Since the constraints for structural design greatly affect plant construction cost, measures to actively mitigate loads are taken to remove the constraints when the economic competitiveness is expected to be improved as a whole. The typical examples are described below.

5.4.1 Thermal Load Mitigation Method

The area near the liquid surface in the reactor vessel is one region where the greatest thermal transient stress may occur in an FR plant. Since the upper part of the reactor vessel is often suspended from the concrete building for support, it is necessary to ensure that the concrete temperature is no greater than 100 °C to maintain the integrity of concrete. On the other hand, the temperature of the upper plenum, which is located between the core outlet and the free sodium surface, reaches 500 °C or higher because high-temperature sodium heated in the core flows into

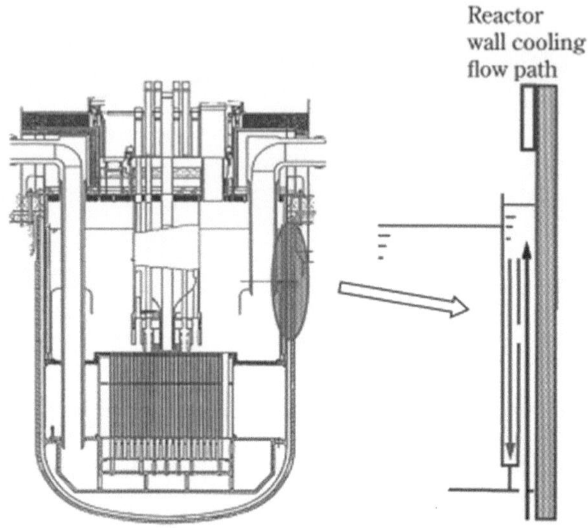

Fig. 5.25 Mitigation of thermal load near the liquid surface in reactor vessel by cooling reactor wall (DFBR)

the plenum. Consequently, a temperature difference equal to or greater than 400 °C occurs within the short distance between the free sodium surface and the top of the reactor vessel. Furthermore, during reactor start-up, the sodium level also rises with the thermal expansion of the sodium itself due to the increase in sodium temperature. In this case, the temperature gradient near the liquid surface becomes steeper, and severe transient stress occurs.

Since this phenomenon occurs repeatedly during start-up, various measures to mitigate the thermal loads are taken to maintain the structural integrity. A typical measure is the mitigation of thermal loads by cooling the reactor wall. An example from the Japan Demonstration Fast Reactor (DFBR) is shown in Fig. 5.25. A portion of low-temperature sodium in the cold plenum located in the lower part of the reactor vessel is sent to the upper part along the reactor wall rather than through the core. Sodium rises and then turns down toward the inside of the reactor vessel across the shroud at a position above the free surface and discharges into the upper plenum at a level somewhat higher than the core. It is necessary to add the shroud to form the flow path, and this increases the diameter of the reactor vessel. However, this method does not directly transfer the effect of the sodium temperature distribution to the vessel wall and can prevent the reduction in material strength by cooling the vessel wall.

Another method to mitigate thermal loads is level control (Fig. 5.26), which was adopted for Monju. When the increase in sodium temperature and the rise in sodium level during start-up are left to chance, the steepest temperature gradient and the maximum stress occur at the completion of the temperature increase process. After the sodium level becomes steady, the temperature distribution becomes flatter

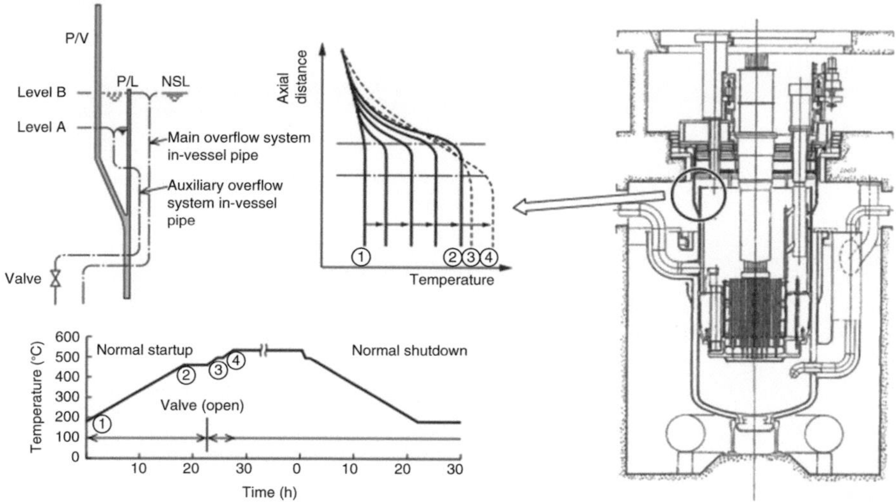

Fig. 5.26 Mitigation of thermal load near the liquid surface in reactor vessel by controlling the liquid level (Monju)

(i.e., temperature homogenization) due to heat conduction in the reactor wall, and thereby the stress decreases. With consideration of this fact, the sodium temperature is increased while maintaining the sodium level at level B in the early stage of the start-up, and at time ②, before the stress reaches the maximum, the temperature is kept at a constant level, and the level is lowered from level B to A by opening valves, as shown in the left figure of Fig. 5.26. Maintaining this condition would reduce the thermal stress as the axial temperature gradient decreases due to temperature homogenization. At time ③, when the stress is sufficiently relaxed, the sodium temperature is increased again while maintaining the sodium level at level A until it reaches the maximum temperature. This procedure allows reduction of maximum stress when the temperature increase is completed. This method requires an overflow system for level control and a bucket structure that separates the reactor wall and upper plenum in the region where the sodium level is controlled; however, the required equipment is simpler than that required for the reactor wall cooling method [10].

Reducing the thermal loads near the sodium surface relieves the constraints on the diameter and wall thickness of the reactor vessel as well as the operating conditions during start-up and shutdown. On the other hand, the addition of equipment to mitigate the thermal loads leads to the increase in the amount of material and the introduction of additional uncertainties. Therefore, it is necessary to compare the advantages and disadvantages before the decision on adoption is made.

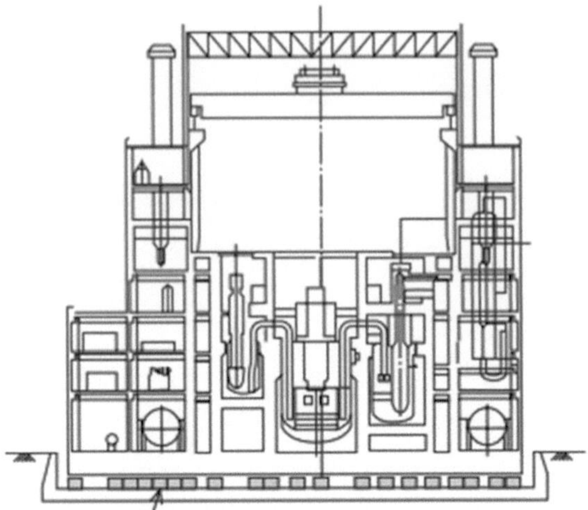

Fig. 5.27 Example of 3-D base-isolated building structure (JSFR)

5.4.2 Seismic Isolation Structure

Harmonization between thermal load design and aseismic design becomes increasingly difficult with the increase in size of a plant in countries such as Japan, which are subject to strong earthquakes. Therefore, it is hoped that the decrease in seismic loads due to the adoption of seismically isolated structure would greatly contribute to improving FR safety and streamlined design. A seismically isolated structure also contributes to design standardization based on different seismic loads of the respective sites.

In order to thoroughly solve the constraints of seismic loads on the design of reactors, etc., research and development is underway on three-dimensional seismic isolation technology that can reduce vertical seismic vibration as well as horizontal vibration [15]. The three-dimensional seismic isolation technology prevents resonance of components and buildings under any assumed seismic condition to realize the seismic isolation function by increasing the period of input seismic vibration. There are two methods employed in three-dimensional seismic isolation technology: 3-D base-isolated building structure, the method of isolating the entire reactor building from seismic vibration, as shown in Fig. 5.27, and the method of combining a horizontally base-isolated building and vertically base-isolated components, as shown in Fig. 5.28 [20].

The use of the seismic isolation method increases the amount of seismic isolation equipment; however, it relieves the constraints of seismic loads on the diameter and wall thickness of the reactor vessel as well as the piping layout and supports.

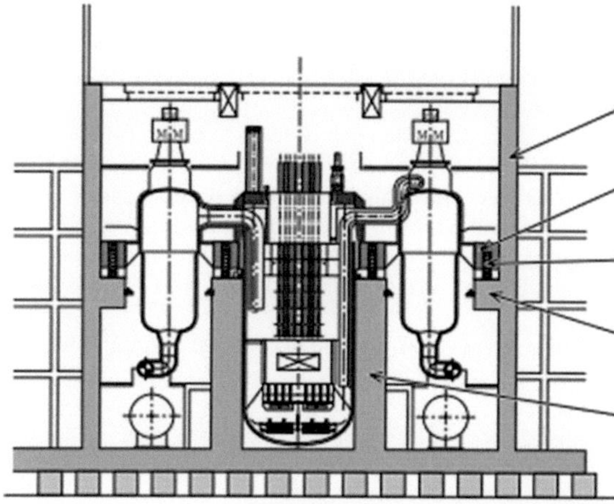

Fig. 5.28 Example of 3-D base-isolated structure with combination of horizontally base-isolated building and vertically base-isolated components (JSFR)

Further Readings

G.A. Webster, R.A. Ainsworth, *High Temperature Component Life Asessment* (Chapman & Hall, London, 1984)

Afcen, *Design and Construction Rules for Mechanical Components of FBR Nuclear Island (RCC-MR)* (AFCEN, Paris, 2002)

British Energy, R5Assessment procedure for the high temperature response of structures. (3) (2003)

Y. Asada, K. Koibuchi, *Kikaikozo kyodo-gaku* (Methodology on Mechanical Structural Strength). Course of Lectures on Integrated Material Strength vol. 8 (Ohmsha, Ltd., 1984) (in Japanese)

K. Miya, G. Yagawa, *Genshiro Kozo Kogaku* (Reactor Structural Engineering), Series in Nuclear Engineering, vol. 1 (University of Tokyo Press, 1977) (in Japanese)

Y. Ando, K. Okabayashi, in *Genshiryoku Puranto no Kozo Sekkei* (Structural Engineering in Nuclear Plants), Series in Nuclear Engineering, vol. 3 (University of Tokyo Press, 1977) (in Japanese)

S. Ishino, *Shosha Sonsho* (Irradiation Damage), Series in Nuclear Engineering, vol. 8 (University of Tokyo Press 1979) (in Japanese)

T. Nagata, et al., Development of Elevated Temperature Structural Design Standard for FBR, Donen Giho (PNC Technical Review), vol 65 (1988), pp. 39–71

G. Yagawa, M. Ichimiya, *Genshiro Kozo Sekkei – Suchikaiseki kara Taishinsekkei made* (Reactor Structural Design – From Numerical Analysis to Seismic Design) (BAIFUKAN Co., Ltd., 1989) (in Japanese)

G. Yagawa, N. Miyazaki, *Yugenyosohou ni yoru Netsuoryoku/ Kuripu/ Netudendo Kaiseki* (Thermal Stress, Creep and Heat Transfer Analysis by Finite Element Method) (SAIENSU-SHA Co., Ltd., 1985) (in Japanse)

N. Ohno, J.D. Wang, Kinematic hardening rules for simulation of ratchetting behavior. Eur. J. Mech, A Solids **13**(4), 519–531 (1994)

H. Wada, et al., in *The Ratchetting Evaluation Methods in Japanese Demonstration FBR Design*, Transactions of the 14th SMiRT, F04/2 (1997)

Y. Takeuchi, *Netsu Oryoku* (Thermal Stress) (Nisshin Shuppan Co., 1975) (in Japanese)
N. Noda et al., *Thermal Stress* (LASTRAN, New York, 2000)
F.K.G. Odqvist, J. Hult, *Kuripu Zuyosa no Riron* (Theory of Creep Strength) (BAIFUKAN Co., Ltd., 1967) (in Japanese)
R.K. Penny, D.L. Marriott, *Design for Creep*, 2nd edn. (Chapman & Hall, London, 1995)
R. Viswanathan (author), F. Masuyama and K. Maruyama (translators), *Ko-on Kiki Buhin no Sonsho Mekanizumu to Jumyo Hyoka* (Damage Mechanisms and Life Assessment of High-temperature Components) (Nikkan Kogyo Shimbun, Ltd., 1993) (in Japanese)
Japan Society of Mechanical Engineers *Standards for Nuclear Power Generation Equipment – Maintenance Standards* (the 2010 supplement versions), (JSME S NA1–2010, 2010) (in Japanese)
Nuclear Standards Committee of Japan Electric Association, *Technical Code for Seismic Design of Nuclear Power Plants*, JEAC4601 (2008)

References

1. Japan Welding Engineering Society, Textbook for seminar on materials, design, manufacture, inspection and maintenance of nuclear structural components (2013)
2. Y. Asada, *Koso teian: Shisutemuka kikaku – sono kihon koso* (Concept Proposal: System based Code – Its Basic Concept), Forum maintenology, vol. 1. (2002) pp. 55–66 (in Japanese)
3. Final Report of the Investigation Committee on the Accident at Fukushima Nuclear Power Stations of Tokyo Electric Power Company, (in Japanese), July, 2012
4. IAEA, *Safety of Nuclear Power Plants: Design Specific Safety Requirements*, SSR-2/1 (International Atomic Energy Agency, Vienna, 2012)
5. K.R. Rao, *Companion Guide to the ASME Boiler & Pressure Vessel Code*, 3rd edn. (ASME Press, New York, 2009)
6. ASME, *ASME Boiler and Pressure Vessel Code, Sec. III, Subsection NB*, New York (2004)
7. Japan Society of Mechanical Engineers, *Standards for Nuclear Power Generation Equipment – Design and Construction Standards, Volume 2, Rules for Fast Reactors*. JSME S NC2–2005 (2005) (in Japanese)
8. ASME, *ASME Boiler and Pressure Vessel Code, Sec. III, Subsection NH*, New York (2004)
9. IAEA-TECDOC-1531, Fast Reactor Database, IAEA (2006)
10. Edited by M. Hori, Kiso Kousokuro Kougaku Henshu Iinkai, *Kiso Kousokuro Kougaku* (Basic Fast Reactor Engineering), Nikkan Kogyo Shimbun, Ltd. (in Japanese) (1993)
11. M. Uesaka, K. Onizawa, N. Kasahara, K. Suzuki, *Genshiro Kouzou Kougaku* (Reactor Structural Engineering), (Ohmsha, Ltd., 2009) (in Japanese)
12. Y. Katto, *Dennetsu Gairon* (General Lecture of Heat Transfer). Yokendo (1964) (in Japanese)
13. ASME,*ASME V&V-10, 2006: Guide for Verification and Validation in Computational Solid Mechanics*, ASME (2006), and ASME, "ASME V&V-10.1-2012: An Illustration of the Concepts of Verification and Validation in Computational Solid Mechanics" (ASME, 2012)
14. ASME,*ASME V&V-20*, 2009*: Standard for Verification and Validation in Computational Fluid Dynamics and Heat Transfer*. (ASME, 2009)
15. Y. Sagayama, K. Okada, T. Nagata, in *Progress on Reactor System Technology in the FaCT Project Toward the Commercialization of Fast Reactor Cycle System*. Proceedings of International Conference on Fast Reactors and Related Fuel Cycles (FR2009), IAEA-CN-176-01-02 (2009)
16. M. Tanaka, et al., *Numerical Simulations of Thermal-Mixing in T-Junction Piping System Using Large Eddy Simulation Approach*. Proceedings of ASME-JSME-KSME Joint Fluids Engineering Conference 2011 (AJK2011-FED), Hamamatsu, Japan, 24–29 July 2011, (Paper ID AJK2011–18009) (2011)

17. M. Tanaka, H. Takita, M. Hideaki, H. Oshima, Investigation on Fluid-Structure Thermal Interaction Related to Eddy Structure on Branch Jet in T-Junction, Nippon Kikai Gakkai Ronbunshu B, **78**(792), (2012–08), 1462–1465 (2012)
18. Maintenance of Prototype Fast Breeder Reactor Monju, The 8th academic lecture meeting held by the Japan Society of Maintenology
19. Japan Atomic Energy Agency, Modes and Sizes of Pipe Break, PNC-TN243 81–06 (1981)
20. K. Inoue, H. Shibamoto, K. Takahashi, S. Ikutama, M. Morishita, K. Aoto, N. Kasahara, T. Asayama, S. Kitamura, Development of elevated temperature structural design standard and three-dimensional seismic isolation technology for Advanced nuclear power plant. J. At. Energ. Soc. Jpn. **48**(5), 333–338 (2006)

Chapter 6
System Conceptual Design (From Function to Mechanism)

Kiyohiko Maeda

Based on functional requirements and function elements discussed in Chap. 3, Chap. 6 presents systems designed to provide mechanisms that achieve the required functions while taking into account the safety measures discussed in Chap. 4 and the structural integrity and economic competitiveness discussed in Chap. 5. It then integrates while ensuring integrity among the respective systems to give a concrete form to the plant concept. The system configuration and main parameters of the plant are determined in this step. This chapter also discusses the typical system configurations of an FR plant.

6.1 System Conceptual Design

Function elements and mechanism elements discussed in Chap. 3 are concretized as structures and then combined for integration as a whole plant. Figure 6.1 shows a schematic of this process.

First, consider the mechanisms and general structure for a core that can demonstrate the expected performance and can load heat generating Pu, fuel breeding ^{238}U, and MA for the reduction of waste. Next, consider the reactor structures needed to support and cool the core. A cooling system is designed to transfer heat from the reactor to steam turbines. In addition, a water-steam system for generating power and power generation equipment is discussed.

Here, overall consistency, such as the balance between heat generated in the core and power generated by turbine, needs to be achieved. Therefore, it is necessary to examine reactor physics, thermal hydraulics, and thermoelectric conversion

K. Maeda (✉)
Japan Atomic Energy Agency, Ibaraki, Japan
e-mail: kiyohiko-maeda@ktj.biglobe.ne.jp

N. Kasahara (ed.), *Fast Reactor System Design*, An Advanced Course in Nuclear Engineering 8, DOI 10.1007/978-981-10-2821-2_6

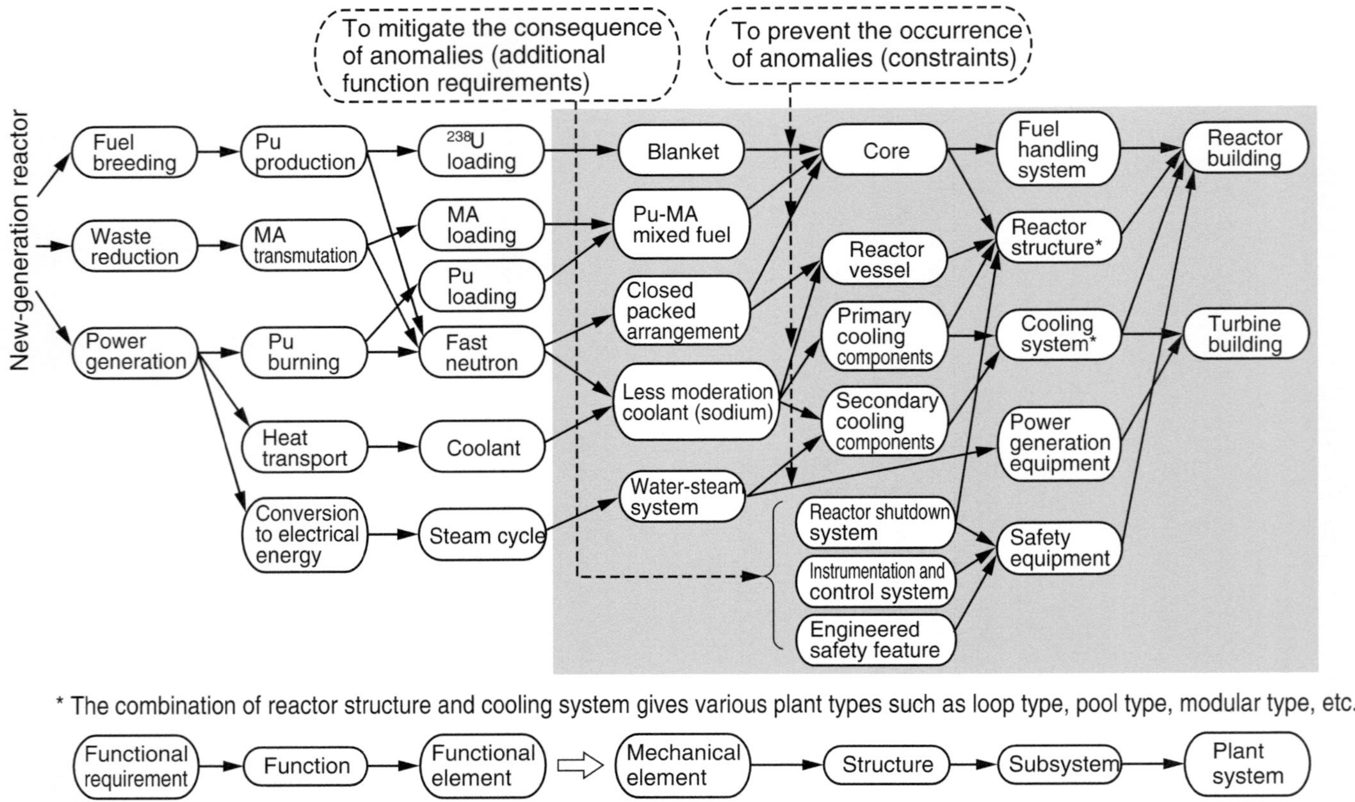

Fig. 6.1 Design flow of a new reactor from functional design to mechanism/structural design

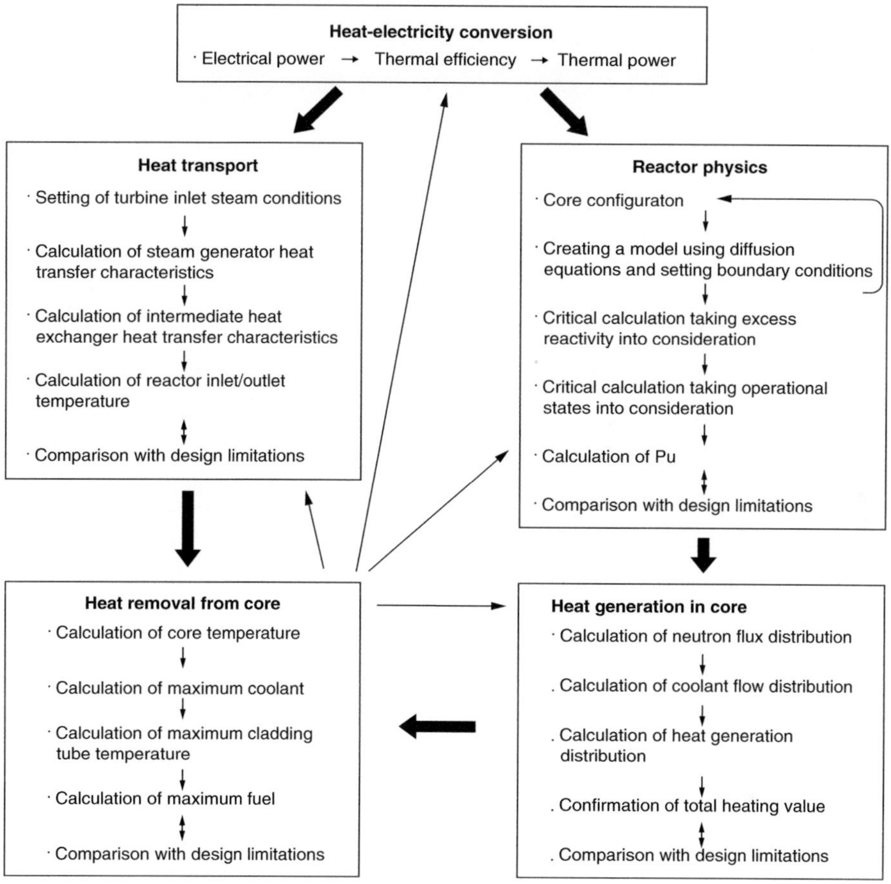

Fig. 6.2 Overall performance design flow

problems, all of which pertain to performance design, in an interrelated manner, as shown in Fig. 6.2, an overall performance design flow.

Before these various mechanisms can be built as part of the plant's systems and structures, they must first meet the constraint of anomaly prevention. For this purpose, the structures are evaluated for their accordance with safety design and structural design policies to be optimized within their respective limits, as illustrated in the schematic of structural design flow in Fig. 6.3.

In addition, a plant requires a fuel handling system for fuel assemblies containing U, Pu, and MA that have a relatively high radiation dose and generate a relatively high amount of heat, as well as blanket assemblies and an auxiliary sodium system for keeping sodium in a highly pure liquid state and for the safe handling of the sodium.

Furthermore, in order to add the function for moderating the consequence of anomalies discussed in Chap. 4, a reactor shutdown system (including control rods),

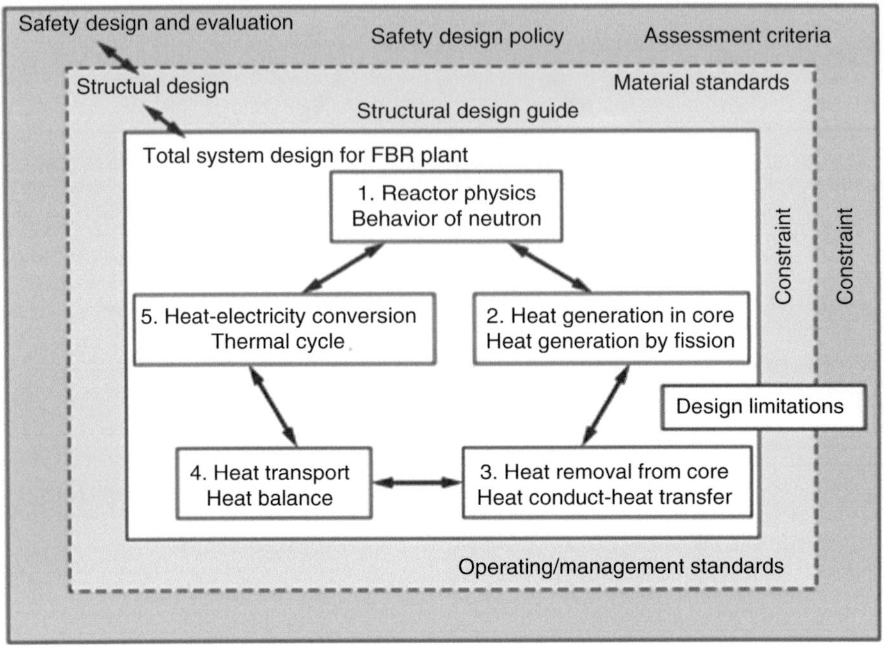

Fig. 6.3 Schematic of structural design flow

a decay heat removal system and a containment facility, which are engineered safety features, an instrumentation and control system for supporting the abovementioned functions, and the like are required.

After the abovementioned systems are designed, a picture of the entire plant needs to be designed in order to verify consistency when the major plant specifications and mechanisms are combined. Consequently, although the mechanisms and structures are not yet finalized, an integrated system in which the overall systems are conceptually combined is designed, and the main plant parameters, such as the inlet and outlet temperatures of components considering heat balance and the like, are calculated for the plant. Furthermore, a conceptual arrangement is designed to incorporate the systems in a reactor building or a turbine building. Based on these results, checks are done to confirm whether the respective system designs are well balanced as a whole, whether the desired plant characteristics can be reasonably achieved, and whether the design is feasible with the level of existing technology and future technological development plans. Problematic parameters or system designs that determine those parameters are corrected as needed. Such evaluations should be implemented as appropriate in accordance with the progress of design to avoid major redesigning.

6.2 Major System Configuration

The systems that constitute a plant can be divided roughly into the following categories:

The core
The reactor structure that constitutes and supports the core
The main cooling system for transferring heat from the core to a steam generator using sodium as the heat transferring medium
The water-steam system for supplying steam for driving a turbine generator and water to the steam generator using water as a heat transferring medium
The reactor shutdown system for shutting down the reactor in the event of an anomaly
The decay heat removal system for cooling the core after reactor shutdown
The reactor containment facility for containing radioactive material released from the core
The cover gas system for covering the surface of sodium coolant
The auxiliary sodium equipment for controlling the purity of sodium and the like
The fuel handling and storage facilities for exchanging new fuel and spent fuel
The instrumentation and control system for monitoring the state of the abovementioned systems

The respective systems are provided with normal or emergency power supply equipment and other auxiliary equipment [1–6].

6.2.1 Core Configuration

In a fast breeder reactor, blanket assemblies are placed around the core fuel region (in axial and radial directions) to absorb neutrons leaking from the core as effectively as possible to convert ^{238}U into ^{239}Pu. Control rods that control the reactor power and, when necessary, shut down the reactor are arranged in the core.

When fuel is evenly placed in a fast reactor core, more power is distributed in the center, and significantly less power is distributed in the peripheral regions. To address this issue, the coolant flow rate is adjusted in accordance with the power generated from each assembly. Furthermore, when the size of the core is increased, the abovementioned characteristics become extremely disadvantageous from both the perspective of safety and economy. Therefore, the core fuel region is divided into two regions such that less reactive fuel assemblies can be placed in the central region and more reactive fuel assemblies can be placed in the peripheral region to reduce the power output in the central region and increase the power output in the peripheral region and thereby flatten power distribution in the core fuel region as much as possible (Fig. 6.4) [1–6].

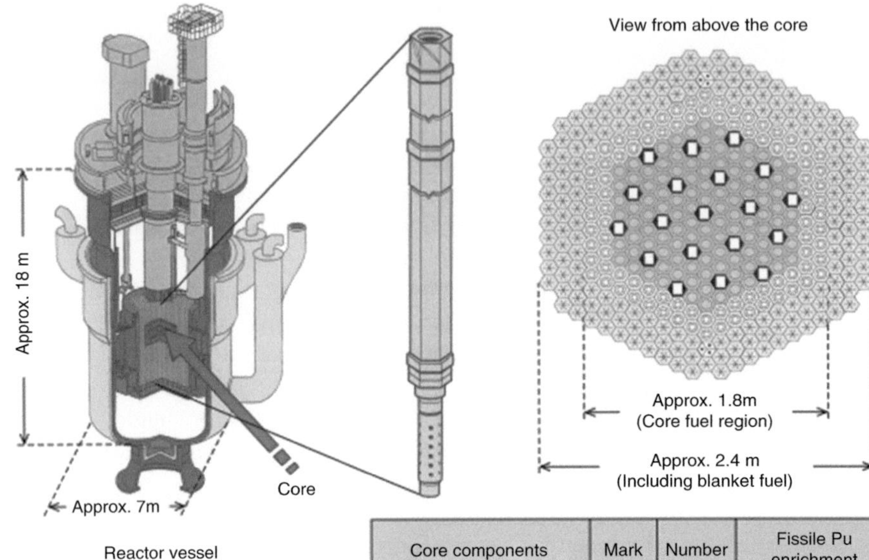

Core components		Mark	Number	Fissile Pu enrichment
Core fuel assembly	Inner		108	Approx. 15 wt%
	Outer		90	Approx. 20 wt%
Blanket fuel assembly			172	(Depleted uraniun)
Control rod assembly			19	-

Fig. 6.4 Example of core configuration (Monju) [6]

6.2.2 Reactor Structure Configuration

A reactor structure is comprised of a core support structure that retains and secures fuel and blanket assemblies, control rod guide tubes and the like inside a reactor vessel to form a core, a core barrel surrounding the core, an in-vessel relay rack that temporarily stores fuel removed from the core during refueling and transfers the fuel to a device for transporting the fuel out of the reactor vessel, a core internal support structure to support the respective equipment, and the like. In addition, an upper-core structure that supports control-rod driving mechanisms and the like from a shield plug, which corresponds to the lid of the reactor vessel, is inserted into the reactor vessel (i.e., to the space above the core).

The core support structure forms the core configuration and distributes the coolant flow rate in accordance with the power distribution of each fuel assembly (Fig. 6.5).

The core barrel is a device for shielding neutrons from the core and restricting the radial displacement and bowing of the core due to thermal expansion of the fuel assemblies, etc., and earthquake motion. Detail of core and fuel design will be described in Chap. 7 [2–5].

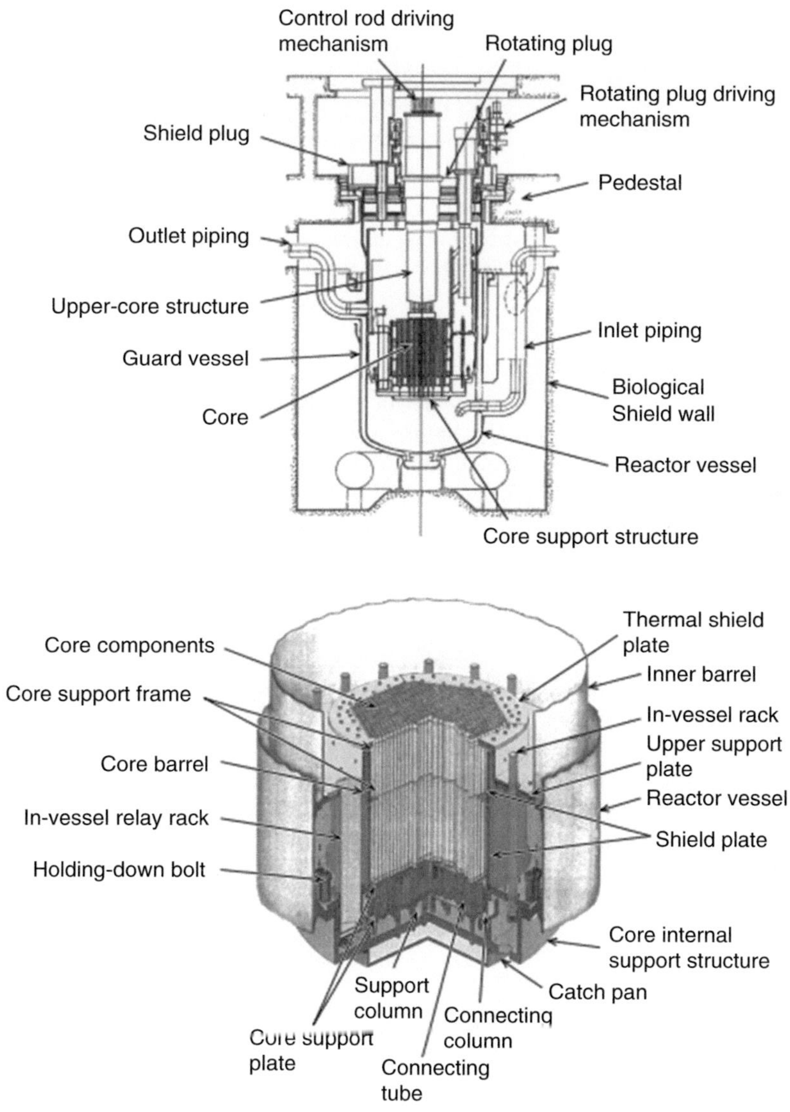

Fig. 6.5 Example of reactor vessel structure (Monju) [1, 3]

6.2.3 Main Cooling System Configuration

The main cooling system is comprised of a reactor vessel, intermediate heat exchangers (IHXs), steam generators, main circulation pumps, and piping systems that connect these members. The main cooling system consists of the primary cooling system, which transfers heat from the reactor vessel to the IHXs, and the secondary cooling system, which transfers heat from the IHXs to the steam generators. A water-steam system comprised of a steam system for supplying generated steam to a turbine generator and a feedwater system for feeding water to the steam generator is connected to the steam generator.

Depending on the arrangement of these components, the primary cooling system is categorized into loop and pool types. From the viewpoint of desired power output, the reactor is categorized into large and modular reactors. Both loop- and pool-type designs were adopted for large reactors. On the other hand, all modular reactor designs were of the pool type. The advantages and disadvantages of the two types are described below [1, 2, 4, 5].

1. Loop type

The loop type is used in primary cooling system configurations. In this configuration, the reactor vessel, IHXs, and circulation pumps are separately installed and connected by piping (Fig. 6.6). The loops required to connect other components are installed around the reactor vessel, in which the reactor core is contained. The required number of loops is determined on the basis of various factors, such as the reactor power, the size of the respective components, total economic competitiveness, and safety considerations.

Advantages of the loop type are as follows. The respective components can be designed in a relatively flexible manner since each component is highly independent, which facilitates development. In addition, since the respective components can be individually installed, the loop type allows for easier seismic design, maintainability, and repairability.

The disadvantages of the loop type are as follows. A relatively large footprint to address thermal stress of piping is disadvantageous in terms of construction cost. In addition, in the event of a piping rupture accident in the primary system, the primary coolant may leak and cause a sodium fire. In order to avoid this, elaborate

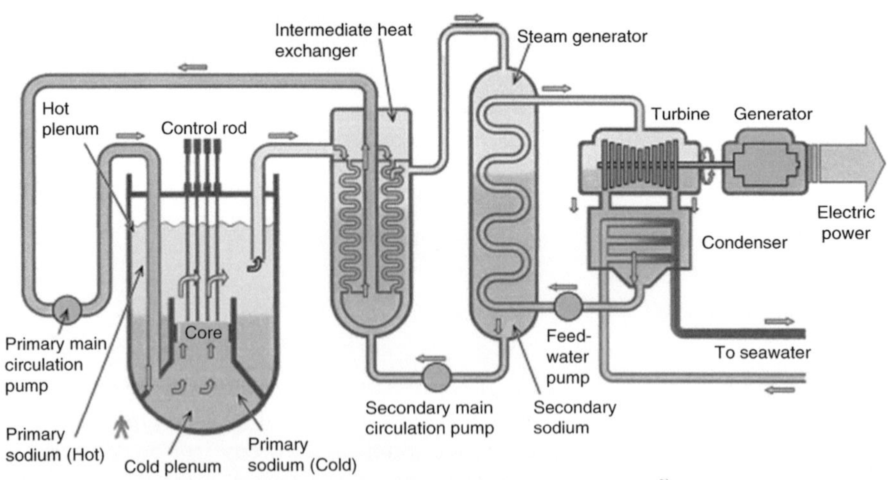

Fig. 6.6 Concept of a loop-type fast reactor [2]

equipment is needed to cope with the leak of the primary coolant (supplying nitrogen atmosphere to the primary system equipment chamber, installing a steel liner in the same chamber, and the like) [1–4]. Note that a major improvement strategy to address the economic disadvantages due to the piping installation space has been suggested by recent design studies, which will be discussed in paragraph 8.2.3.

Examples of the loop-type reactors include Monju, E. Fermi, BN-350, SNR-300, and CRBR. Figures 6.7 and 6.8 show the typical system configurations. The loop-

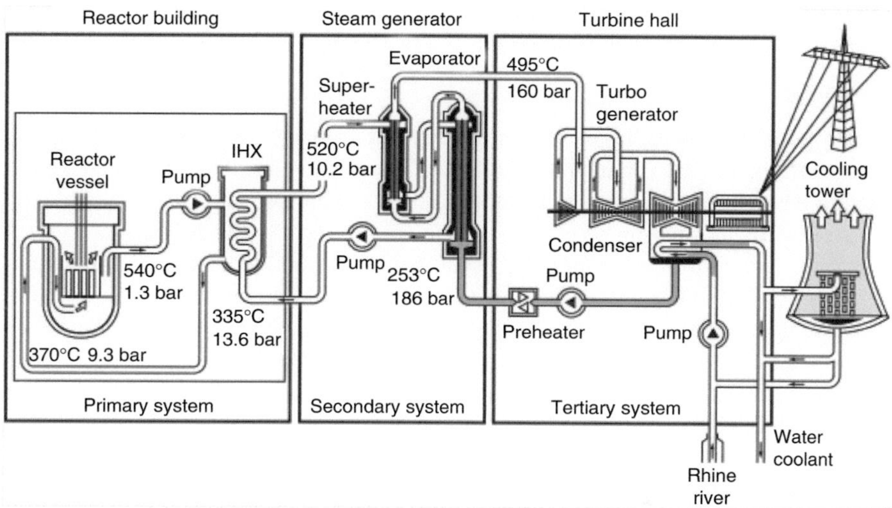

Fig. 6.7 Example 1 of cooling systems of a loop-type fast reactor (SNR-300, Germany) [1]

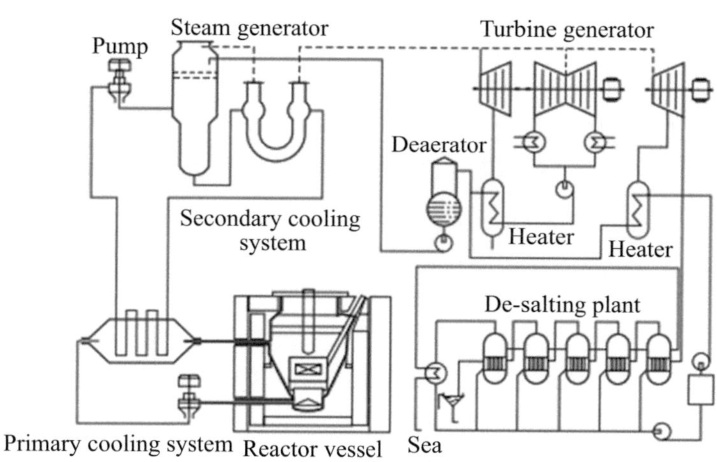

Fig. 6.8 Example 2 of cooling systems of a loop-type fast reactor (BN-350,Kazakhstan) [1]

type design is used in experimental as well as prototype reactors for its ease of development.

Because of the high boiling point of the coolant, there is no need for special pressurization. As a result, the primary cooling systems have a simple configuration in which an IHX and a circulation pump are connected by piping. In many cases, the circulation pump is placed in the downstream of the IHX (cold leg) with the aim of achieving a low-temperature design of the pump, which is an active component. However, in some cases, the circulation pump may be placed in the upstream of the IHX (hot leg) to improve the heat transfer characteristics of the IHX (due to the more complicated primary coolant flow path) because placing the IHX on the pump discharge side eliminates the constraint for pressure drop on suction pressure [1–6].

2. Pool type

The pool type (tank type) is another type of primary cooling system. In this configuration, the IHX and the circulation pump are placed in the main vessel (reactor vessel), in which the reactor core is contained, and are separated by partition walls and piping structure to form the flow paths inside the main vessel (Fig. 6.9). The number of components (number of circuits) to be placed in the main vessel is determined in a manner similar to that for the loop type.

Advantages of the pool type include the following. Compact arrangement of components can be achieved, which is advantageous from the standpoint of construction cost. The shape of the main vessel is relatively simple, which makes the main vessel less likely to be damaged. Even if the partition inside the main vessel fails, since there is sodium on both the inside and outside of the partition structure except the section of the main vessel that is in contact with sodium, the primary

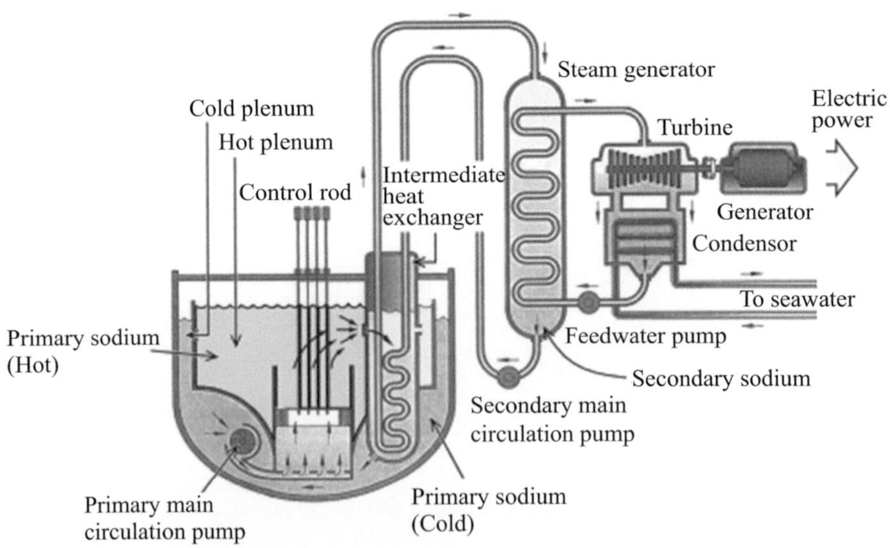

Fig. 6.9 Concept of a pool-type fast reactor [2]

coolant is unlikely to leak into the atmosphere to cause sodium fire. Therefore, measures against the leak of primary coolant can be implemented relatively easily. Furthermore, because of the large amount of coolant in the main vessel, heat capacity effects can be used to remove decay heat, and a rapid temperature change in the coolant is less likely to occur.

Meanwhile, the pool type has the following disadvantages. Since the respective components are placed inside the main vessel, their shapes and arrangement are constrained, which requires careful attention to design. Close attention must also be paid to ensuring shielding performance, to avoid significant activation of the secondary coolant, and the maintainability and repairability of components, structural members, and the like in the main vessel. Due to its large size (generally, 20 m in diameter), transporting the main vessel is quite difficult during reactor construction. As a result, more manufacturing work has to be done on site, which requires special attention to ensure quality. It is also difficult to secure seismic resistance. Furthermore, the internal structures in the main vessel, which form the flow paths for the primary coolant, undergo relative displacement due to thermal expansion and the like. To address this issue, an advanced flow evaluation, temperature distribution evaluation, and structural design are necessary [1, 2, 4, 5].

Examples of the pool-type reactors include PFR, Phenix, Super Phenix, and BN-600. Figures 6.10 and 6.11 show the typical system configurations. The pool-type configuration is often used in reactors which are designed as demonstration or commercial reactors after a prototype reactor phase for its economic advantages.

In all these reactors, large main vessels are placed to make contact with the coolant on the low-temperature side (i.e., core inlet side) so that the main vessel does not become the subject of high-temperature design. In all pool-type reactors, internal structures form the flow paths. Taking this feature into account, pumps are

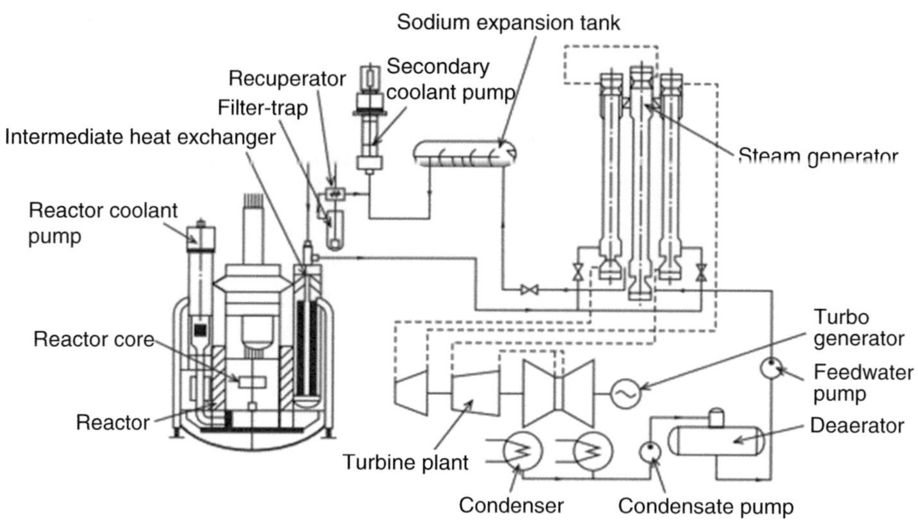

Fig. 6.10 Example 1 of cooling systems of a pool-type fast reactor (BN-600, Russia)

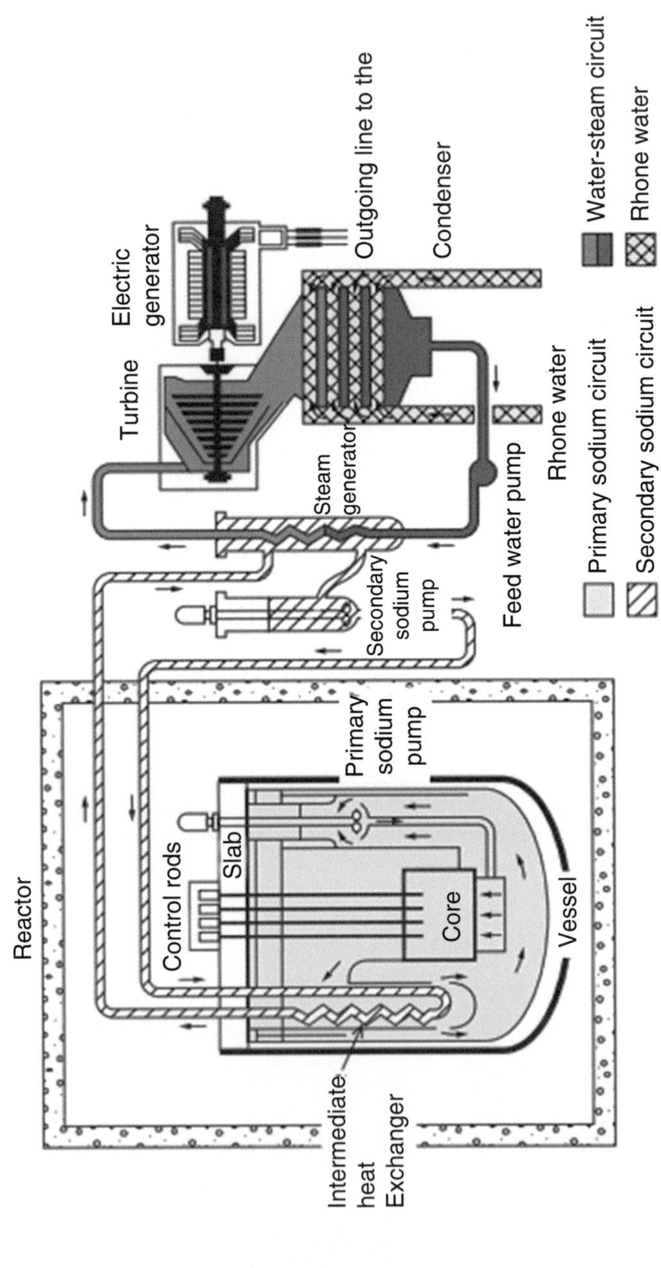

Fig. 6.11 Example 2 of cooling systems of a pool-type fast reactor (Phenix, France) [1]

placed immediately upstream of the core inlet (cold leg) to limit the sections that are subjected to the discharge pressure from the pumps [1, 2, 4, 5].

3. Modular type

As opposed to large-scale reactors, which mainly focus on the advantage of scale resulting from large-size components and reduction of cycle cost through increased fuel burnup, a modular-type reactor is a plant design based on the concept of mass production efficiency by reducing the size of a single unit and mass-producing facilities of the same type. The modular-type reactors also aim to establish unique safety logic by avoiding the core characteristics, which are relatively disadvantageous in large-scale FRs, by reducing the size of the core. Since the components are small and can be easily transported, more manufacturing work can be done in a factory during power plant construction, and thereby on-site work can be reduced. As a result, the components may become more reliable, and the plant can be constructed in a shorter period of time. In addition, restrictions on plant site selection are reduced. On the other hand, in terms of the amount of energy obtained per unit area of the site, the modular-type reactors are considered to be disadvantageous. The modular type is defined by the size of a plant. In previous design examples, pool-type reactors, which are economically advantageous, are typically used for modular-type plants aiming for reduced size because difficulties resulting from the use of large components associated with the pool-type reactors can be avoided.

Figures 6.12 and 6.13 show the ALMR reactor as an example of the modular-type reactor [1, 2, 4].

6.2.4 Configurations of Fuel Handling and Storage Facilities

Fuel handling and storage facilities are comprised of equipment used for operations such as carrying in new fuel, loading new fuel into the core, removing spent fuel from the core, transferring spent fuel to a fuel storage facility, and storing spent fuel in the spent fuel storage facility until it is transported to off-site facilities. Figure 6.14 shows a system configuration related to fuel handling and storage facilities.

New fuel is temporarily stored in a new fuel storage rack before being loaded into the core. The new fuel is transferred to the reactor vessel by fuel transfer equipment via a fuel transfer machine, such as an under-floor transfer car, and is then loaded into an in-vessel transfer machine. Two fuel assemblies (one new fuel assembly and one spent fuel assembly) can be loaded into the in-vessel transfer machine. A new fuel assembly is loaded at a prescribed position in the core (position from which a spent fuel assembly is removed) by a fuel exchange machine via the in-vessel transfer machine.

Spent fuel is pulled out from the core by the fuel exchange machine and transferred to the in-vessel transfer machine in the reverse order of that for new fuel. The spent fuel is removed from the in-vessel transfer machine to the outside of

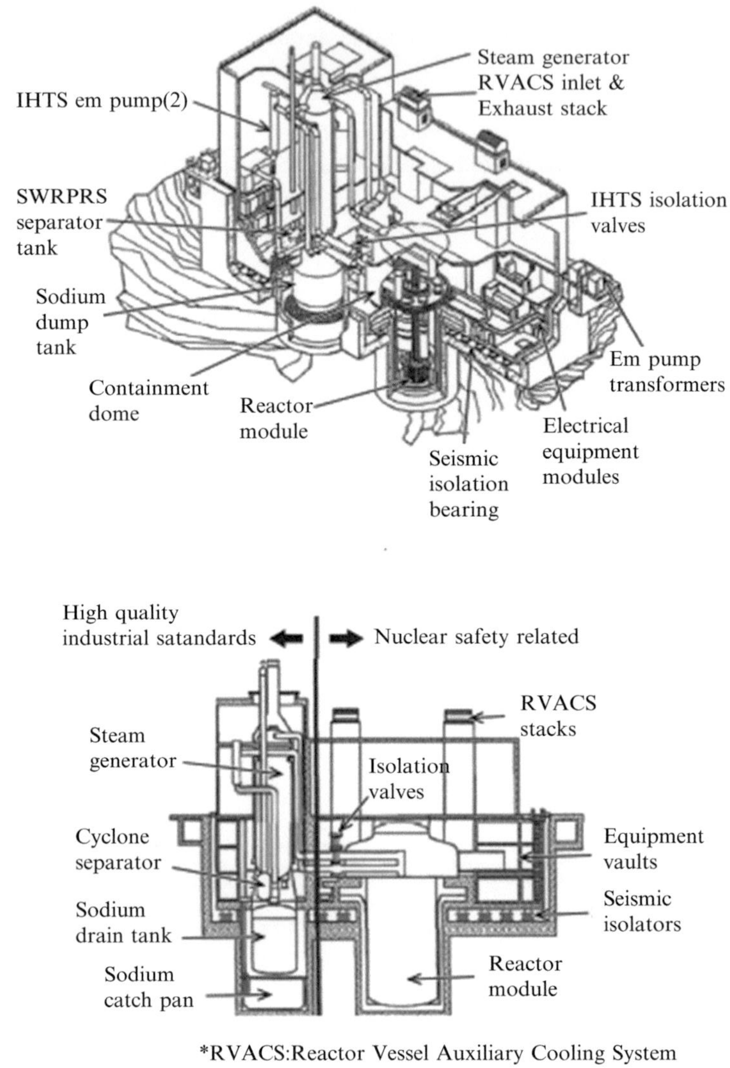

Fig. 6.12 A modular-type fast reactor [1]

the reactor vessel by the fuel transfer equipment. Immediately after removal from the reactor vessel, sodium is still attached to the spent fuel. Because of the significant decay heat, the spent fuel is temporarily stored and cooled inside an ex-vessel fuel storage equipment (under sodium) before the sodium attached to it is cleaned in fuel cleaning equipment. The spent fuel is then transferred to a fuel pool where it is stored until transport to off-site facilities. Between the reactor vessel and the ex-vessel fuel storage equipment, fuel is handled while being stored in a fuel pot in which sodium is placed in order to better cool the fuel.

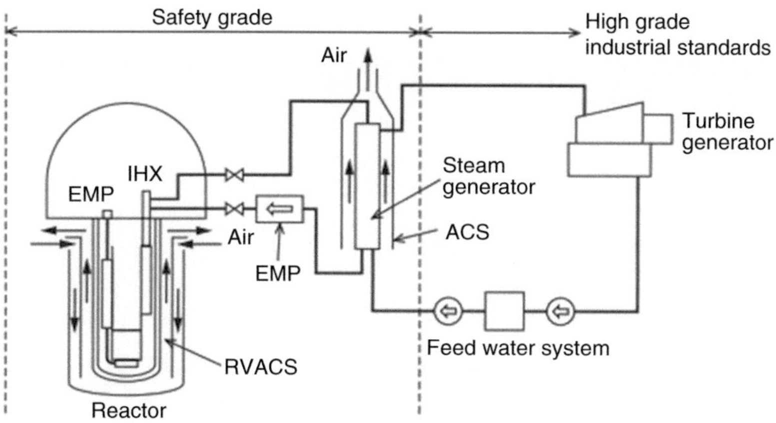

Fig. 6.13 Example of cooling systems of a modular-type fast reactor (ALMR, USA)

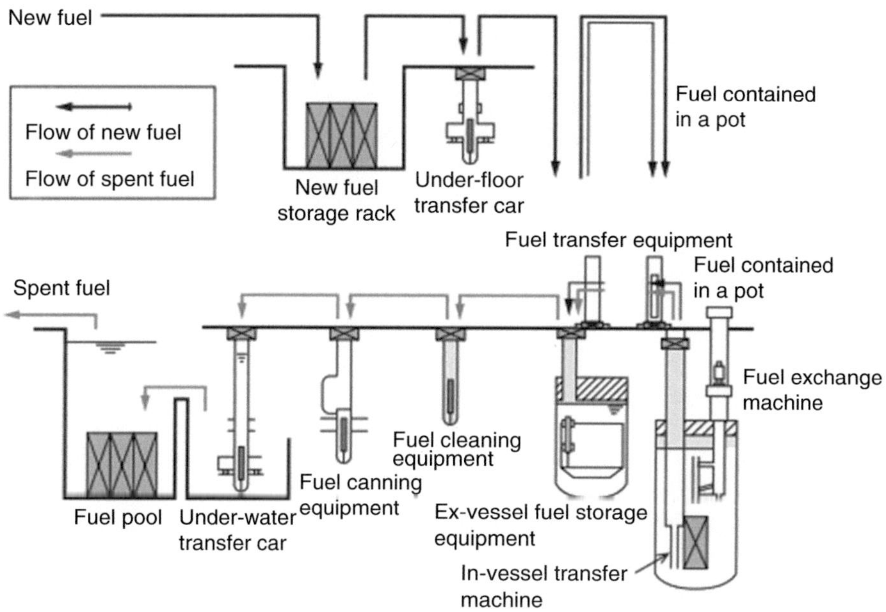

Fig. 6.14 System configuration of fuel handling and storage facilities (Monju) [7]

Since new and spent fuel is carried into and discharged from the core through the containment vessel boundary, travelling fuel transfer equipment (Fig. 6.15) was adopted for Monju, and the containment vessel was designed to be temporarily opened while refueling.

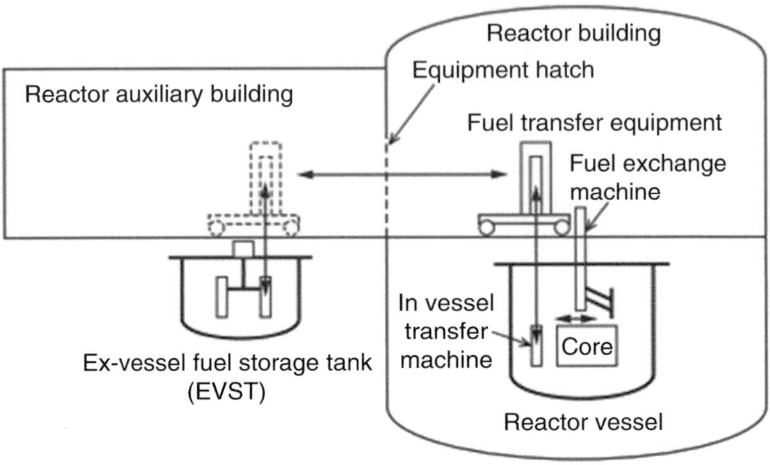

Fig. 6.15 Example of travelling fuel transfer equipment (Monju)

In Monju, spent fuel is sealed inside a highly airtight can by fuel canning equipment when the spent fuel is stored in the fuel pool in consideration of radioactive material leak from failed fuel and the like.

The storage capacity of ex-vessel fuel storage equipment is determined as follows: the number of stored assemblies waiting to be cooled (waiting to be transferred to the fuel pool) plus the number of stored assemblies waiting to be loaded into the core (number of fuel assemblies to be replaced at a time). In addition, urgent evacuation of all fuel from the core (the number of fuel assemblies in one core) in case of emergency is taken into consideration.

In smaller reactors, the number of fuel assemblies to be replaced at a time is small. Because of this, spent fuel may be stored and cooled in sodium in a fuel rack, which is placed on the outer periphery of the core inside the reactor vessel (e.g., Joyo and E. Fermi) [5, 6]. When spent fuel is temporarily stored inside the core, the system for cooling the core also cools the spent fuel. As a result, cooling of the spent fuel and containment of radioactive material can be done in a highly reliable manner. However, sodium attached on spent fuel assemblies must be cleaned during a refueling period if there is no under-sodium storage equipment outside the reactor vessel. Consequently, the refueling period is prolonged when a large number of fuel assemblies need to be replaced. In addition, there are other challenges such as the increase in the size of the reactor vessel, influence on neutron instrumentation, and more complex fuel transfer equipment. Application of this design to large reactors is thus considered difficult.

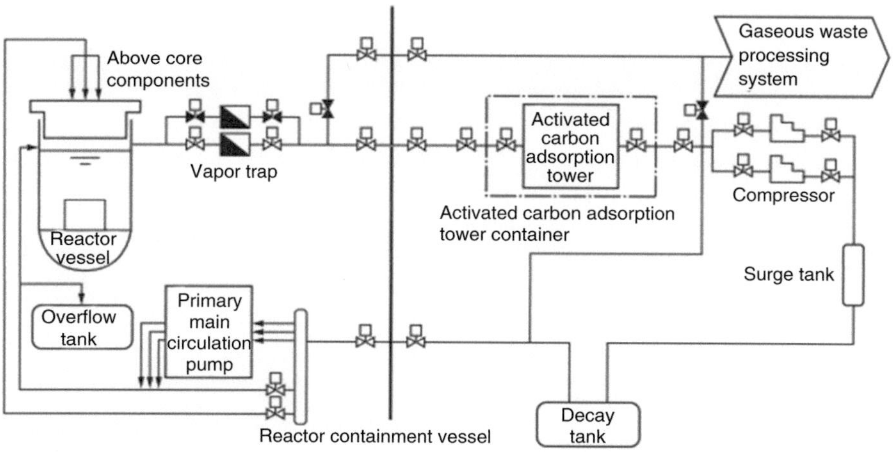

Fig. 6.16 Example of cover gas system (Monju) [3]

6.2.5 Auxiliary Sodium Equipment Configuration

1. Cover gas system configuration

An inert gas (typically, argon gas) is provided above the liquid level in a component that contains sodium (gas space) in order to inactivate the surface of sodium, which is chemically active, and to prevent a negative pressure in a sodium system. This gas is referred to as cover gas. An example of the cover gas system is shown in Fig. 6.16.

Generally, to ensure the purity of cover gas, the pressure in a cover gas system is set higher than the pressure of the atmosphere outside component/piping at every point in the system so that outside gas never flows into the cover gas system in the event of the failure of the cover gas system piping [2–4].

In Monju, cover gas is cleaned and recirculated to reduce the amount of gaseous waste generated. The cover gas cleanup system used in Monju includes an activated carbon adsorption tower and a decay tank, both of which use the decaying of radioactive material. A liquefaction-distillation method, which uses the difference in boiling points between cover gas and FP gas, is another cleaning method.

2. Other Auxiliary Sodium Equipment Configuration

Auxiliary sodium equipment in a sodium-cooled FR includes the following: an overflow system for keeping the sodium level in a reactor vessel at a prescribed level, a sodium purification system for maintaining the purity of sodium, and a charge/drain system that temporarily receives sodium during maintenance and repair of sodium components. Figure 6.17 shows an example of auxiliary sodium equipment design [1–4].

As a method for keeping a constant sodium level, the overflow method is typically used. In this method, an overflow tank with a capacity that can

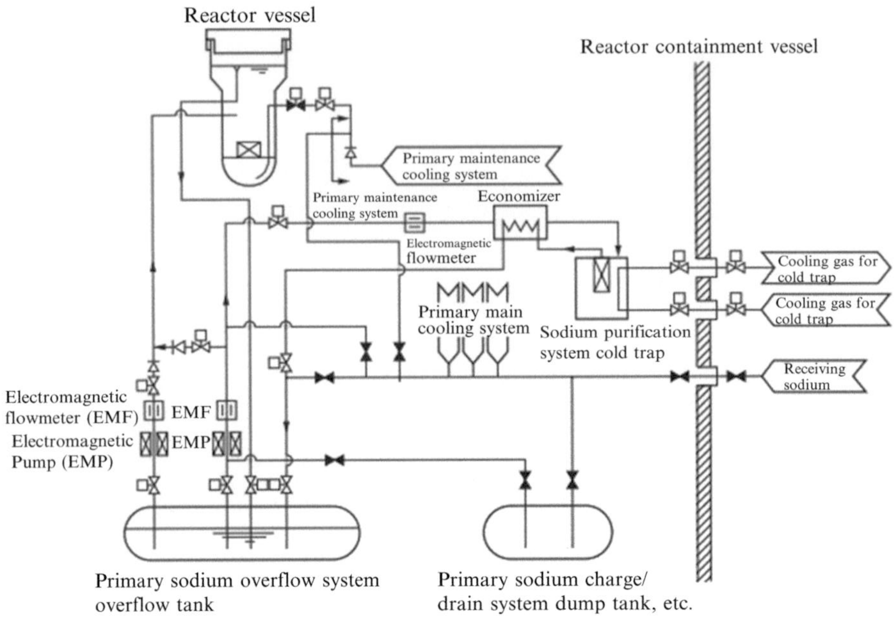

Fig. 6.17 Example of auxiliary sodium equipment (Monju) [3]

accommodate the thermal expansion of sodium, which is a major cause of level change, is prepared. In order to simplify the system, instead of providing an overflow system, there is an arrangement in which a large space is secured above the free liquid level of coolant so that the space accommodates the thermal expansion of coolant. In this arrangement, attention must be paid to liquid level changes inside the system caused by a change in operation state.

A cold trap that uses the precipitation of impurity with a decrease in temperature is typically used as the device for controlling the purity of sodium. To improve the thermal efficiency, an economizer (i.e., dump heat exchanger) is usually provided at the inlet/outlet of the cold trap to recover heat.

Draining of sodium from a sodium component is achieved for the most part by gravity. Charging is typically performed through the pumping up of sodium by an electromagnetic pump and the pushing up of sodium by a difference in cover gas pressure. Before charging sodium, the relevant sodium components need to be preheated to a prescribed temperature by an electric heater or heated gas.

6.2.6 Safety System Configuration

1. Shutdown system configuration

In general, the shutdown system of an FR uses control rods. In addition to a main shutdown system that has both normal reactor control function and primary shutdown function, a backup shutdown system that only has a reactor shutdown function is typically provided from the standpoint of ensuring the multiplicity of the shutdown system for higher reliability. These multiplexed shutdown systems are made as diverse as possible by using mutually different disconnecting structures, acceleration mechanisms, and the like to improve the reliability of the reactor shutdown function.

Various highly passive systems have been suggested as other backup shutdown systems. Such systems include the following:

A system in which a temperature-sensitive magnet that uses the Curie point is employed to allow for dropping the control rod when the holding force of the magnet is reduced solely by an increase in coolant temperature without depending on a control signal (self-actuated shutdown system (SASS))

A system that uses gas expansion caused by an increase in coolant temperature (gas expansion mechanism (GEM))

Systems in which neutron absorbers are dropped to insert the core due to the decrease in hydraulic force at a reduced coolant flow rate or by melting a low-melting-point metal [1–4].

2. Decay Heat Removal System Configuration

In the decay heat removal system of an FR, coolant is, even if no special pressure is applied, always in a liquid phase ranging from the normal operation mode to the decay heat removal state in the event of an accident. Therefore, the configuration can be simpler compared to that of a light-water reactor (LWR). In addition, should a coolant boundary be damaged, a loss of coolant caused by decompression boiling of the coolant does not occur, thereby eliminating the need for emergency injection of the coolant. Figures 6.18 and 6.19 show examples of measures to prevent the leak of primary coolant (double-wall vessel and piping method) using these characteristics [1–6].

The measures to prevent coolant leak using the double-wall vessel and piping method are attained by providing double walls in the vessel and piping in the primary system. The entire region that is in contact with the coolant is double walled, and the gap between the inner vessel/inner piping and the outer vessel/piping is made airtight so that if the inner vessel or piping that constitutes the

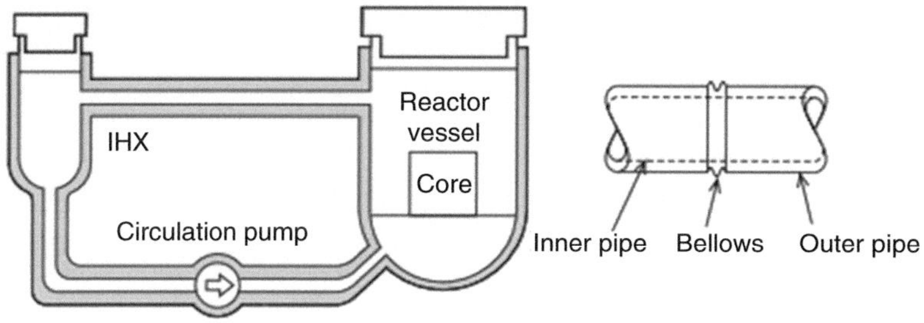

Fig. 6.18 Measures to prevent coolant leak using double-wall piping and vessels

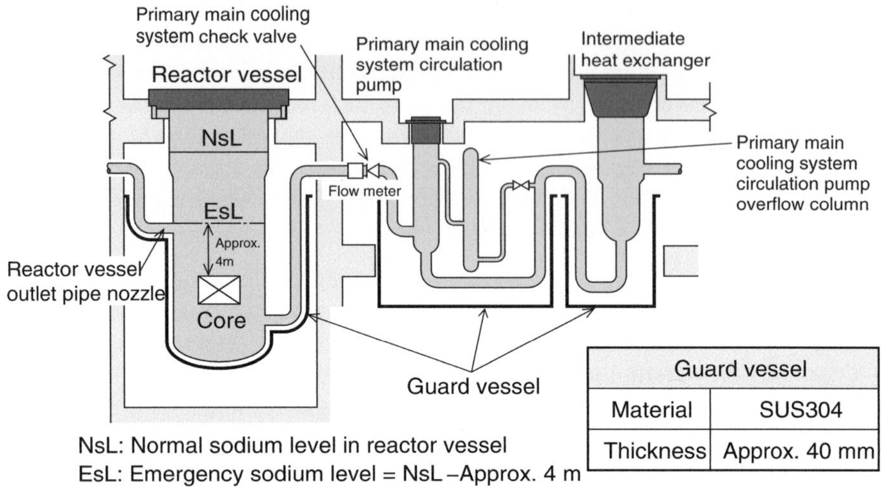

Fig. 6.19 Measures against sodium leak using high-elevation layout of primary cooling system piping and guard vessel

boundary is damaged, the outer vessel or piping keeps the leaking coolant and maintains a coolant level necessary for decay heat removal (Fig. 6.18). The piping layout is not restricted, unlike the high-elevation horizontal layout, which is described below. Therefore, a compact piping layout is applicable. In addition, detection of sodium leak is made easier because of the limited gap space, which alleviates the need for significant measures against burning (fire) of leaking sodium. However, sufficient technical considerations, such as the placement of bellows on the outer vessel/piping, should be given to absorbing the difference in thermal expansion between the inner vessel/piping and the outer vessel/piping, particularly, the transient expansion difference during transients. Furthermore, attention must be paid to the structure that supports the inner vessel/piping from the outside of the outer vessel/piping.

In the high-elevation cooling system piping layout and guard vessel method that are employed as a measure against coolant leak (Fig. 6.19), if high-elevation piping breaks, coolant flows out, and once the liquid level is reduced to the opening level, coolant is no longer discharged by internal pressure unless the pump is operated at a high speed. Therefore, the high-elevation piping is designed to be placed at a prescribed level (referred to as SsL) or higher so that the static liquid level of coolant after completion of the coolant leak becomes equal to or higher than the level necessary for decay heat removal (referred to as EsL: the level at which the liquid level inside the reactor vessel is sufficiently above the upper end of the outlet nozzle) (Fig. 6.20). With regard to components and piping placed at a lower level, the static liquid level after the coolant leak is designed to be maintained at a level necessary for decay heat removal by providing guard vessels (outer vessels) and receiving the leaking coolant in the guard vessels (Fig. 6.21). This is based on the

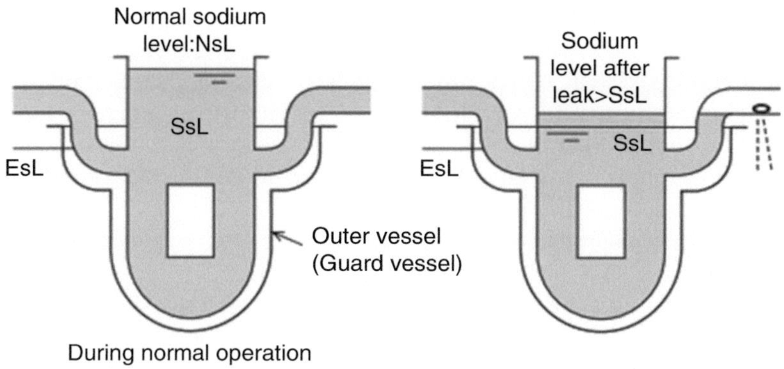

Fig. 6.20 Concept of maintaining sodium level by means of high-elevation layout of cooling system piping (after a pipe break at a high elevation)

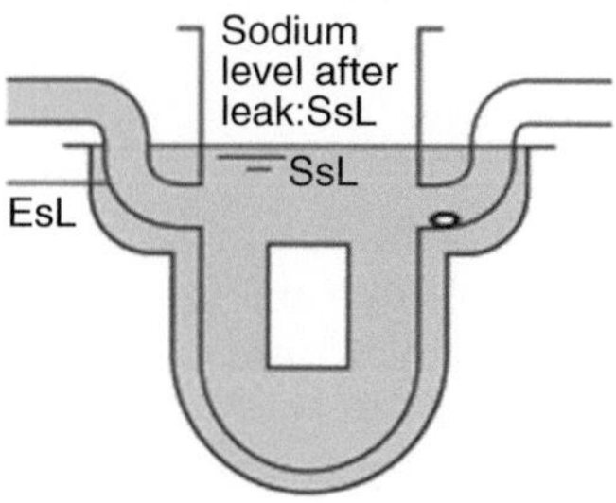

Fig. 6.21 Concept of maintaining sodium level by means of guard vessel (after a pipe break at a low elevation)

principle that coolant leak stops when the coolant levels inside and outside the damaged piping (internal and external pressures) become equal to each other. Although complex structure of the vessels and piping can be avoided, the piping layout is constrained to the high-elevation horizontal layout. As a result, sufficient attention must be paid to this constraint (see Paragraph 8.3.3). In addition, the main vessels (reactor vessel, IHX, and the like) and the guard vessels need to be structurally independent of each other in terms of support and handling of thermal expansion to avoid a common cause failure.

In the high-elevation cooling system piping layout and guard vessel method, the volume of the free space in the guard vessel is limited to less than an allowable amount of leak to ensure that the liquid level inside the guard vessel is increased to SsL due to a sodium leak that does not reduce the liquid level in the reactor vessel below the allowable limit (EsL) when damage occurs in the lower section, thereby stopping the leak using the difference between internal and external pressures. Furthermore, to allow decay heat removal operation of the circulation pumps even when the liquid level inside the reactor vessel is reduced to EsL, the shaft length, discharge pressure during low-speed operation, and the like of the pumps must be examined.

In the double-wall vessel/piping method and the high-elevation cooling system piping layout and guard vessel method, heat insulator for the vessel and piping of the main cooling system is typically provided on the outer surface of the outer vessel/piping or the guard vessel. In the guard vessel method, a certain amount of gap can be secured within the volume limit to allow volumetric inspection as an in-service inspection in a relatively easy manner. In the double-wall vessel and piping method, however, volumetric inspection is often difficult.

In an FR, the coolant temperature is maintained at approximately 200 °C even when the reactor is shut down, and in most previous reactors, the air is used as the final heat sink. This is a difference from most LWRs, which have relied on seawater as the final heat sink. However, it should be noted that seawater is often used to cool a component cooling water system and an emergency generator, which are important to safety, in FRs. Like LWRs, diverse final heat sinks are sought as a result of the lessons learned from the Fukushima accident.

Figure 6.22 shows examples of typical paths to show at which location decay heat removed from the core is guided to the final heat sink. The examples include the following: a direct reactor auxiliary cooling system (DRACS), in which the heat is removed directly from the reactor vessel; a primary reactor auxiliary cooling system (PRACS), in which the heat is removed from the primary cooling system; an intermediate reactor auxiliary cooling system (IRACS), in which the heat is removed from the intermediate cooling system; and a steam generator auxiliary heat removal system (SGAHRS), in which the heat is removed from the steam generator.

Generally, a mode in which the heat is removed from a section closer to the core can reduce the number of related components up to the final heat sink and is considered to be more reliable. However, to select a heat removal mode, the heat removal performance, simplicity of the total system, operational reliability, economic competitiveness, and the like must be assessed in a comprehensive manner. Multiple modes may be combined.

In most FRs, coolant is circulated by an emergency drive motor (pony motor) installed in a main circulation pump. Since sodium has a relatively large thermal expansion coefficient and a high boiling point, a great temperature difference can be obtained between hot and cold legs, allowing for significant natural circulation. As a result, decay heat can be removed by natural circulation in most FRs. In order to take advantage of heat removal by natural circulation, the difference in elevation of heat exchangers is important. Figure 6.23 shows the elevations of components in Monju [1–5].

3. Containment facility configuration

The containment facility contains radioactive material released from the core in the event of an accident to prevent the radioactive material from spreading to the environment. The containment vessel in an FR is very large (approximately 50 m in diameter in Monju) since it contains component/piping of the primary main cooling system and auxiliary primary sodium equipment. In the event of a coolant leak accident, neither eruption of high-pressure steam nor decompression boiling of

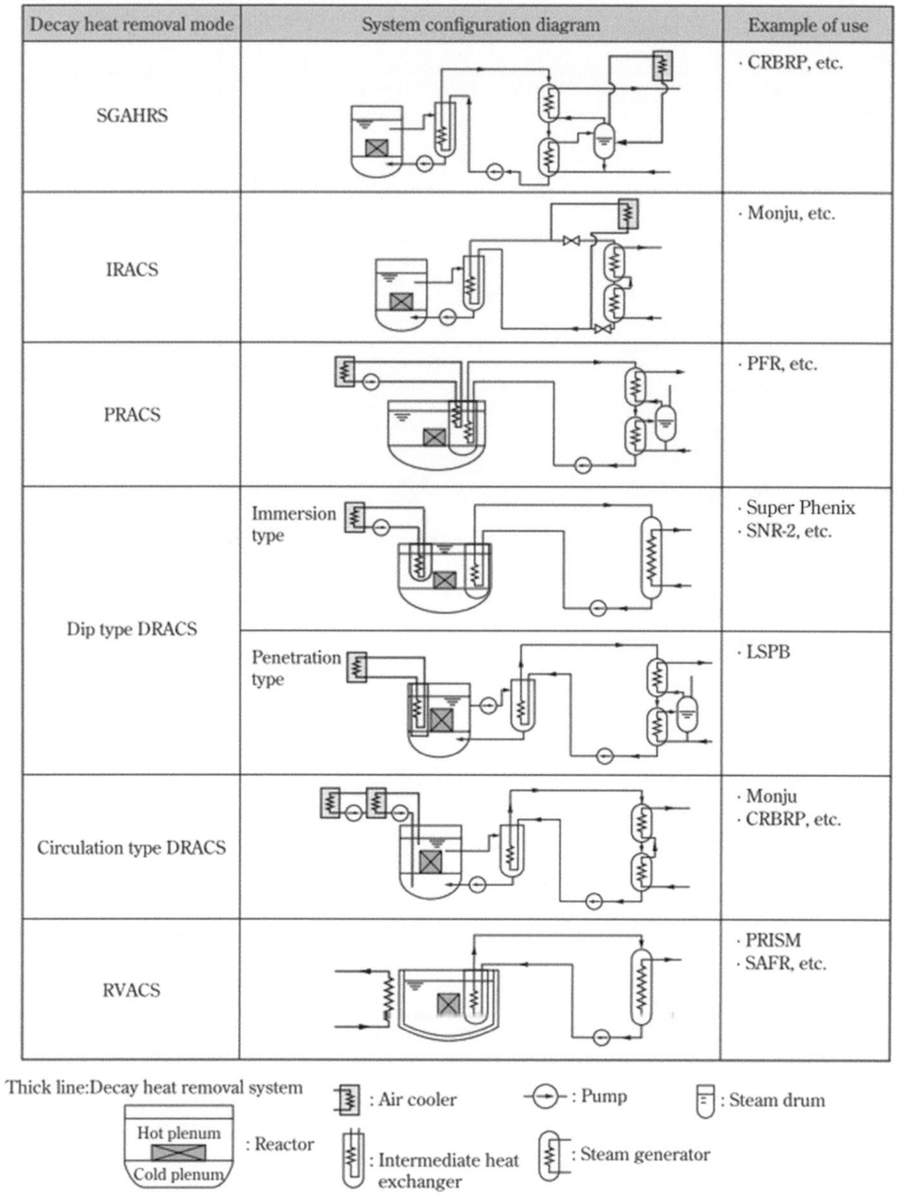

Fig. 6.22 Examples of decay heat removal modes [2]

coolant would occur. Therefore, there is no major cause for pressurization of the containment vessel, and the vessel is generally not required to be highly pressure resistant. Because of this feature, many types of containment vessel methods are employed in FRs. These include the steel containment vessel method, the steel-

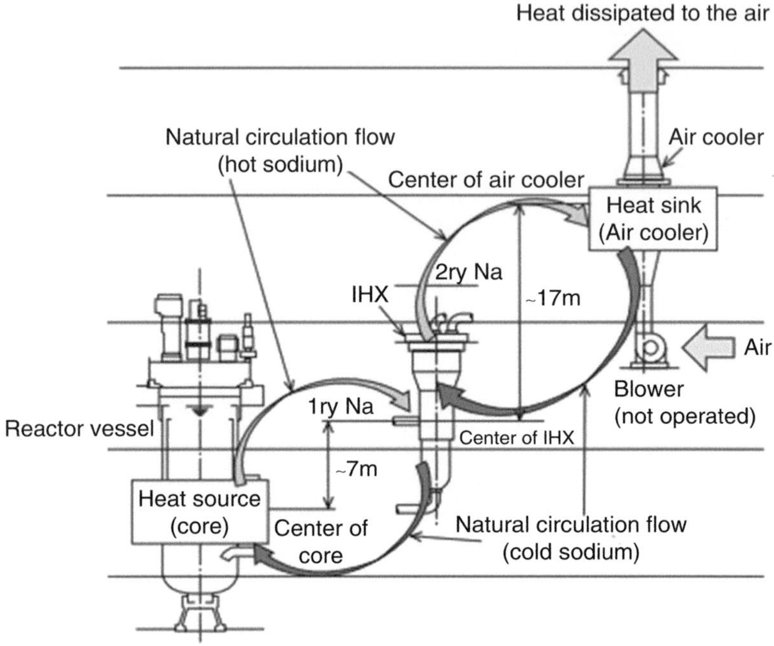

Fig. 6.23 Example of the elevations of cooling components (Monju) [7]

lined concrete building containment method, and the steel dome safety vessel method.

The steel containment vessel generally has hemispherical or semielliptical mirrors that are attached to the top and bottom sections of the cylindrical body. Like a PWR, a shield wall is provided on the outside, and the gap between the shield wall and the containment vessel is kept at a negative pressure as an annulus space. Exhaust air is cleaned by a high-efficiency filter and is then discharged from a stack (Figs. 6.24 and 6.25).

Although the thickness of the containment vessel required for the internal pressure is thin, close attention must be paid to securing rigidity and seismic resistance since the vessel is quite large in size. To ensure sufficient rigidity and seismic resistance, a stiffening ring may be placed on the outer surface of the steel containment vessel. The steel containment vessel method is disadvantageous in terms of component layout (space tends to be wasted). This method is used in Joyo, Monju, E. Fermi reactor, CRBR, and the like.

The steel-lined concrete building containment structure relies on concrete for its structural strength. A steel lining is attached to the outside or inside of the concrete structure to make the structure airtight. The gap between the concrete building and the steel lining is kept at a negative pressure, and a filter is provided for filtering the exhaust air to reduce the release of radioactive material (Fig. 6.26).

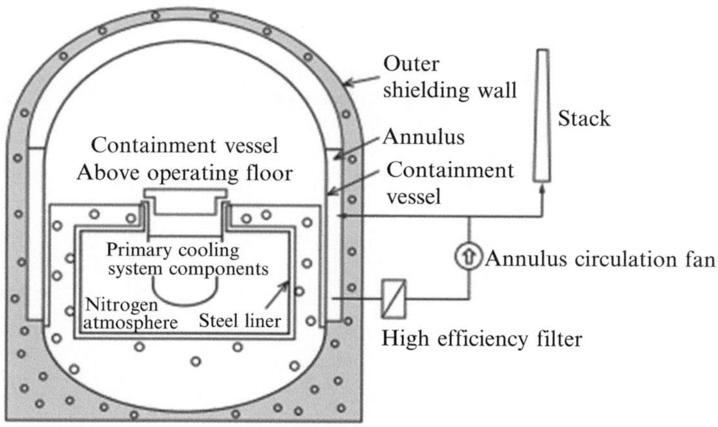

Fig. 6.24 Concept of containment system (Monju)

The steel-lined concrete building containment structure can be flexibly designed in accordance with a component layout, and the containment structure can be designed to accommodate the building design. As a result, a compact plant design can be achieved. SNR-300 is a typical example of this structure.

A method of placing a steel dome on the upper section of a reactor vessel (main vessel), placing a concrete containment vessel on the dome, and placing a safety vessel (guard vessel) on the outside the lower half of the reactor vessel to form a containment boundary may be used instead. The structure of the lower half uses the reactor vessel auxiliary cooling system (RVACS), in which decay heat of the core is removed by a cooling medium or outside air via the safety vessel, which is a decay heat removal system in the event of a major accident. Containment facilities of Super Phenix and the modular ALMR are examples that use this system (Figs. 6.27 and 6.28). In systems where the guard vessel (safety vessel) of the main vessel also functions as the containment vessel, decay heat removal can be performed in a more reliable manner. However, the reliability of containment performance must be considered thoroughly [1–6].

In the containment facility, penetration for piping and electric wiring, a hatch for carrying in/out the components, an air lock for personnel entry/exit, and the like are provided. All these penetrations have strict airtight structures through the use of a double-seal structure, the welding of a bellow seal structure, and the like. In addition, these structures allow for checking a local leak rate. Many methods are available for the transportation of new and spent fuel assemblies through a containment vessel as described in Sect. 8.3.

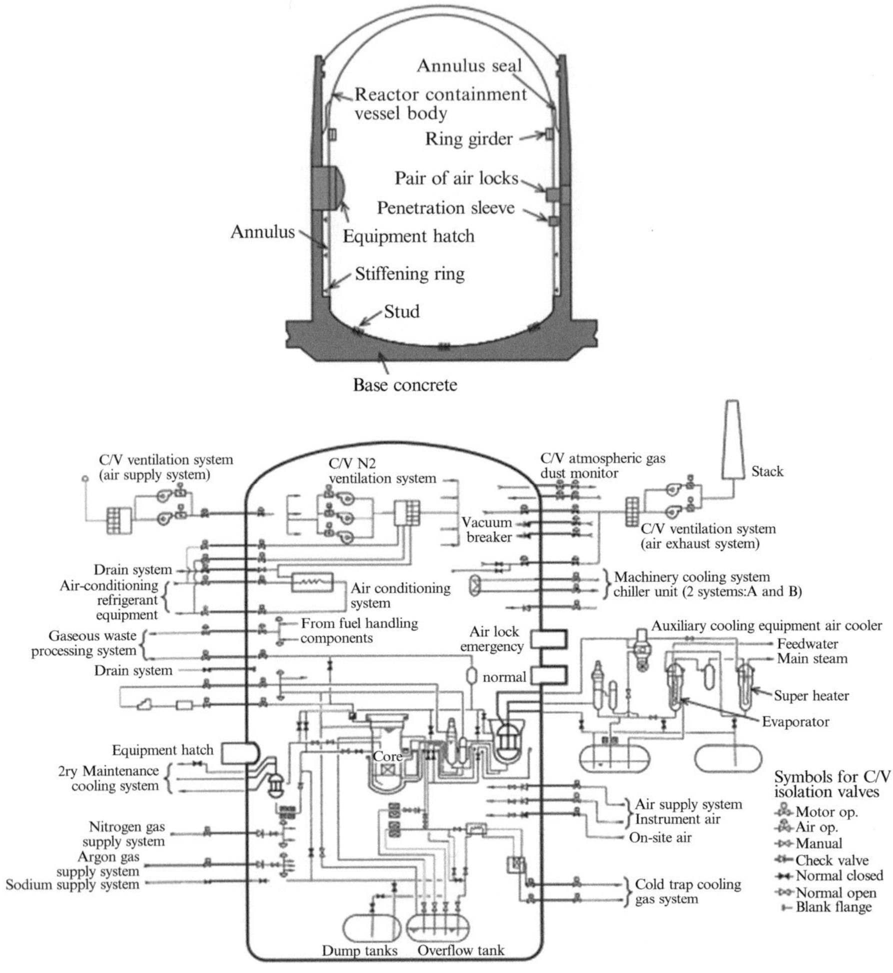

Fig. 6.25 Example of boundary arrangement of steel containment vessel (Monju) [3]

6.2.7 Instrumentation and Control System

Instrumentation and control systems that provide necessary accuracy and reliability are installed for the functions required from the viewpoint of operating a reactor system with the highest performance.

The instrumentation and control systems can be classified into those pertaining to “the safety protection system” that is essential to ensuring reactor safety, those required for normal operation, and those provided to improve safety. The safety protection system detects anomalies of the reactor system and generates a signal to scram the reactor and/or actuate engineered safety features if necessary in case of emergency in order to ensure reactor safety. Table 6.1 shows examples of reactor trip (emergency shutdown) signals [3].

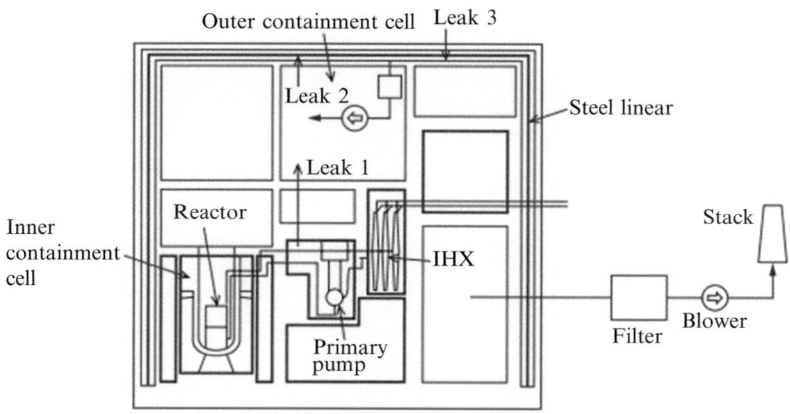

Fig. 6.26 Steel-lined concrete containment building (SNR-300)

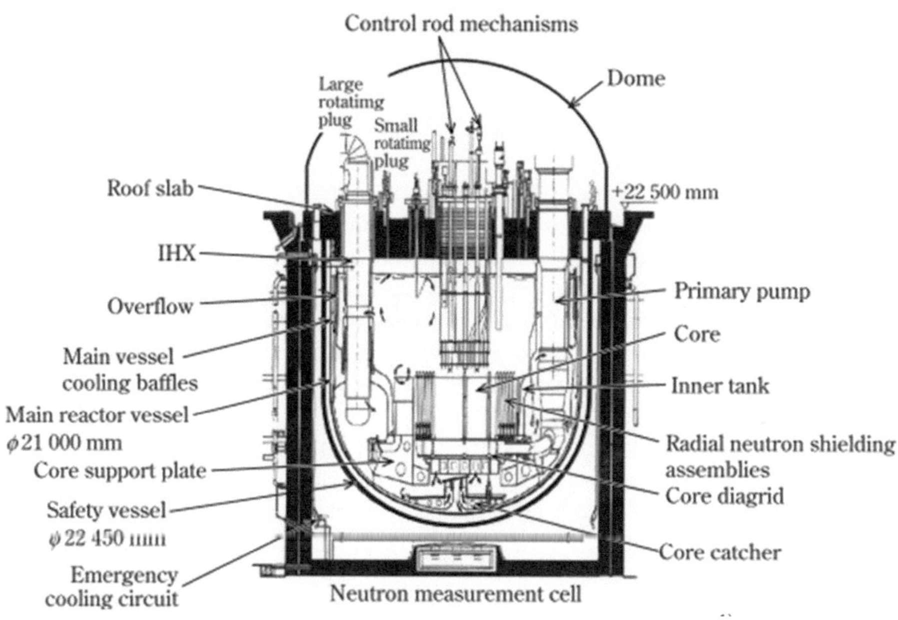

Fig. 6.27 Containment system of Super Phenix [1]

The safety protection system is provided with multiplicity or diversity and independence considering separation of systems, including the power supply, to prevent the loss of function due to a single failure [2–5]. In case of actuation logic "x/y," y shows number of detectors (channels), and x shows required number of detecting abnormal condition signal to activate reactor trip system.

The instrumentation and control system components necessary for normal operation need to maintain required accuracy that constrains the error in the set output and cannot have adverse effects on safety, including the safety improvement equipment.

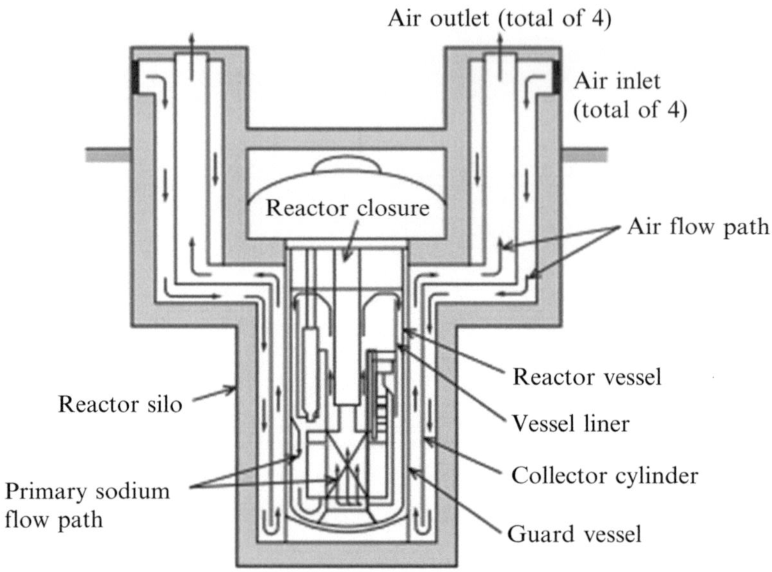

Fig. 6.28 Example of containment facility concept of a modular reactor (ALMR, USA) [1]

In addition to the neutron instrumentation, other instrumentation and control systems are provided to monitor the in-core behavior and detect anomalies. Such systems include a fuel assembly outlet temperature and flow (flow velocity) meter, a fuel failure detection system (FFD), and a fuel failure detection and location system (FFDL). The fuel assembly outlet temperature and flow (flow velocity) meter detects cooling anomalies (local flowpath blockage, etc.) in each fuel assembly. The FFD is used for early detection of fuel failure inside the core. The FFDL is used to identify the failed fuel assembly.

Sodium, which is used as coolant, has a very high boiling point and remains in liquid form within the service temperature range with no need of pressurization. Therefore, measurement and control of the coolant pressure are not particularly required. The temperature, flow rate, liquid level, and purity are the major measurement items for the coolant.

The control systems, which perform one of the functions of the instrumentation and control systems, are used to control the reactor power, coolant flow rate, the water-steam system, and the turbine system. However, in all FRs, it is extremely simple to control the reactor power during normal operation due to the fact that the power coefficient is negative, coolant is circulated in a single-phase liquid state, and special pressurization is unnecessary. In many FRs, the reactor power and the coolant flow rate are controlled to be approximately proportional to each other to keep the coolant temperature approximately constant with the aim of mitigating thermal transient stresses caused by temperature changes.

FR operation costs tend to be lower than construction costs. In this cost allocation, a baseload operation at the maximum power is more economically

Table 6.1 Examples of reactor trip signals (Monju) [3]

Reactor trip signal	Detector	Actuation logic	Interlock
Source range neutron flux high	Source range neutron flux detector	1/2	Manual block when (P-A) set value is reached or exceeded
Wide range neutron flux high			Manual block when (P-B) set value is reached or exceeded
(a) Lower set value	Wide range neutron flux detector	2/3	
(b) Higher set value	Wide range neutron flux detector	2/3	
Power range neutron flux high	Power range neutron flux detector		Manual block when (P-B) set value is reached or exceeded
Rate of change in power range neutron flux high	Power range neutron flux detector	2/3	
Reactor vessel sodium level low	Reactor vessel sodium level gauge	2/3	Manual block when (P-A) set value is reached or exceeded
Reactor vessel outlet sodium temperature high	Reactor vessel outlet sodium temperature detector	Each loop 2/3	
Intermediate heat exchanger primary-side outlet sodium temperature high	Intermediate heat exchanger primary-side outlet sodium temperature detector	Each loop 2/3	
Primary main cooling system circulation pump rotational frequency low	Primary main cooling system circulation pump rotational frequency detector, power range neutron flux detector	Each loop 2/3	Manual block when (P-A) set value is reached or exceeded
Primary main cooling system circulation pump rotational frequency high	Primary main cooling system circulation pump rotational frequency detector, power range neutron flux detector	Each loop 2/3	
Primary main cooling system flow rate low	Primary main cooling system circulation pump rotational frequency detector, wide range neutron flux detector	Each loop 2/3	Manual block when (P-A) set value is reached or exceeded
Secondary main cooling system circulation pump rotational frequency low	Secondary main cooling system circulation pump rotational frequency detector, power range neutron flux detector	Each loop 2/3	Manual block when (P-A) set value is reached or exceeded

(continued)

Table 6.1 (continued)

Reactor trip signal	Detector	Actuation logic	Interlock
Secondary main cooling system flow rate low	Secondary main cooling system circulation pump rotational frequency detector, wide range neutron flux detector	Each loop 2/3	Manual block when (P-A) set value is reached or exceeded
Evaporator outlet sodium temperature high	Evaporator outlet sodium temperature detector	Each loop 2/3	
Turbine trip	Main steam stop valve position detector	When 2 valves closed	Automatic block when (P-C) set value is reached or exceeded
Normal bus voltage low	Normal bus voltage undervoltage relay	Each bus 2/3	
Failed fuel detection	Delayed neutron detector	Each loop 2/3	
Reactor containment vessel isolation			
Earthquake acceleration large			
(a) Horizontal acceleration	Horizontal acceleration detector	2/3	
(b) Vertical acceleration	Vertical acceleration detector	2/3	
Manual operation		1/2	

advantageous than adjusting power in accordance with the load of the transmission system. Under this concept, Monju uses a control system that operates the reactor at a predetermined constant power. In other words, the reactor power is adjusted to correspond to the set power, and the flow rates of the primary coolant, the secondary coolant, and the feedwater are adjusted to be approximately proportional to the set power. During a reactor trip, the reactor is scrammed (emergency shutdown), and at the same time, the drive motors of the primary and the secondary system circulation pumps are switched from the main motors to the small pony motors that are operated by the emergency power supply, and the final heat sink is switched to a cooler that is cooled by air.

However, when FRs are used as a main power source, load following operation that can follow the changes in system load is needed. In response, studies on load following characteristics of FR systems are underway [2–4].

6.2.8 Power Supply Equipment

The majority of equipment in a plant is operated by electricity. The power supply equipment can be divided into the normal power supply equipment for normal operation and the emergency power supply equipment for safety functions required in the event of an accident. There is an operation power supply that is mainly used for supplying power to the normal and emergency power supply equipment independently, and there is a control power supply that is primarily used for measurement and control. The operation power supply usually uses alternating current, whereas the control power supply generally uses direct current.

The emergency power supply equipment needs to be fed electricity from an emergency generator and a battery to ensure power supply to the safety equipment in the event of an accident. In addition, multiplicity and system separation must be taken into account. During normal operation, a portion of power generated by the main generator is supplied via a station service transformer as the power for the operation of on-site components.

During reactor shutdown or in the event of an accident, power is received via a start-up transformer and is supplied to the facility if an off-site power supply is available. If no off-site power supply is available, the emergency generators are started to feed power to the emergency bus. Furthermore, to ensure the measurement and control functions during the time period until the emergency generator is started, DC power supply equipment called an uninterruptible power supply for supplying power from a battery is also prepared.

The emergency power supply equipment including the emergency generator and batteries must have multiplicity in accordance with the redundancy of the safety equipment to which power is supplied. Also, to prevent these systems from losing function due to a single failure, system separation must be ensured.

From the lessons learned through the accident at the Fukushima Dai-ichi Nuclear Power Plant of Tokyo Electric Power Co., the importance of securing a power supply in the event of an accident was recognized. As a result, the following measures to reinforce the power supply systems are required by the regulatory standards set by the Nuclear Regulation Authority. Figure 6.29 shows a concept of power supply equipment [2, 3].

1. The connection to off-site transmission systems shall be enhanced via multiple lines. In addition, the loss of off-site power supply for an extended period of time in the event of an accident shall be taken into consideration.
2. The emergency AC power equipment shall have necessary power supply capability (power generation capability) in the event of the loss of off-site power supply. Power supply equipment shall be designed and arranged such that its function as safety equipment is not lost by earthquake, tsunami, or other common cause by implementing multiplicity and system separation corresponding to the redundancy of the safety equipment. Multiplicity and system separation shall also be ensured in auxiliary components required for achieving the safety functions.
3. The emergency AC power supply equipment shall maintain its function and maintain 7 days' worth of fuel supply in the event of a major accident.
4. The emergency DC power supply equipment shall be provided with batteries that can supply power for 8 hours without load shedding after the occurrence of a major accident and for a subsequent 16 hours with load shedding.

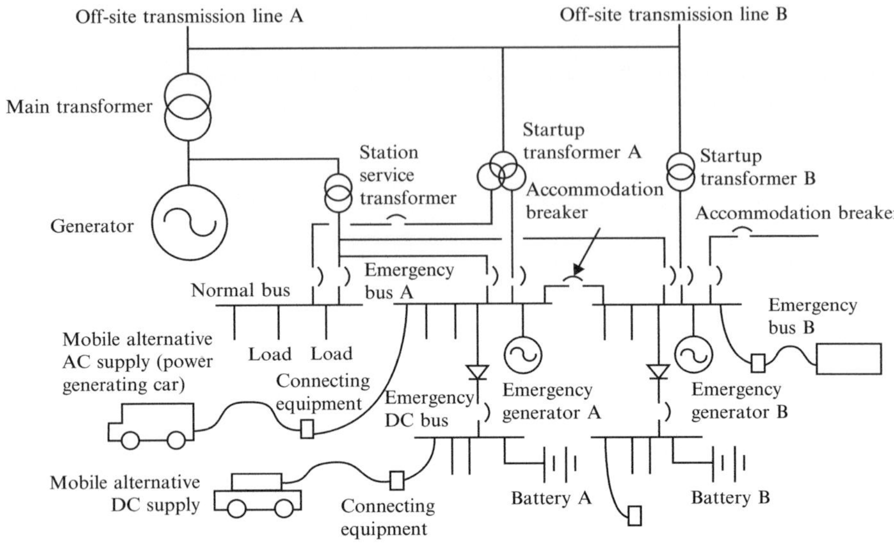

Fig. 6.29 Concept of power supply equipment

5. Alternative (AC and DC) power supply equipment in case of a loss of emergency power supply function shall be prepared. In addition, connection equipment for the emergency bus shall be prepared for the use of mobile alternative power supply equipment.
6. Diversity and positional redundancy of alternative power supply equipment shall be taken into consideration to prevent the loss of function due to a common cause with emergency power supply equipment.
7. In case of a major accident, power interchange from other power supply systems or emergency power supply equipment of other reactor facilities may be assumed despite the principle of system separation if safety can be improved.

6.3 Various Factors of System Integration

6.3.1 Factors Determined by Heat Balance

When designing the respective systems described above, mutual consistency as a system needs to be ensured. Heat balance is a representative example of such consistency.

In the design flow shown in Fig. 6.30, heat balance must be achieved from heat generation by nuclear reactions in the core to thermoelectric conversion.

A clear feature of the heat balance of a sodium-cooled system is that the temperature and pressure of the steam for driving turbine generators can be made

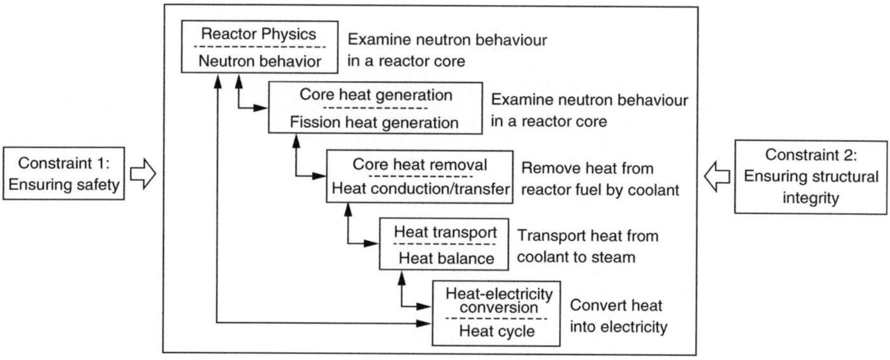

Fig. 6.30 Design flow to satisfy functions under constraints

Table 6.2 Comparison of plant systems between fast reactor and light-water reactor [1]

Specifications	Fast reactor	Light-water reactor
Breeding ratio (conversion ratio)	1.1–1.3	0.6
Burnup (average when discharged)	80,000–100,000 MWd/t	40,000–45,000 MWd/t
Reactor vessel outlet coolant temperature	500–550 °C	280–330 °C
Difference between core outlet and inlet temperatures	130–150 °C	5–30 °C
Maximum coolant operating pressure	0.5–1 MPa	7–16 MPa
Steam conditions (at turbine inlet)	483 °C/12.4 MPa - 500 °C/ 14.7 MPa	270 °C/6.0 MPa- 280 °C/6.5 MPa
Thermal efficiency	40–45%	32–35%

Representative values are listed for the respective reactor types that have been continuously improved

higher than those for a light-water reactor. In other words, the coolant can be recirculated in liquid phase at a high temperature with no special pressurization. Therefore, the coolant temperature can be increased to a level that can be allowed by high-temperature design of fuel and cooling systems, resulting in superior thermal efficiency compared to conventional LWRs. (Table 6.2 and Fig. 6.31)

6.3.2 Factors Determined by Balance Between Functional Requirements and Safety and Structural Integrity Constraints

The relevant factors include the following: maximum coolant and cladding tube temperatures, burnup, breeding ratio, coolant pressure, and the number of loops.

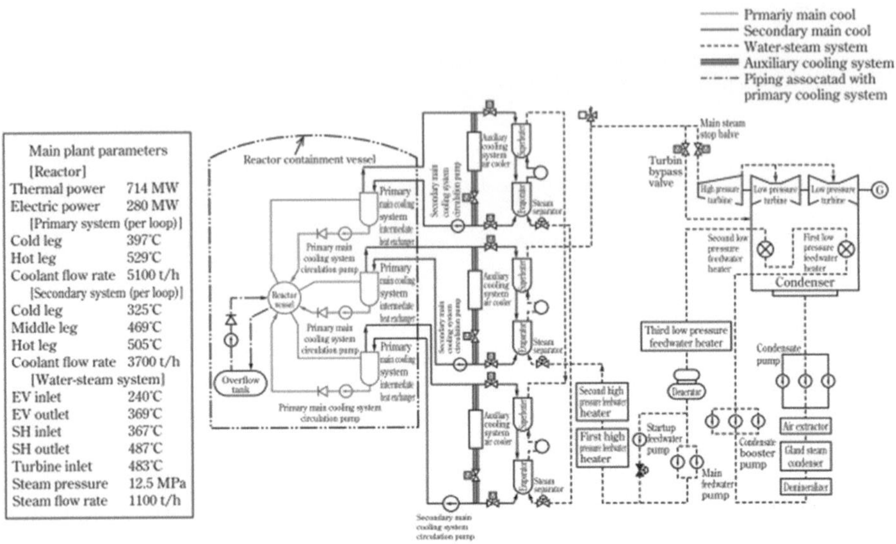

Fig. 6.31 Example of heat balance in fast reactor (Monju) [3]

However, when seeking a high-temperature design allowing for high-thermal efficiency, special attention must be paid to the fuel and structural design problems due to higher temperatures (higher thermal expansion, impact on material characteristics by high temperature, and the like). Since heat is removed by single-phase liquid recirculation, only sensible heat (difference in temperature) can be used for heat removal. Combined with the relationship with specific heat of sodium (much smaller than that of water), the difference in coolant temperature between the inlet and outlet sides of the heat exchanger needs to be relatively large compared to a case that uses water, which can use latent heat from boiling and condensation. As a result, thermal transient must be taken into account because a large change in temperature may occur during a transient operation.

In addition, when the fuel cladding temperature increases as a result of the exploration of high fuel burnup in order to reduce cycle cost, creep behavior and the like become noticeable. To address this issue, high-temperature and high-burnup designs must be optimized.

Here, burnup is determined by the balance between reactivity reduced by core burnup and excess reactivity, the balance between FP accumulation in the fuel cladding tube and gas plenum volume and pressure-resistant design, and the balance between neutron fluence of fuel material and the irradiation resistance of the material. Table 6.3 shows the main system parameters of the respective FRs determined as described above.

With regard to the number of loops, the smaller the number is, the more advantageous the construction cost is in general. This, however, leads to an increase in the size of circulation pump and heat exchanger, making development more

Table 6.3 Main system parameters of various fast reactors

	Japan		France		Russia		India	USA
	Monju	Demonstration reactor	Phenix	SPX	BN-600	BN-800	PFBR	CRBR
Electric power (MWe)	280	1500	255	1242	600	870	500	380
Reactor outlet/ inlet temp. (°C)	529/397	550/395	560/395	542/395	535/365	547/354	547/397	535/388
Main steam temp. (°C)/ pressure (MPa)	483/12.5	495/18.0	510/16.8	487/17.7	505/13.7	490/13.7	490/16.7	482/9.81
Cladding material	Modified 316	ODS	Cr17Ni13 Mo2.5Mn1.5TiSi	Same as the left	Cr16Ni13 Mo2.5Mn1.5TiSi	Same as the left	Cr15Ni15MoTi (CW)	316(20% CW)
Burnup ave./ peak (MWD/t	80,000/ 94,000	150,000/ 200,000	125,000/170,000	90,000/ 113,000	72,000/120,000	66,000/ 98,000	77,000/ 113,000	50,000/ 74,200
Reactor vessel material	304	316FR	316	316	Cr18Ni9	Cr18Ni9	316LN	304
Steam generator material	2.25Cr1Mo/ 321	Mod.9Cr1Mo	2.25Cr1Mo/ 321H	Ni33 Cr21 TiAlMn	2.25Cr 1Mo/Cr18 Ni9	10Cr 2Mo VNB/ 10Cr 2Mo VNB	Mod.9Cr 1Mo/ Mod. 9Cr1Mo	2.25Cr 1Mo/ 2.25Cr 1Mo

difficult. Meanwhile, in terms of safety, it is advantageous from the perspective of multiplicity to have more loops. In Monju, which is a prototype reactor, the number of loops was determined in consideration of the engineering extrapolation for increasing the size of pumps and heat exchangers of the experimental reactor Joyo as well as from the viewpoint of mitigating influences on the core in case of a component failure and coping with a single failure.

6.3.3 Factors for Achieving Physical Consistency Among Systems

Factors for achieving physical consistency among systems include the following: overall layout, liquid levels, approximate size, weight, and support method through the building of the core and main components.

For example, in the primary cooling system, since the coolant pressure drops along with the coolant circulation, differences in level are generated if a free liquid level is present along the flow path. Considering the level differences, the reactor cover gas pressure and liquid levels during rated operation in the respective components are determined, and the pressure loss during the rated flow condition is limited. Figure 6.32 shows an example of the relationship between pressure and liquid level at the respective positions of the primary cooling system in Monju.

When supporting the respective components, the components need to be supported so as to support their own weight in a stable manner. In addition, attention must be paid to preventing the local application of an excessive load to

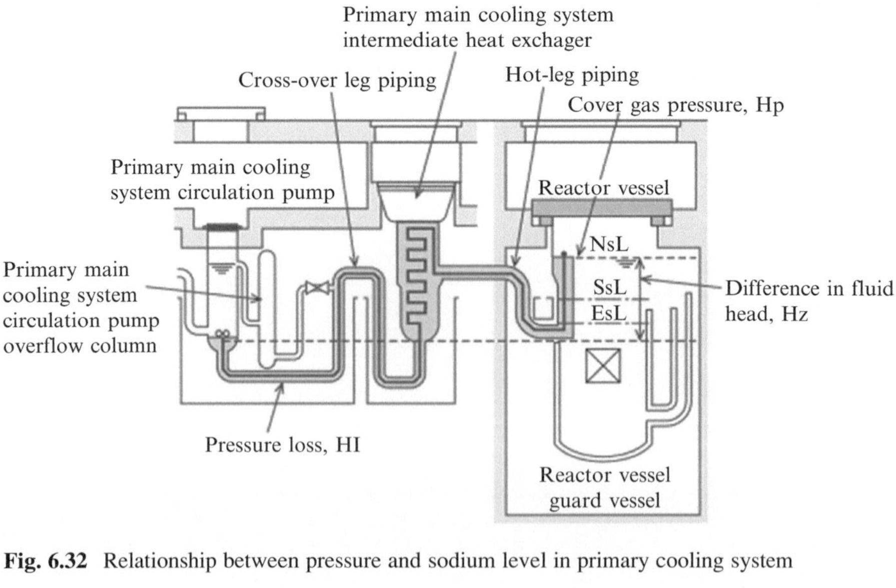

Fig. 6.32 Relationship between pressure and sodium level in primary cooling system

an attachment section of the support structure and the occurrence of significant displacement of components during seismic motion.

The layout of the components in the secondary main cooling system is determined with the aim of securing the pressure difference between the primary and secondary coolant in the IHX. In other words, the installation levels of the components in the secondary cooling system are set to be above a prescribed level to set the pressure of the secondary cooling system higher than that of the primary cooling system in all operation modes using the head pressure of the coolant, thereby preventing activated primary coolant from leaking to the secondary cooling system even if a heat transfer tube of the IHX is damaged. In addition, the difference in levels between the primary and secondary cooling systems also needs to be taken into account since the difference determines the natural circulation heat removal capability (Fig. 6.23). It is advantageous to be able to handle the secondary coolant as nonradioactive material from the safety and economic standpoints. To achieve this, the pressure difference between the primary and secondary systems needs to be maintained to prevent radioactive material from mixing into the secondary cooling system. In addition, the neutron flux from the core needs to be sufficiently shielded in the proximity of the IHXs, where the secondary coolant approaches nearest to the core, to prevent the secondary coolant from being activated. Particularly, in a pool-type reactor, which is discussed below, the core and the IHXs are located near each other. Consequently, shielding the IHXs is an important design factor.

6.3.4 Factors Determined by Cost and Technological Level

Factors determined by cost and technological level are typically related to material characteristics. From the energy efficiency standpoint, it is preferable to increase the temperature as much as possible. However, the maximum service temperatures and the like are limited due to constraints in the high-temperature characteristics of the fuel and cooling system materials as well as in structural integrity. Those involved in design make an effort to design the structure so as to obtain maximum performance while trying to achieve a balance between high performance and economic competitiveness. Furthermore, from a broader perspective, the power generation cost of an FR is determined by the facility construction cost and the operation cost, including the nuclear fuel cycle cost. As a result, it is necessary in design to strike a balance between the reduction of construction costs and operation costs.

A result of optimization assessment on the main steam temperature and pressure is shown below as an example (Fig. 6.33). Although the thermal efficiency is improved with an increase in temperature, the pressure on the water side also needs to be increased to avoid unstable flow on the water side. As a result, both thermal stresses and pressures are increased with an increased temperature. Since material strength decreases with an increase in temperature, construction and maintenance costs increase due to constraints on structural integrity.

Fig. 6.33 Optimization of main steam temperature and pressure

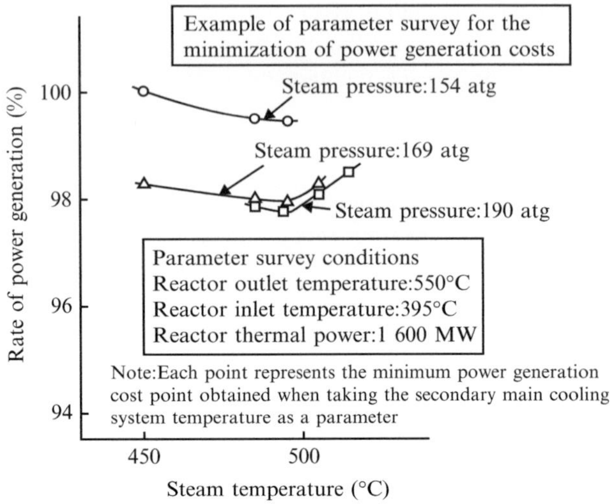

Further Readings

Power Reactor and Nuclear Fuel Development Corporation, Application for reactor establishment license: Experimental Fast Breeder Reactor, Joyo

Nuclear Safety Commission, "Kosoku zoshokuro no anzensei no hyoka no kangaekata (Philosophy in Safety Evaluation of Fast Breeder Reactors)" (in Japanese), November, 1980

Nuclear Regulation Authority, "Jitsuyo hatsudenyo gensiro oyobi sono fuzokushisetsu no ichi kozo oyobi setsubi no kijun ni kansuru kisoku" (Regulation on the Standards for the Location, Structure and Equipment of a Commercial Nuclear Power Reactor and its Associated Facilities) (in Japanese), July, 2013

References

1. IAEA, Tecdoc-1531, Fast Reactor Database Update (2006)
2. Japan Atomic Energy Relation Organization, "FBR koho sozai shu" (Collection of Public Relations Material on FBR, No.2) (in Japanese)
3. Power Reactor and Nuclear Fuel Development Corporation, "Kosoku zoshokuro Monju hatsudensho secchi kyoka shinseisho" (Application for reactor establishment license) (in Japanese): Prototype Fast Breeder Reactor, Monju
4. IAEA, Tecdoc-1691,Status of Fast Reactor Research and Technology Development
5. M. Hori (supervising editor), "Kiso kosokuro kogaku (Basic Fast Reactor Technology)" (in Japanese), Power Reactor and Nuclear Fuel Development Corporation (1993)
6. Japan Atomic Energy Agency's brochure
7. Peer Review Committee for Safety Measures for Monju, "Kosoku zoshokuro Monju no anzen kakuho no kangaekata" (Safety assurance strategy for the Fast Breeder Reactor) (in Japanese), Monju (2014)

Chapter 7
Core and Fuel Design (From Mechanism to Structure)

Nobuo Nakae, Toshikazu Takeda, and Hiroyuki Ohshima

Chapter 7 deals with core and fuel design. Concepts, procedures, and/or processes involved in determining the structure and specifications of core and fuel are mainly described in this chapter as well as typical design results. These matters are closely related to the result of conceptual system design described in Chap. 6, but they have less relation with the type of reactor (loop type, pool type). The conceptual design provides fundamental plant parameters such as the type of reactor, thermal output, coolant material, fuel type, and inlet and outlet coolant temperatures. The sequence of design work should maintain the constraints such as reactor safety and reactor component integrity described in Chaps. 4 and 5. The relationship among each design level is shown in Fig. 7.1, and the area surrounded by a dotted line corresponds to the main function of core and fuel design. The minor actinide (MA) content of plutonium-MA fuel is determined by both nuclear design and thermal hydraulic design. The former assures core criticality and the latter assures the required linear heat generation rate (LHR). The cladding, which contains fuel pellets, should be designed to maintain its integrity during normal operation and anticipated operational occurrences. Each fuel pin is wrapped by a (spacer) wire to maintain sufficient coolant flow area, and the fuel pins are closely packed for effective neutron utilization. Each fuel pin bundle is installed into a wrapper tube (duct) to form the core fuel assembly. The core is composed of several components such as the core fuel assembly, whose function is heat generation by fission; the

N. Nakae (✉)
Nuclear Regulation Authority, Tokyo, Japan
e-mail: nobuo_nakae01@nsr.go.jp

T. Takeda
University of Fukui, Fukui, Japan
e-mail: t_takeda@u-fukui.ac.jp

H. Ohshima
Japan Atomic Energy Agency, Ibaraki, Japan
e-mail: ohshima.hiroyuki@jaea.go.jp

N. Kasahara (ed.), *Fast Reactor System Design*, An Advanced Course in Nuclear Engineering 8, DOI 10.1007/978-981-10-2821-2_7

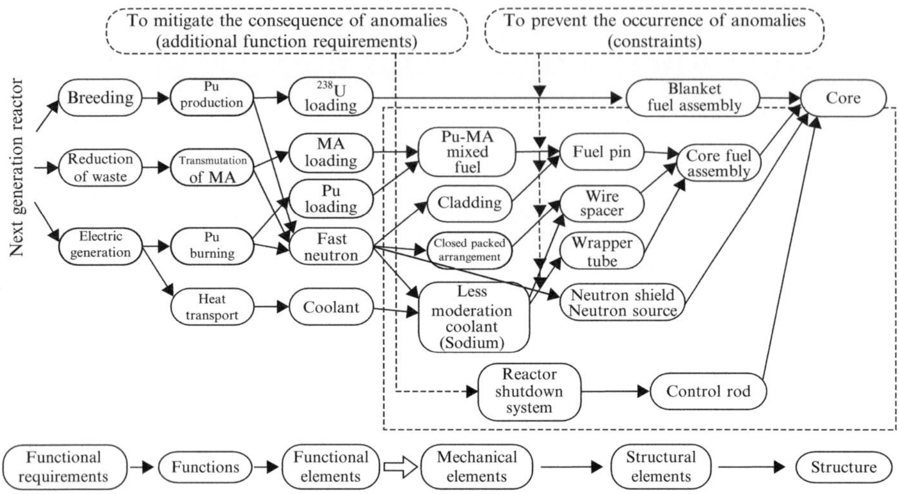

Fig. 7.1 Thinking process diagram for core and fuel

blanket fuel assembly, whose function is plutonium buildup; the control rod assembly, whose function is to keep criticality and control reactivity; the neutron shield assembly; and the neutron source assembly. These components should also be closely packed.

The main purpose of core and fuel design is to determine specifications for maximum performance that ensure safety and structural integrity. Required conditions also involve the assurance of nonproliferation and economy.

Core design is defined as design that assures the control of reactor power level and the removal of heat from a reactor core. Thus, core design is composed of nuclear design, thermal hydraulic design, and plant dynamics analysis. Fuel design is defined as design that assures the integrity of fuel with the function of heat generation. Core design and fuel design are very closely connected. Fuel specifications are based on the results of core and fuel design. Fuel fabrication is based on the fuel specifications. Therefore, the results of core and fuel design have a strong influence on fuel fabrication.

This chapter introduces the procedure and basic concept of nuclear design, thermal hydraulic design, and plant dynamics analysis. The outline of reactor physics and thermal hydraulic engineering are also introduced in the first half of this chapter, and fuel design follows. Fuel irradiation behavior, trend of fuel design, and fuel technical standards are introduced as parts of fuel design. The development of cladding for the improvement of fuel burnup and reliability, fuel physical property measurement, and fuel irradiation testing are introduced as fuel development topics in the second half of this chapter.

7.1 Core Components (Fuel Assembly, Control Rod Assembly, Neutron Shield Assembly, Neutron Source Assembly, Surveillance Assembly)

Figure 7.2 shows the configuration of the Monju core. The core is composed of fuel assemblies, control rod assemblies, neutron shield assemblies, neutron source assemblies and surveillance assemblies. Two kinds of fuel assemblies are considered for the fast breeder reactor (FBR). One is a core fuel assembly and the other is a blanket fuel assembly.

The core fuel assembly is also called driver fuel because the fuel is used for reactor operation. The fissile content, which corresponds to the plutonium content, is high, and the core fuel assemblies produce almost all heat in the core. The core fuel pellets typically consist of plutonium and uranium mixed oxide (MOX), and depleted or natural uranium is used.

The blanket fuel assemblies are distributed in the surroundings of the reactor core, and ^{238}U in the blanket fuel pellets is converted to ^{239}Pu by absorbing neutrons leaking from the core. The blanket fuel plays an important role in fuel breeding. The blanket fuel pellets typically consist of depleted uranium dioxide (UO_2) and sometimes natural uranium dioxide.

The control rod assembly has the functions of shutting down the reactor and controlling the reactor power level, in other words, reactivity. Accordingly, the control rod pellets often consist of boron carbide (B_4C) containing boron with high neutron absorption capability.

The neutron shield assembly reflects neutrons leaking from the core and protects surrounding components from fast neutrons and gamma rays leaking from the core. It is made of stainless steel.

The neutron source assembly has the role of safety reactor startup. Californium is often used as the neutron source.

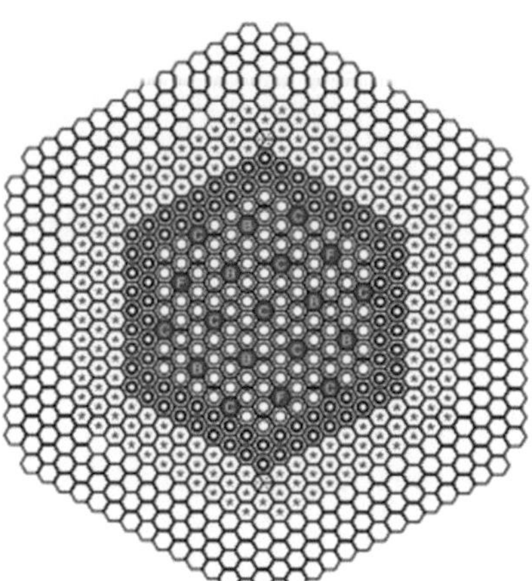

Core component		Symbol	Number
Core fuel assembly	Inner core		108
	Outer core		90
Blanket fuel assembly			172
Control rod assembly	Fine control rod	F	3
	Coarse control rod	C	10
	Backup control rod	B	6
Neutron source assembly			2
Neutron shield assembly			316
Surveillance assembly			8

Fig. 7.2 Configuration of Monju core [1]

The surveillance assembly is fabricated by partially changing the inner structure of neutron shield assembly to contain surveillance test samples of the reactor vessel and inner core structure. It has a role in the periodical inspection of changes in material properties due to neutron irradiation, aging, etc.

7.2 Core and Fuel Structure

Core structure is determined by the amount and arrangement of core components shown in Fig. 7.2.

Fuel structure is shown in Fig. 7.16. Fuel pellets are installed into a cladding tube together with other components such as plenum spring and tag gas capsule. The upper and lower ends of the cladding tube are capped with plugs and welded. The fuel pins are assembled with a constant distance by wire wrapping, and the fuel bundle is inserted in a wrapper tube. The wrapper tube is attached with an entrance nozzle on the bottom side and with a handling head on the top side.

The structure of core and fuel should be determined in order to satisfy the fundamental requirements such as fuel breeding, radioactive waste reduction and electric generation, and design requirements such as reactor safety and the integrity of components.

7.3 Blanket Fuel and Breeding Characteristic

Breeding characteristic, which is one of the fundamental requirements of FR, is attained by converting ^{238}U in blanket fuel pellets into ^{239}Pu through absorption of neutrons leaking from the core. Blanket fuel assemblies should be arranged in the surroundings of the core and its configuration in a way that is very similar to a person wearing a blanket. This is the reason that fuel for breeding is called blanket fuel. Two types of blanket fuels are designed. One is radial blanket fuel and the other is axial blanket fuel. The radial blanket fuel assemblies surround the core fuel assemblies. The axial blanket pellets are installed above and below the core fuel pellets. The arrangement of blanket fuel in Monju is shown in Fig. 7.3.

In the high performance core shown in Fig. 7.3, the radial blanket fuel assemblies are replaced by reflector assemblies, the height of axial blanket fuel is shortened with the increase in core fuel height, and the height of the control rod absorber region is increased. The purpose of such change is to extend operation cycle length and to achieve high burnup in consideration of plutonium supply and demand and the reduction of the amount of spent fuel.

The breeding ratio is calculated as the sum of the ratios of the amounts of fissile material newly produced in core fuel, radial blanket fuel, and axial blanket fuel to the amount of fissile material consumed in the entire core region. The fractions calculated with the Monju core are approximately 0.6, 0.4, and 0.2 for core fuel,

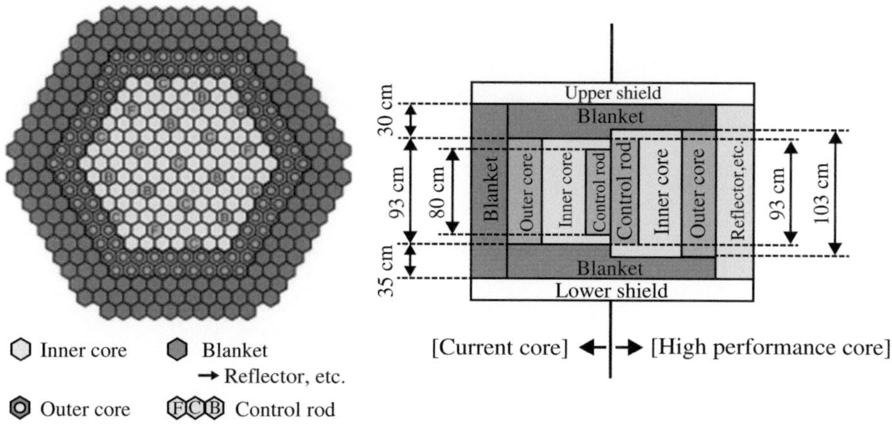

Fig. 7.3 Core structure and arrangement of Monju [2]

radial blanket fuel, and axial blanket fuel, respectively, resulting in the total breeding ratio of approximately 1.2. The breeding ratio depends on fuel burnup. Thus, the power distribution between core and blanket fuel regions varies with the increase in fuel burnup. The power shared by the blanket fuel region changes from 9 % at the beginning of cycle to 11 % at the end of cycle. The breeding ratio strongly depends on core design.

7.4 Brief Flow of Core Design

The process from the setting of electric power to the determining of fuel irradiation condition consists of two paths. One is controlled by limitations and the other is an iteration path. The first path is to produce the result under the limitations of safety and component integrity, and the second path is to check whether or not the result satisfies the limitations, and if the result does not satisfy the limitation, another calculation should be carried out with changed fuel specifications and fuel irradiation conditions. A brief flow of core design is shown in Fig. 7.4. The flow in this figure indicates the result in the case that the result satisfies the limitations required. The values shown in Fig. 7.4 refer to the result of design equivalent to Monju.

7.4.1 Determination of Thermal Output

Electric power is set to be 280 MW. Thermal output is calculated to be 714 MW by assuming that the power generation (or thermal) efficiency, a fraction of electric power to thermal output, is 39.2 %. The power generation efficiency depends on coolant outlet temperature, and higher efficiency is obtained with higher outlet temperature. The power generation efficiency of the FR is higher than that of the

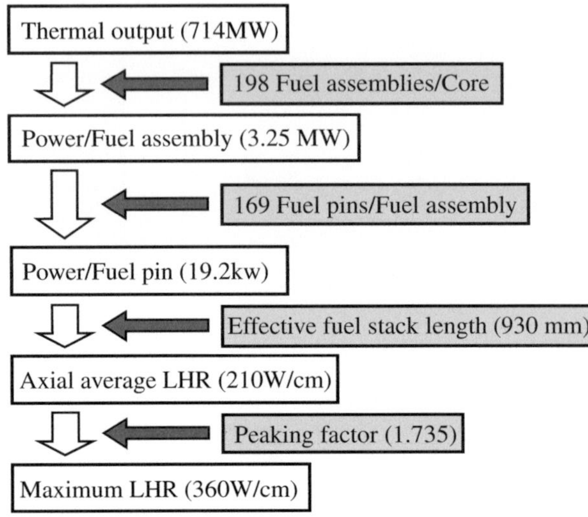

Fig. 7.4 Brief flow of core design

light water reactor (LWR) since the outlet temperature is much higher in the FR than in the LWR.

7.4.2 Determination of the Number of Fuel Assemblies

We are now considering the number of fuel assemblies that can supply the thermal output determined above (714 MW). The LHR, which is defined as the thermal power per unit length of fuel pin, is important since high LHR tends to cause fuel melting. The typical value of LHR for oxide fuel is believed to be 350–450 W/cm. When a higher thermal output of one fuel assembly is required, the number or total length of fuel pins in the assembly should be increased. A larger fuel assembly would be needed in this case. It is necessary to select the size of fuel assembly to allow for easy handling of the fuel assembly. In the Monju core, the thermal output per fuel assembly is set to be 3.25 MW. Thus, the number of fuel assemblies in the core becomes 198 ($714 \times 0.9/3.25 = 198$). In this estimation, the power shared by the blanket region is taken into account. It is approximately 10 %. The arrangement of fuel assembly in the core should be closely packed because the fission cross sections of fissile materials are fairly small in a fast neutron energy spectrum. To this end fuel assemblies are arranged in a hexagonal pattern. Control rod assemblies are also installed in the core of FR.

7.4.3 Determination of the Number of Fuel Pins

Next, we determine the number of fuel pins in a fuel assembly. It is geometrically determined that the number of fuel pins in a fuel assembly is 1, 7, 19, 37, 61, 91,

127, 169, 217, or 271 because fuel pins should be closely packed in a fuel assembly (specifically, in a hexagonal pattern). If the number of fuel pins is small, thermal power per fuel pin would be high, increasing the risk of fuel melting. Thus, it is reasonable that the number of fuel pins is set to be from 127 to 217. The number for Monju was determined to be 169. The thermal output per fuel pin becomes 19.2 kW (3250/169 = 19.2).

7.4.4 Determination of Fuel Stack Length and LHR

Linear heat generation rate is a very important parameter in fuel design. If the fuel stack length is determined, LHR is easily set. If the fuel stack length is set to be short, LHR becomes high, increasing the risk of fuel melting. If the fuel stack length is set to be long, the size of fuel assembly increases and may cause difficulty in fuel handling. Thus, the fuel stack length should be determined very carefully by considering both effects on fuel melting and fuel handling. The fuel stack length for Monju was set to be 93 cm, and the average LHR becomes 210 W/cm (19,200/93 = 206). It is well known that power distribution should exist in the core, and the spatial power distribution is taken into account by use of the power peaking factor. There are three peaking factors, including the fuel assembly power peaking factor, which is expressed by the ratio of the maximum to the average power of a fuel assembly; the local power peaking factor, which is the ratio of the maximum to the average of axially integrated power for a selected fuel assembly; and the axial power peaking factor, which is the ratio of the maximum power along the axial power distribution to axially averaged power for a selected fuel pin. The maximum LHR can be calculated by multiplying the product of these three factors (i.e., total peaking factor) and the average LHR. If the total peaking factor is 1.735, the maximum LHR is calculated to be approximately 360 W/cm ($206 \times 1.735 = 357$).

7.4.5 Determination of Fuel Burnup

Fundamental fuel irradiation conditions include LHR, burnup, fast neutron fluence, and temperature. A procedure to determine fuel burnup is described below. In case of Monju, the average fuel assembly burnup is set to be 80 GWd/t. The burnup can be calculated from the average LHR (W/cm), fuel pellet diameter (cm), fuel pellet density (g/cm^3), and fuel residence time in the core (day). The pellet diameter of FR fuel should be smaller than that of LWR fuel in order to achieve higher burnup. The density of FR fuel should be lower than that of LWR fuel in order to reduce fuel swelling under a higher burnup condition. In the Monju design, the pellet diameter and density are 0.54 cm and 9.35 g/cm^3, respectively. The pellet density of 9.35 g/cm^3 corresponds to 85 % of the theoretical density (TD). The evaluation of theoretical density is based on crystal structure and lattice constant. The crystal structure of MOX is a fluorite-type face-centered cubic. Lattice parameter depends on the plutonium content of MOX, and it is approximately 5.7×10^{-8} cm for the MOX fuel used in Monju. Therefore, the theoretical density of the Monju fuel is calculated to be approximately 11 g/cm^3.

To achieve an average burnup of 80 GWd/t-HM (71 GWd/t-oxide) under the condition cited above, the fuel residence time in the core is calculated to be 724 days.

$$
\begin{aligned}
t(d) &= \frac{71(\mathrm{GWd/t}) \times \pi\left(\frac{0.54}{2}\right)^2(\mathrm{cm}^2) \times 1(\mathrm{cm}) \times 9.35(\mathrm{g/cm}^3) \times 10^{-6}(\mathrm{t/g})}{210(\mathrm{W/cm}) \times 1(\mathrm{cm}) \times 10^{-9}(\mathrm{GW/W})} \\
&= 724(d)
\end{aligned}
\tag{7.1}
$$

To continue reactor operation for this period without any fuel exchange, a very large control rod worth is needed to control the excess reactivity at the beginning of cycle (BOC). Accordingly, this period should be divided into shorter periods, and a part of the fuel loaded in the core should be exchanged at each period. In case of Monju, this period is divided into five periods, and one fifth of the fuel is exchanged at each cycle. This type of fuel exchange is called a 5-batch dispersed exchange procedure. The actual cycle length for Monju was set at 148 days. Therefore, the fuel residence time in the core is 740 days.

Burnup is expressed by several different units. They include GWd/t-HM or GWd/t-oxide, which corresponds to integrated heat generation per unit weight of fuel (heavy metal or oxide); at%, which corresponds to the ratio of fissioned atoms to the total atoms in the core; and fissions/cm^3, which in turn corresponds to the number of fissions occurring per unit volume. The relationship between the burnup in GWd/t-HM and the burnup in at % is expressed by Eq. (7.2). The relationship between the burnup in fissions/cm^3 and other burnup units, such as GWd/t-HM and at %, depends on fuel types, such as oxide and metal, fuel composition, and fuel density. It is recommended that readers derive these relationships, including Eq. (7.2).

$$
1\mathrm{at}\% = 10\mathrm{GWd/t\text{-}HM} \tag{7.2}
$$

7.4.6 Determination of Fast Neutron Flux and Plutonium Content

It is necessary to determine the amount of fissile material that can continue the operation with a thermal output of 714 MW for 148 days, one cycle. The main fissile material is plutonium in case of the FR. Plutonium content survey calculation should be made by the use of nuclear design code. The determination of core neutron flux distribution and plutonium content is based on the nuclear calculation result. In case of Monju, the maximum fast neutron flux obtained is 3.6×10^{15} n/cm ^{2}sec, and the plutonium content obtained is approximately 20 wt% for the inner core and approximately 30 wt% for the outer core. Two different plutonium contents were set in the Monju core in order to flatten the radial power distribution. Lower plutonium content was set for the inner core because of higher neutron flux in the inner core and higher plutonium content for the outer core.

7.4.7 Determination of Coolant Flow Rate

Heat balance between generation and removal should be maintained in thermal hydraulic design. The determination of coolant flow rate is based on the following equation:

$$\begin{aligned}&(\text{Core thermal output}) = (\text{Coolant specific heat}) \times (\text{Coolant flow rate})\\&\times (\text{Difference between core inlet and outlet temperatures})\end{aligned} \tag{7.3}$$

Coolant flow rate distribution should be determined in order to balance heat generation with heat removal in each region. The majority of coolant is sent through the fuel assemblies with large thermal output. The ratio of flow in the core fuel region to the flow in the blanket region is set at 9:1 since the thermal output of the core fuel region is 90 % of the total thermal output and that of the blanket region is 10 %. In case of Monju, the inner core fuel region has five flow areas, the outer core fuel region has three flow areas, and the blanket fuel region has three flow areas to match the flow rate to thermal output in each region.

7.4.8 Determination of Basic Fuel Specifications

Basic fuel specifications except for fuel pellet diameter and fuel pellet density include the inner and outer diameter of cladding, cladding thickness, the gap between fuel pellet and cladding, smeared density, plenum length, arrangement pitch of fuel pins in fuel assembly, and the dimensions of each part. These specifications are determined by fuel design to ensure fuel integrity. Nuclear design and thermal hydraulic design are described in Sect. 7.5, and fuel design is described in Sect. 7.6.

7.5 Core Design

7.5.1 Activities Related to Core Design (What Is Core Design?)

The mission of core design is to determine core specifications such as the amount of fissile material loaded in core, plutonium content, operation length per cycle, the number of fuel exchange batches, the number and arrangement of control rod assemblies, boron (^{10}B) enrichment, LHR, burnup, core inlet and outlet coolant temperatures, coolant flow rate and its distribution, and smeared fuel density. The specifications determined are provided to fuel design as part of the fuel design conditions. Core design also has a mission to provide the results to plant design as design conditions and to take the balance between plant and core design. The determination of core and fuel specifications corresponds to the determination of the structure of core and fuel.

Nuclear design and thermal hydraulic design should be performed in order to confirm core validity, in other words, to ensure safety and structural integrity of the core. Moreover, plant dynamic analysis is carried out in order to confirm that plant would be recovered to a stable condition via the inherent safety functions even if a disturbance, such as ramp and/or step-type power change and 50 % loss of load, were to occur. Inherent safety means that reactor safety should be ensured without any active functions. Plant dynamic analysis is included in core design. The flow of core design is shown in Fig. 7.5.

7.5.2 *Trend of Core Design*

The fast breeder reactor and its related fuel cycle which is called "the FR cycle system" should be developed because the FR cycle system is considered to be the most reasonable system in nuclear fission energy utilization systems. When we develop the FR cycle system, we have five viewpoints: safety, economy, sustainability, nonproliferation, and the reduction of environmental impact. The trend of core design from these five viewpoints is shown in Fig. 7.6. Small core size, low power level, and low burnup are preferred from the viewpoint of safety. However,

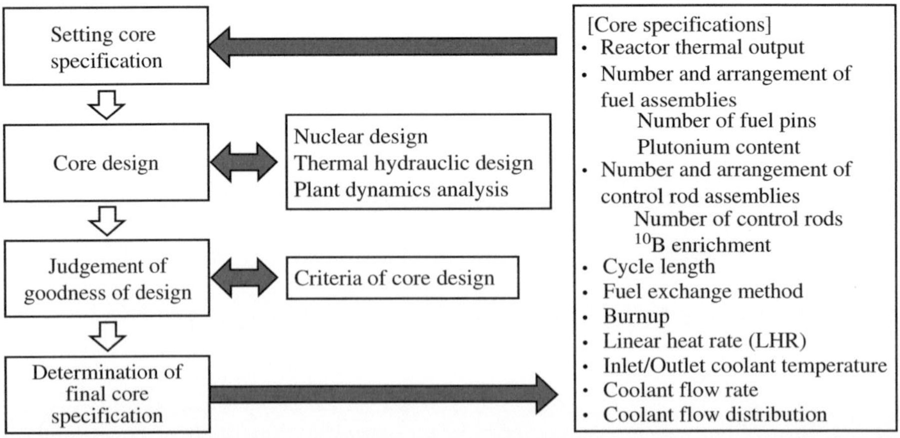

Fig. 7.5 Flow of core design

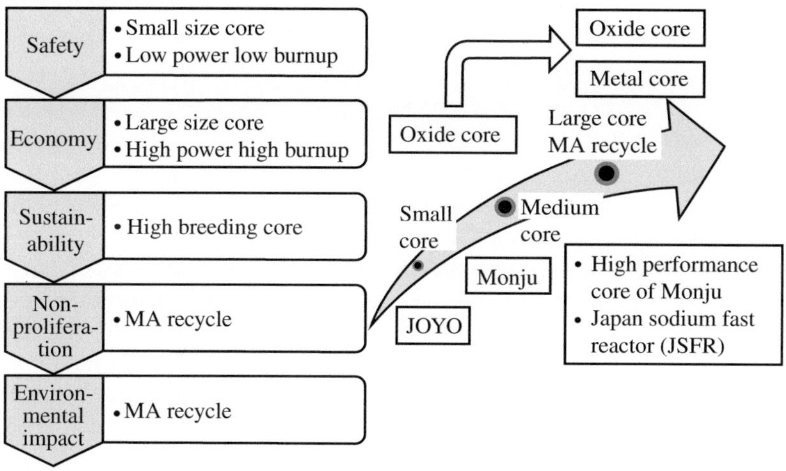

Fig. 7.6 Trend of fast reactor core design

large core size, high power level, and high burnup are preferred from the viewpoint of economy. High breeding cores tend to be selected from the viewpoint of sustainability. Minor actinide (MA) recycle should be studied from the viewpoints of nonproliferation and the reduction of environmental impact. High breeding core and MA recycle should be regarded as the common targets. However, core size, power level, and burnup should be determined very carefully in consideration of the balance between safety and economy. As long as safety is confirmed, it is best for us to choose large core size, high power, high burnup, and MA recycle as promising future FR core features.

7.5.3 Fundamentals of Reactor Physics for Nuclear Design

Nuclear power reactors including the FR generate electricity by converting thermal power produced by fission. Neutrons play a very important role in fission. It is not too much to say that reactor nuclear characteristics are determined by neutrons. Reactor physics can be understood as a kind of physics related to the behavior of neutrons which contribute to fission. This is called neutron physics, and it deals with probability of reactions between nuclei and neutrons (i.e., nuclear reaction cross sections), neutron energy dependence of nuclear reaction cross sections, critical state, reactivity, etc. The nuclear reaction cross section is defined as the probability of a reaction between neutron and nucleus. The number of reactions per unit time and unit volume, R (cm^{-3} s^{-1}), is proportional to the neutron flux, Φ (cm^{-2} s^{-1}), which is the number of incident neutrons per unit time per unit area, and the number density of a nucleus existing in a system, N_0 (cm^{-3}). The relationship is expressed as $R = \sigma\phi N_0$ with the proportional constant (σ). (See Fig. 7.7.) Then, σ is expressed as follows:

$$\sigma = \frac{R/N_0}{\phi} \tag{7.4}$$

where the denominator of Eq. (7.4) is the number of incident neutrons per unit time per unit area and the numerator is the number of reactions per one target nucleus per unit time. The constant σ corresponds to the number of reactions per one target nucleus per unit time when one neutron enters into a nucleus per unit time per unit area. Since the dimension of constant σ is square centimeter, σ is called the reaction cross section. This cross section changes depending on the size of target nucleus and an incident neutron energy. The neutron energy dependence of absorption cross section (σ_a), capture cross section (σ_c), fission cross section (σ_f), and so on are experimentally determined for various nuclei.

Figure 7.8 shows the neutron energy dependence of fission cross section for ^{235}U and ^{239}Pu. Fission cross section is inversely proportional to the neutron velocity in low energy regions, because the fission reaction is proportional to the time in which a neutron passes neighborhood of the nuclei and the time is inversely proportional to the neutron velocity. The fission cross section of ^{235}U decreases with increased

Fig. 7.7 Neutron flux and fission reaction

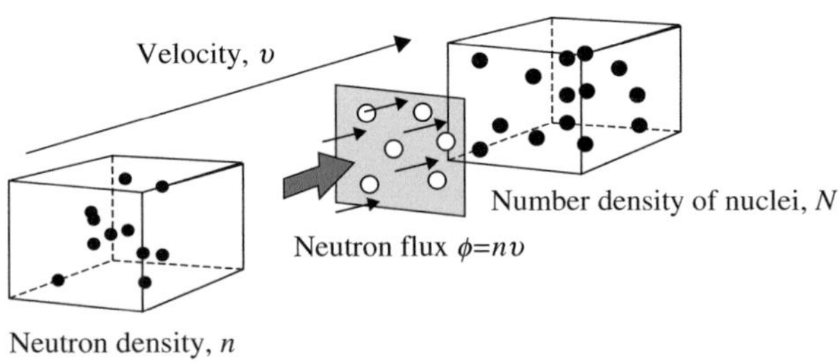

Fig. 7.8 Neutron energy dependence of fission cross section [3]

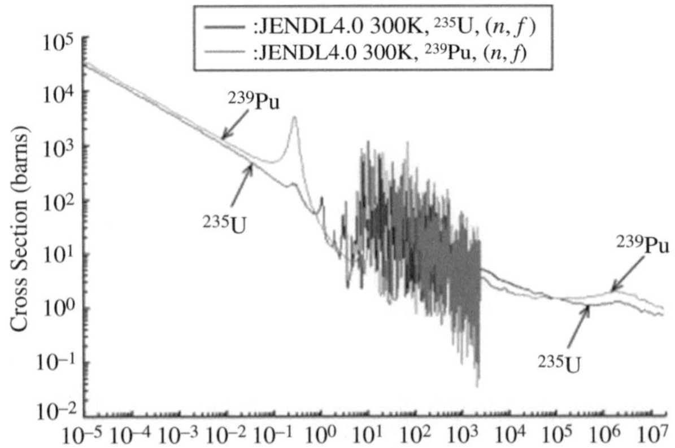

neutron energy in the range higher than a few keV, while that of ^{239}Pu is kept at almost constant. Since the average neutron energy in a FR core is a few hundred keV, the fission cross section of ^{239}Pu is approximately one four hundredth of that in the thermal neutron region ($E \cong 0.025$ eV). Therefore, fuel with high plutonium content should be needed in a FR core. Resonance absorption peaks appear in intermediate energy regions (from 1 eV to a few keV). The resonance capture cross section of ^{238}U plays an important role in neutronic behavior.

Next, criticality and reactivity are explained. The four-factor formula is sometimes referred to in explaining critical conditions. Criticality is achieved when the effective neutron multiplication factor (k_{eff}) becomes unity. Neutrons generated by fission are categorized into two groups. One is neutrons absorbed in the system, and the other is neutrons leaking from the system. The fraction of neutrons that are absorbed in the system is denoted as P. The neutron multiplication factor in an infinite system is denoted as k_∞. The effective neutron multiplication factor (k_{eff}) is expressed as follows:

$$k_{\mathrm{eff}} = k_\infty \cdot P = \eta \cdot \varepsilon \cdot p \cdot f \cdot P \tag{7.5}$$

where η is the number of neutrons generated by fission when fissile material absorbs one neutron, called the neutron regeneration rate, ε is a factor expressing the effect of fission of ^{238}U caused by fast neutron, p is the probability of escape from the resonance absorption, and f is the fraction of neutrons absorbed by fuel, called the fuel utilization efficiency. If the number of neutrons per fission is denoted as ν, the neutron generation rate (η) is expressed as $\nu\sigma_f/\sigma_{aF}$, where σ_f is the fission cross section and σ_{aF} is the fuel absorption cross section. The value of ν is almost constant (approximately 2.9 for ^{239}Pu and approximately 2.5 for ^{235}U) with neutron energy of less than approximately 1 MeV, and it gradually increases at higher energy regions. The value of η changes depending on neutron energy, and it varies by nuclides. η is strongly related to breeding characteristic, and the condition for breeding is expressed by $\eta \geq L+2$, where L is the number of neutrons leaking from a system and those absorbed by coolant and structure. This condition can be achieved by using the combination of plutonium fuel and fast neutrons. The value of η is approximately 2.45 if typical mixed oxide fuel is used together in a sodium cooled fast reactor (SFR).

When the effective neutron multiplication factor k_{eff} is less than unity, the number of neutron decreases with time and the chain reaction diminishes. In this situation the reactor remains in a subcritical state. When k_{eff} is greater than unity, the chain reaction rate increases and reactor enters a supercritical state. Reactivity (ρ) is an index that indicates the level of deviation from criticality. It is expressed by the following equation:

$$\rho = (k_{\mathrm{eff}} - 1)/k_{\mathrm{eff}} = 1 - 1/k_{\mathrm{eff}} \tag{7.6}$$

When $\rho > 0$, the reactor is supercritical. When $\rho = 0$, the reactor is critical. When $\rho < 0$, the reactor is subcritical. The reactivity coefficient to a reactor state x is expressed by the following equation assuming that k_{eff} is close to 1.0:

$$\alpha_x = d\rho(x)/dx = \left(dk_{eff}/dx\right)/{k_{eff}}^2 \cong \left(dk_{eff}/dx\right)/k_{eff} \quad (7.7)$$

The reactor state x includes temperature, density, and the void fraction.
The following reactivity coefficients are important for the FR.

1. Doppler coefficient

When heat generation increases due to enhanced chain reaction, fuel temperature rises significantly, especially in the case of ceramic fuel because of its low thermal conductivity. Resonance absorption of ^{238}U is enhanced by the temperature increase to cause negative feedback of reactivity. This phenomenon is called the Doppler effect. Since neutron flux having energy of around 1 keV causes the maximum effect on the Doppler coefficient, the Doppler effect becomes more significant with a softened neutron spectrum.

2. Expansion effect of fuel

If fuel expands in an axial direction, the number density of fuel decreases relatively, and neutrons tend to be absorbed by materials other than fuel or neutrons that tend to leak from the core. This should decrease reactivity.

3. Deformation and expansion effect of structural materials

If the core support plate is expanded by temperature increase, the radial dimension of core increases and the number density of fuel decreases. It should decrease reactivity. The expansion effect of cladding is similar to that of coolant sodium, as described below.

4. Sodium void effect

If void is generated in coolant sodium or sodium temperature increases, the following changes should occur and affect reactivity. In this case reactivity tends to be positive in the middle of the core. The third effect is the most important.

1. Neutron absorption by sodium decreases. This causes a positive effect.
2. The mean free path of neutrons increases due to the reduction in sodium density, and this increases the probability for neutrons to escape out of the system. This causes a negative effect.
3. Since the probability of neutron scattering decreases, the neutron spectrum becomes harder. It causes a positive effect.

The neutron transportation theory is one of the fundamental theories for nuclear design calculations. The neutron transportation theory is described by the Boltzmann transport equation because neutron transportation is, in principle, quite similar to the diffusion of gases. The neutron diffusion equation is practical for use in nuclear calculations due to its simplicity. This type of procedure is very useful and rational, especially for the FR because the neutron mean free path in case of the FR is much longer than the fuel pin diameter and/or the dimension of coolant subchannel and no local distortion or peaking of neutron flux exists in almost all neutron energies. However, since fission occurs in all neutron energies in case of the FR, the effect of neutron spectrum on nuclear characteristics is significant, and a detailed treatment of the effect is very important. Especially in cases where a variety of nuclides, such as minor actinides, exist in a core, the effect of neutron energy spectrum on reactivity should be appropriately taken into account to estimate the change in reactivity due to the change in core conditions. Three-dimensional diffusion or perturbation calculation codes with increased energy group numbers have recently become more common, thanks to enhanced computer capabilities. Correction of diffusion calculation results is sometimes carried out using the results obtained by multidimensional transport calculation code especially for systems having extremely heterogeneous configuration and/or significant change in neutron distribution. The Monte Carlo calculation has been used recently.

The flow of reactor physics calculation for nuclear design is shown in Fig. 7.9. Core parameters such as the macroscopic neutron fission cross section (Σ_f), macroscopic neutron capture cross section (Σ_c), and macroscopic neutron absorption cross section (Σ_a) are calculated by the use of specific nuclear properties, including the microscopic neutron fission cross section (σ_f), microscopic neutron capture cross section (σ_c) and microscopic neutron absorption cross section (σ_a), core component properties including neutron source intensity, and core structure. These parameters should be set in consideration of the dependence of these parameters on neutron energy spectrum. The necessary plutonium content can be calculated by the use of multigroup and multidimensional neutron diffusion equation, which is shown as the basic equation in the figure, and the criticality condition. In this calculation the parameters cited above are used as input data, and excess

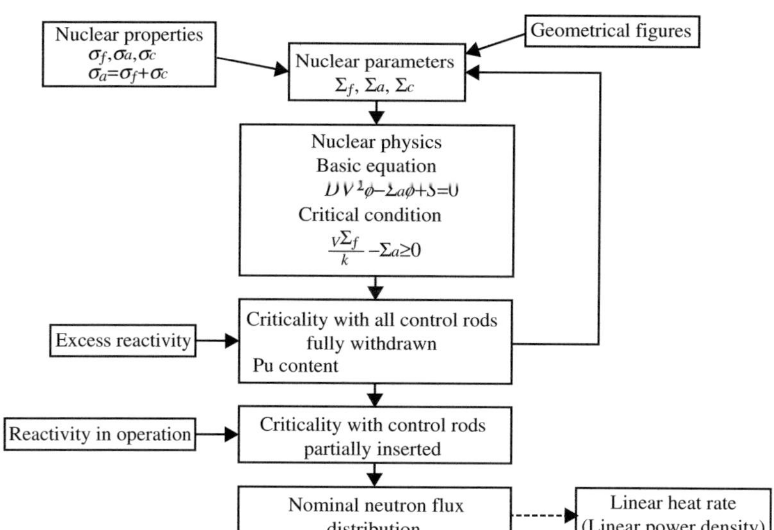

Fig. 7.9 Reactor physics calculation flow for nuclear design

reactivity, which corresponds to the reactor reactivity with all control rods fully withdrawn, is used as the boundary condition. If the calculated plutonium content deviates from the assumed value, the calculation is repeated with changed plutonium content. The final plutonium content is determined by iterative calculation. Next, the control rod insertion level at which criticality would be achieved is calculated by the use of reactivity during operation as the boundary condition. Neutron diffusion equation should be solved in this state to obtain neutron flux. The number of fissions per unit length of fuel can be calculated by the products of neutron flux, the amount of fissile materials per unit volume, and the microscopic fission cross section. This number can be converted into the LHR using the energy generated per fission.

7.5.4 Nuclear Design

1. Fundamental requirements for nuclear design

The fundamental requirements of nuclear design are summarized as follows: (1) the reactor can continue steady fission, (2) the reactor can be shut down whenever it is demanded, (3) reactivity should inherently decrease whenever abnormal event occurs, and (4) power distribution in core should be flattened.

In order to satisfy these requirements, the following six items should be performed in nuclear design: (1) the division of reactivity control equipment functions and the determination of reactivity that has to be controlled by each piece of equipment; (2) the confirmation of required functions of the main shutdown system consisting of fine control rods and coarse control rods and the backup shutdown system consisting of backup shutdown rods; (3) the calculation of reactivity coefficients, which are needed for reactor control; (4) the confirmation of inherent safety characteristics; (5) the confirmation of flattened power distribution; and (6) the determination of fuel exchange pattern and method.

2. Procedure to be used in fast reactor nuclear design (Flow of nuclear calculation)

The flow of reactor physics calculation for nuclear design is described in Sect. 7.5.3. An example of nuclear calculation flow for FR is explained using Fig. 7.10 in this section. First, determine what nuclear data library (a set of cross section and other nuclear data) should be used. The latest nuclear data, JENDL-4, which is developed by the Japan Atomic Energy Agency (JAEA), is currently used as shown on top of this figure. It is known that JENDL-4 can generally provide more accurate nuclear characteristics than JENDL-3.3. A cross section set called the group constant having approximately 100 energy groups is calculated and produced, and the group constant is shown as UF-LIBJ33 in the figure. Previously, an effective cross section in a unit lattice system was calculated from JENDL-4 after one-dimensional neutron spectrum calculation was performed using the SLAROM code. Currently, however, the SLAROM-UF code, which divides a resonance

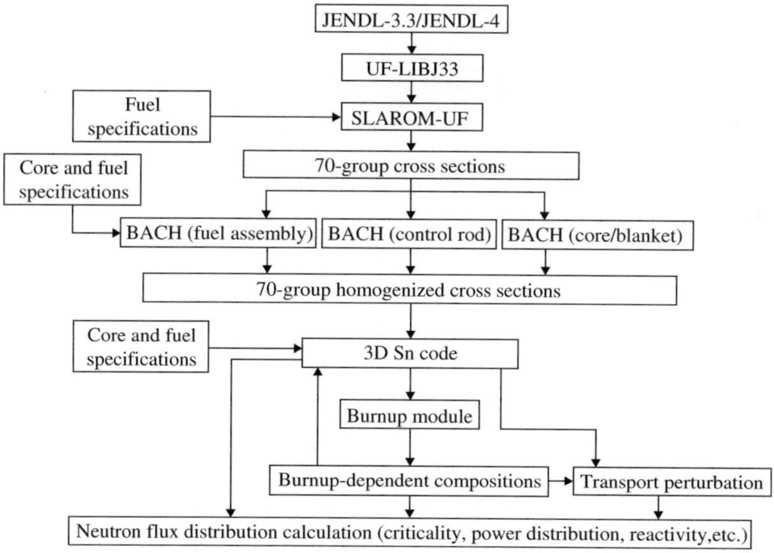

Fig. 7.10 Example of nuclear calculation flow [4]

energy region less than 50 *keV* into 100,000 groups, is often used for the one-dimensional calculation in order to handle the resonance effect in more detail. As a result 70-group cross sections are obtained. It is important to consider the self-shielding effect in the course of multigroup effective cross section calculation because fuel is surrounded by cladding and a fuel bundle is surrounded by wrapper tube. This effect is called the double heterogeneous effect. Then, the average cross section of a fuel assembly is obtained from the multigroup effective cross sections. BACH code, which can treat heterogeneous effect in hexagonal-shaped fuel assembly by transport theory, is used because the fuel assembly of FR has hexagonal shape. In this way 70-group homogenized cross sections are obtained. The core calculation is next started mainly by the use of the code based on transport theory, 3D Sn code shown in the figure. TRITAC code for three-dimensional XYZ geometry and NSHEX code for hexagonal-Z geometry have been developed in Japan as calculation codes based on the transport theory. Though calculation code based on a diffusion calculation code such as CITATION is frequently used for a large-scale FR where neutron leakage is insignificant, the calculation code based on transport theory should be used in order to obtain accurate local neutron flux near control rods. It is expected to use a calculation code based on transport theory for a small-scale fast reactor like Joyo (Japanese experimental FBR) where neutron leakage is significant. Nuclear characteristics such as criticality, power distribution, and reactivity (control rod reactivity worth, temperature coefficient of fuel, power coefficient, and so on) can be calculated using these codes. The burnup effect on the composition is also included in the figure.

The ratios between calculated and measured (C/E) criticality (effective multiplication factor) for various fast reactor cores are shown in Fig. 7.11. The horizontal

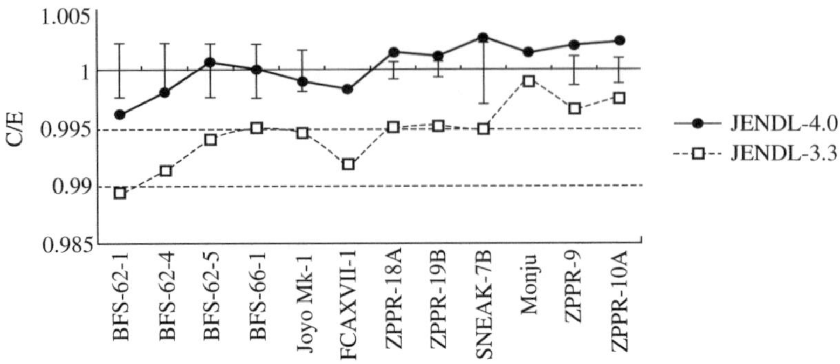

Fig. 7.11 C/E values of criticality (effective multiplication factor) [5]

axis in the figure shows various criticality experiments, and the vertical axis is the C/E values. Underestimation larger than 0.5 % of the effective multiplication factor in BFS and FCA fast reactor cores is found with JENDL-3.3. However, JENDL-4 provides good results for all types of fast reactor cores. The criticality experiments should accompany experimental error and the range of the error is shown as bars in Fig. 7.11.

3. Nuclear design items

Nuclear design items include reactor shutdown margin, reactivity insertion rate, excess reactivity, power coefficient, power distribution, stability, fuel burnup, and breeding ratio. Nuclear design should confirm the satisfaction of fundamental requirements by clarifying design principle and the means to achieve them. Design principle and the means are summarized in Table 7.1.

4. Nuclear design results

Table 7.2 shows the results of nuclear design for the Monju core. It is concluded that nuclear design results satisfy the fundamental requirements as summarized below:

- Fuel exchange pattern is the five-batch dispersal method.
- Two different fissile plutonium contents are set for the inner and outer cores, respectively, by considering the flattening of radial power distribution.
- The breeding ratio is obtained to be 1.2, which satisfies the design objective.
- The maximum LHR is determined to be 360 W/cm, which satisfies the limit to prevent fuel melting.
- The required control rod reactivity worth of regulating rods and backup rods is determined and the mission of these rods is clarified.
- Reactivity coefficients of structure temperature, coolant temperature, and void reactivity are positive; however, the gross power coefficient is negative.
- Prompt neutron lifetime and effective delayed neutron ratio are obtained, and they satisfy the design objectives.

Table 7.1 Nuclear design principle and means for achieving the principle [1]

	Nuclear design principle	Means for achieving principle
Reactor shutdown margin	To design so that reactor shutdown is done with one rod stuck and subcriticality is kept under low-temperature conditions	Reactor shutdown margin is kept more than $0.01\Delta k/k$
		Shutdown function by both main and backup shutdown systems
Reactivity insertion rate	To design so that maximum reactivity insertion rate is controlled in order to keep integrity of reactor coolant boundary against reactivity insertion accident and to keep integrity of reactor vessel components for maintaining core coolability	Maximum reactivity insertion rate is limited to be no greater than 8×10^{-5} $\Delta k/k$/sec
Excess reactivity	To design so that excess reactivity should compensate reactivity changes due to power change from low- to high-temperature conditions, burnup, and operational margin	Excess reactivity should include reactor shutdown margin
Power coefficient	To design so that the power coefficient has negative value to suppress the changes in power when power is changed	Negative reactivity feedback characteristics
		Doppler coefficient is negative and global power coefficient including temperature coefficients of fuel, core structure, coolant, and core supporting plate is also negative during all operational periods
Power distribution	To design so that power distribution during normal operation and anticipated operational occurrence does not cause any condition that specified acceptable fuel design limit is exceeded	Flattening of power distribution (2 core regions)
		Limitation of the maximum LHR (360 W/cm)
		Fuel exchange by dispersal method
Stability	To design so that power fluctuation during normal operation and anticipated operational occurrence does not cause any condition that specified acceptable fuel design limit is exceeded	Negative Doppler and fuel temperature coefficients (self-regulation characteristics)
		Limitation of local distortion of neutron flux distribution by control rod insertion
		Small neutron absorption effect by fission products
Fuel burnup	To design so that average fuel assembly burnup satisfies the design target	Determination of cycle length and fuel exchange pattern
Breeding ratio	To design so that the target breeding ratio is achieved	Determination of the blanket fuel regions

Table 7.2 Results of nuclear design for the Monju core [1]

Fuel exchange pattern		Control rod worth	Initial core	Equilibrium core
Core fuel assembly	5 batches	Regulating rod (one rod stuck)	About 0.07$\Delta k/k$	About 0.07$\Delta k/k$
Blanket fuel assembly	5 batches	Backup rod	About 0.05$\Delta k/k$	About 0.06$\Delta k/k$
Fissile plutonium content		Reactivity coefficient		
Initial core (inner/outer)	About 15 wt%/about 20 wt%	Doppler	- (4.4~7.6) $\times 10^{-3}$ T dk/dt	
		Fuel temperature	- (3.0~3.9) $\times 10^{-6}$ $\Delta k/k$/°C	
Equilibrium core (inner/outer)	About 16 wt%/about 21 wt%	Structural material temperature	+ (6.0~14) $\times 10^{-7}$ $\Delta k/k$/°C	
		Coolant temperature	+ (1.0~28) $\times 10^{-7}$$\Delta k/k$/°C	
		Core support plate temperature	- (10~12) $\times 10^{-6}$$\Delta k/k$/°C	
		Power coefficient	- (8.4~11) $\times 10^{-6}$$\Delta k/k$/MW	
		Maximum void reactivity	+ (1.1~1.9) $\times 10^{-4}$$\Delta k/k$	
Breeding ratio				
Initial core	About 1.2	Delayed neutron lifetime	0.36~0.45 μs	
Equilibrium core	About 1.2			
Average fuel burnup				
Initial core	About 13,000 MWd/t	Effective delayed neutron generation rate	0.0032~0.0039	
Equilibrium core	About 80,000 MWd/t			
Linear heat rate				
Core averaged	About 210 W/cm			
Maximum	360 W/cm			

7.5.5 *Thermal Hydraulic Design*

1. Flow of thermal hydraulic design

A flow of thermal hydraulic design is shown in Fig. 7.12. Coolant flow rate should be determined by the use of thermal power, core inlet and outlet coolant temperatures, etc. in order to achieve a balance between heat generation and heat removal. Appropriate coolant flow regions and necessary flow rates should be determined in order to unify the fuel assembly outlet temperatures and achieve effective heat removal. The temperature distribution in a fuel assembly is calculated by subchannel analysis to confirm that the maximum cladding surface temperature is within the design limitation.

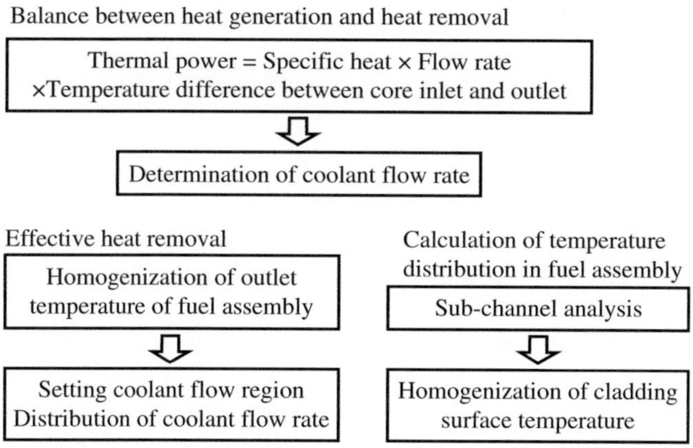

Fig. 7.12 Flow of thermal hydraulic design

Once the fuel cladding surface temperatures are determined, the fuel pellet center temperature can be estimated by calculation of the temperature gap between the inner and outer surfaces of cladding, the temperature gap between cladding inner surface and fuel pellet outer surface, and the temperature difference within a fuel pellet.

2. Outline of thermal hydraulic engineering

Thermal hydraulic design is performed in order to get the most economical parameters in consideration of heat generation and removal conditions and hardware specifications. Therefore, thermal hydraulic design should be carried out based on the neutron flux distribution that is obtained as a result of nuclear design. It is very important to calculate the temperature distribution of each component as accurately as possible by clarifying heat removal characteristics of coolant against heat generation by fission. Here, phenomena which are focused on from the viewpoint of thermal hydraulic behaviors in core and their evaluation procedures are reviewed, and the concept and flow of thermal hydraulic design are explained.

Fuel pins are closely packed because the fission cross sections of FR are lower than those of LWR. Therefore, the pitch between fuel elements (pins) is narrow. The mechanism employed to hold the distance among the pins is called a wire-wrapped spacer. A wire is wrapped spirally along the fuel pin to prevent pin-to-pin contact. The wire-wrapped spacer has the advantage of reducing pressure loss (drop) and enhancing the coolant mixing effect in comparison with a grid-type spacer. Sodium coolant enters into a fuel assembly from the entrance nozzle, receives heat from fuel pins, rises spirally around fuel pins, and flows out from the exit at the top of the fuel assembly. The fuel pins (pin bundle) are installed in a wrapper tube to keep necessary flow rate and to protect the fuel pin bundle mechanically. The focus in the in-core thermal hydraulic design should be placed on the following points:

1. Fuel assembly pressure loss and temperature distribution during normal operation
2. Thermal hydraulics in low flow rate conditions such as natural circulation (mixed convection)
3. Fuel deformation due to high burnup

Evaluation of the pressure loss in a fuel assembly is indispensable in obtaining appropriate flow distribution for all flow rate regions for homogenizing the fuel assembly outlet temperatures and setting the discharge pressure of the primary pumps. In Joyo and Monju fuel assembly designs, pressure loss characteristics were clarified by thermal hydraulic experiments with full-scale mock-ups, and the experimental results were used as data for design. Temperature distribution in a fuel assembly is determined by the balance between the axial and radial distributions of heat generation in fuel assembly, and the heat removal characteristics which depend on the coolant flow distribution in subchannels and the mixing effect (including wire spacer and turbulence effects). The characteristics of temperature distribution are confirmed by experiments using liquid sodium because thermal conductivity differs significantly between water and sodium.

Flow distribution is explained here. An FR core is composed of core fuel assemblies, blanket fuel assemblies, control rod assemblies, and neutron shield assemblies located in the periphery of the core. Since the amounts of heat generation in these components are different due to their compositions and loading positions, it is very important to appropriately determine flow distribution by dividing flow rate regions in order to remove heat generated in the core effectively under the thermal restrictions. This is one of unique FR issues and is quite different from LWR. Appropriate flow distribution is realized by providing high- and low-pressure plenums under the core region in the reactor vessel. Coolant flows into the core fuel assemblies having higher heat generation from the high-pressure plenum, and coolant flows into the blanket fuel assemblies having lower heat generation from the low-pressure plenum. A mechanism designed to control the flow rate of each assembly is also installed in the lower part of each assembly. The mechanism is the combination of orifice holes set at the entrance nozzle and the flow rate control device set at the connection tube, into which the entrance nozzle is inserted. The mechanism is shown in Fig. 7.16.

A design concerning flow distribution in the core is carried out by the following three steps:

1. Flow rates in core/blanket fuel assemblies are determined so that the maximum cladding temperatures in all assemblies satisfy the thermal limitation and are homogenized, based on the evaluation of cladding temperatures using the power distribution in the core.
2. Flow rate regions are set by grouping fuel assemblies whose required flow rates are similar.
3. The flow rate control devices are designed so as to homogenize the pressure losses from lower to upper plenums in the reactor vessel in all flow rate regions.

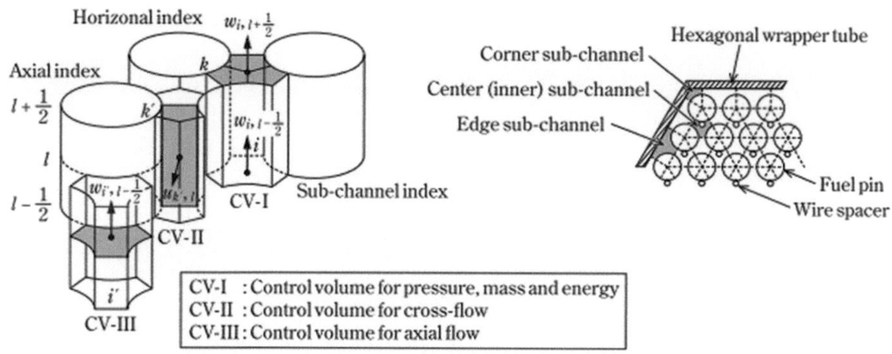

Fig. 7.13 Definition of control volume in subchannel analysis

It is necessary to avoid complicated design of the flow rate control device when the flow rate regions are determined. The pressure loss coefficients at various structure parts which are needed for designing the flow rate control device should be evaluated by mock-up water tests and/or numerical simulations. Attention should be paid to the possibility of cavitation, which may occur as the coolant passes through the orifice holes at the entrance nozzle with higher flow velocity and exert an influence on the structural integrity.

A representative method for analyzing the thermal hydraulic behavior of a whole fuel assembly is subchannel analysis. The definition of "subchannel" is a unit flow channel which is surrounded by fuel pins or a unit flow area which is surrounded by fuel pin(s) and the wrapper tube wall. Subchannels are schematically shown in Fig. 7.13.

The outline of the subchannel analysis is explained as follows: Each subchannel as a unit flow channel is axially divided into control volumes. Conservation equations of mass, momentum and energy of coolant, and energy conservation equation of fuel pin are three-dimensionally solved based on the control volumes. The coolant flow rate at the inlet of the fuel assembly, which is estimated in the in-core flow distribution analysis, and the heat generation distribution of fuel pins are used as input data for the analysis. The coolant velocity and temperature distributions and the cladding and fuel temperature distributions are calculated by these procedures. Models of cladding surface heat transfer coefficients, pressure loss, and turbulent mixing are used in the subchannel analysis. Parameters applied to the models are evaluated through the data obtained by thermal-hydraulic experiments using a full-scale model for each design. However, detailed numerical simulation has recently been applied instead of mock-up experiments because of the difficulty involved in conducting experimental measurements and the high cost of experiments. This trend is growing with the progress of computer power and analysis technology. If numerical simulation is applied to substitute for experiments, the accuracy of the simulation should be evaluated in advance.

In FR, a high-temperature low-pressure system, the adoption of a decay heat removal system operated exclusively by natural circulation without using any active component has been studied based on the safety design requirements. In low flow rate conditions, such as natural circulation, buoyancy force becomes dominant rather than inertia force as the force exerting on coolant. Therefore, it is expected that radial temperature distributions both in the core and in fuel assemblies are flattened due to redistribution of coolant flow. The flattening is accelerated by heat transfer among fuel assemblies. It is also probable that sodium flow in gaps between fuel assemblies has an influence on temperature distribution in the core. Several types of analytical methods have been examined in this field. The simplest method is one-dimensional flow network model, in which coolant flow channels are modeled by a combination of one-dimensional branch elements and the flow rate, flow direction, and temperature of each element are calculated so as to satisfy conservation laws. The three-dimensional model is somewhat complex, and it models the whole core with combining subchannel analysis methods and taking into account the inter-wrapper flow.

Bundle-duct interaction (BDI) may occur at very high burnup because of the expansion of fuel pin bundle due to the swelling of structural materials by fast neutron irradiation. BDI is remarkable when expansion of fuel pin bundle is larger than that of duct (i.e., wrapper tube). When BDI is remarkable, thermal integrity of cladding might be degraded due to the bowing of fuel pins along the wrapping wire pitch and the reduction in distance between fuel pin and duct. To confirm the design feasibility, a detailed evaluation of the temperature distribution is required under conditions where fuel pins are deformed by considering the influence of swelling and the thermal effect on cladding deformation. Since the deformation of structural material and the coolant temperature distribution affect each other, combined analysis of thermal hydraulics and fuel deformation is usually carried out in thermal hydraulic evaluation with deformed fuel bundle.

In other aspects, local flow channel blockage and local boiling due to the entry of foreign material are evaluated from a safety perspective because the fast reactor has features, such as closely packed fuel pins and high LHR.

3. Fundamental requirements and principle of thermal hydraulic design

Fundamental requirements on thermal hydraulic design are summarized as follows: (1) thermal characteristics should be determined through a balance of axial and radial heat generation distribution, which are determined by nuclear design, and heat removal characteristics determined by thermal hydraulics in fuel assemblies, and (2) consideration should be given to the thermal output; heat transfer surface area; coolant flow rate; pressure; engineered safety factor (hot spot factor), which is explained later; over power coefficient, which is also explained later; fuel temperature; cladding temperature; etc. when thermal characteristics are set.

The principles of thermal hydraulic design are summarized as follows: (1) the specified acceptable fuel design limits shall not be exceeded in normal operational states and anticipated operational occurrences by the use of functions of the reactor cooling system, intermediate cooling system, instrumentation and control system, and safety protection system, and (2) fuel integrity shall be maintained until a specified burnup is reached.

4. Specified acceptable fuel design limit

The definition of "specified acceptable fuel design limit" is a limit within which the degree of fuel damage is acceptable from the viewpoint of reactor safety and reactor operation can be continued. Specific principles in Monju design are to prevent coolant boiling, to limit the maximum fuel temperature to less than the melting point (2650 °C), and to limit the cladding temperature during anticipated operational occurrences to less than 830 °C. Sodium coolant boiling should be avoided because the reactor power may increase due to the insertion of positive sodium void reactivity and thereby cause excessive increase in cladding temperature due to the degradation of heat transfer properties on cladding surface. The maximum fuel temperature is typically limited to less than the fuel melting point in order to prevent the loss of cladding integrity due to volume expansion of fuel. While the LWR limits the heat flux on cladding surface, the FR limits the cladding temperature itself which influences the cladding mechanical property because the FR is a high-temperature system and coolant sodium has very good heat transfer characteristics. The maximum cladding temperature during normal operation is limited to maintain fuel integrity until target burnup is reached based on creep strength test data and irradiation test data, and it is often set at a temperature between 650 °C and 700 °C. The maximum cladding temperature during anticipated operational occurrences is limited to prevent burst failure due to abrupt increase in cladding temperature and is determined based on out-of-pile rapid heating test data using irradiated cladding materials.

5. Temperature distribution

Temperature distribution from coolant to fuel center can be calculated by the use of several equations that can be given by solving the heat balance and heat conduction equations. Figure 7.14 is a schematic of radial temperature distribution

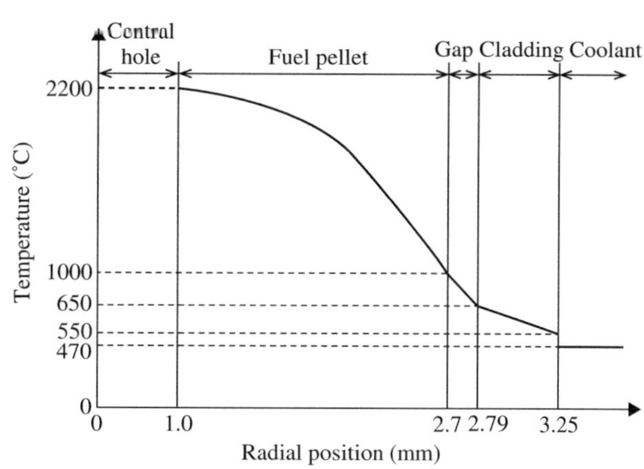

Fig. 7.14 Schematic diagram of fuel pin radial temperature distribution

of Monju fuel pin at the point of maximum heat generation, which corresponds to the axial center of fuel pin.

The values shown in this figure are nominal ones for which no uncertainty was taken into account. However, a number of uncertainties related to temperature evaluation need to be taken into account for design evaluation. For example, the uncertainty of LHR is taken into account as the over power coefficient. Moreover, fabrication tolerances and the uncertainties of physical properties should also be taken into account. These uncertainties are treated as hot spot factors and divided into two types: one is the multiplication term by which systematic errors are treated cumulatively and the other is the statistic term by which random errors are treated statistically. The former includes the power distribution and the latter includes the fuel pellet fabrication error. Hot spot factors are introduced in order to estimate the maximum values in evaluating temperatures under various heat generation and removal conditions. Fabrication tolerances, modelling errors, and measurement errors for physical properties should be taken into account in the design. The hot spot factor with fuel pellet, which is applied to the temperature difference between fuel center temperature and coolant inlet temperature, was set at 1.2 in Monju design. Since the over power coefficient in Monju was set at 1.16, the nominal temperature difference should be multiplied by 1.4 ($=1.16 \times 1.2$) to obtain the hot spot temperature difference in over power condition.

6. Thermal hydraulic design result

The results of thermal hydraulic design for Monju are shown in Table 7.3. The LHR is a particularly important parameter in core design.

7.5.6 Plant Dynamics Analysis

The outline and results of plant dynamics analysis are explained below.

In the fast reactor, greater coolant temperature difference between reactor vessel inlet and outlet can be set because the core power density in the FR is higher than in the LWR, and coolant sodium has very good heat transfer characteristics and a high boiling point. In addition the heat transport system from reactor core to turbine generator inherently has a delay time due to sodium transport in pipes and components and the heat capacity of sodium and structural materials. Therefore, the plant output level should be controlled in accordance with the value designated by the power demand master, and the flow rate of the main cooling system is controlled in proportion to the designated power level in order to keep plant stability and prevent significant deviation from a thermal equilibrium state. The reactor vessel outlet sodium temperature should be controlled so as to keep the degree of superheat of the main steam constant.

Reactivity (ρ), which is explained in Sect. 7.5.3, is a barometer of the deviation from criticality. If reactivity is added in a core, the effective neutron multiplication factor (k_{eff}) exceeds unity, meaning that the neutron balance shifts to positive. In

Table 7.3 Results of thermal hydraulic design [1]

Thermal output	About 714 MW	Hot spot factor	
		Core fuel assembly	
		Core fuel pellet	About 1.20
		Cladding	About 1.26
		Coolant	About 1.25
		Blanket fuel assembly	
		Blanket fuel pellet	About 1.27
		Cladding	About 1.32
		Coolant	About 1.32
Gross flow rate of primary loop	About 15.3×10^6 kg/h	Linear heat rate of fuel pin	
		Core average	About 210 W/cm
		Maximum (peak)	360 W/cm
Reactor vessel inlet temperature of primary coolant	About 397 °C	Maximum fuel temperature	
		Steady state	2350 °C
		Over power	About 2600 °C
Reactor vessel outlet temperature of primary coolant	About 529 °C	Maximum cladding midwall temperature	
		Steady-state core fuel assembly	675 °C
		Steady-state blanket fuel assembly	700 °C
Power peaking factor		Maximum coolant temperature	
Fuel assembly	About 1.42	Core fuel assembly	About 659 °C
Axial	About 1.21	Blanket fuel assembly	About 696 °C
Local	About 1.01		
Over power coefficient	About 1.16		

this case the reactor nuclear kinetics model shown in Fig. 7.15 sends a signal warning that the amount of neutrons is excessive and the power control system works to insert control rods into the core via the control rod driving mechanism (CRDM). Thus, the change in reactivity is used to control the reactor power by use of information on the neutron balance.

Response analysis is performed by computer, which contains plant simulation models. Characteristics of the main components and functions of the control system, safety protection system, etc. are installed in the plant simulation models. The outline of plant simulation models is shown in Fig. 7.15.

Plant dynamics analysis shows that the reactor can be easily returned to a steady-state condition without any reactor trip even if the following disturbances occur during normal operation with over 40 % power: (1) ± 10 % step change in power, (2) ramp change in power of ± 5 %/min, and (3) 50 % loss of load.

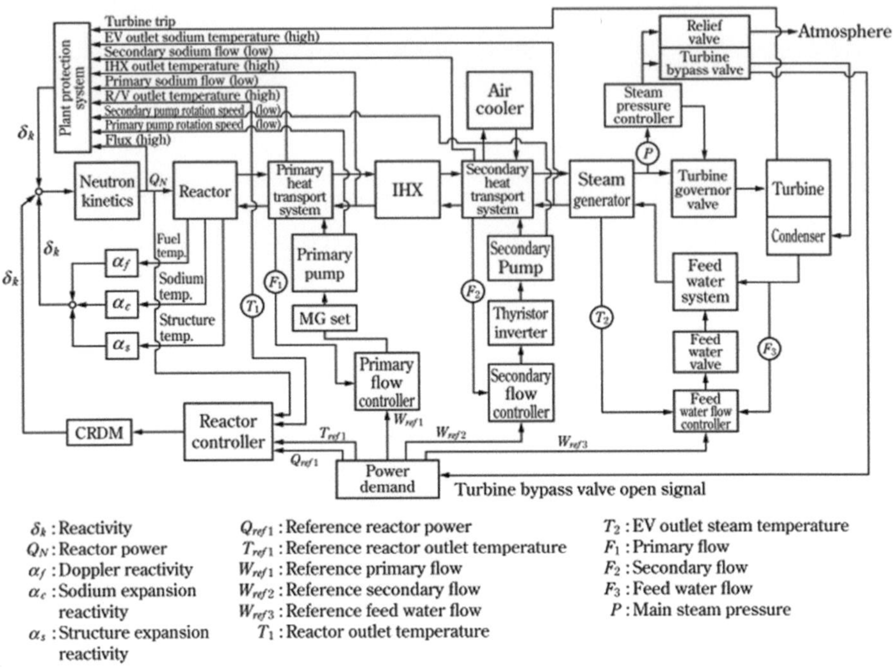

Fig. 7.15 Schematic diagram of plant dynamics analysis models [1, 6]

7.6 Fuel Design

7.6.1 Characteristics of FR Fuel

1. What is fuel design?

The purpose of fuel design is to determine specifications such as pellet diameter, density, cladding inner radius, cladding thickness, plenum length, dimensions of fuel assembly, etc., based on the information provided from core design including the LHR, burnup, temperature, etc.; and fuel specifications are supplied to fuel fabricators. Although fuel and cladding materials are very closely related to fuel design, they are determined at the stage of conceptual design, because the target burnup, LHR, and temperatures should be determined by the performance of fuel and cladding materials. That is, conceptual system design is based on the performance of fuel and cladding materials. In the Monju design, MOX fuel is used for the fuel material, and 316 SS equivalent stainless steel, which contains a very small amount of Nb, Sn, and Ti as additives, is used for the cladding material. Cladding swelling is a phenomenon which is caused by the production of alpha particle (helium atom) by (n, α) reaction and the displacement of atoms by fast neutrons. The positions of atoms directly displaced become vacancies, which form a void by

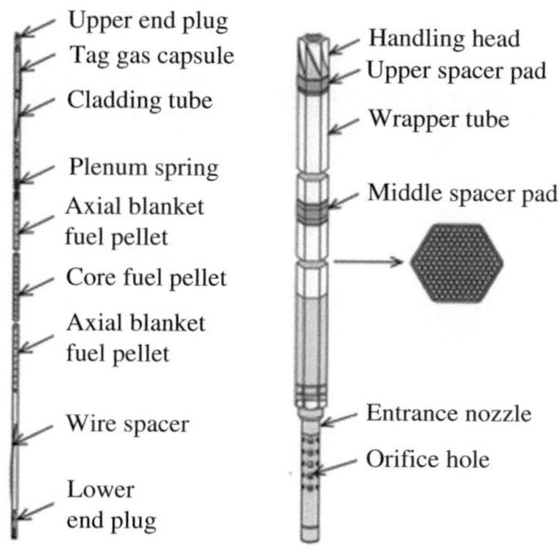

Fig. 7.16 Structure of core fuel assembly (*right* side) and the fuel pin (*left* side)

gathering together. Such voids cause an expansion in the volume of material. This series of the process is called void swelling.

2. Fuel structure

The structure of the Monju core fuel assembly and fuel pin are shown in Fig. 7.16, which is prepared based on the "Application for reactor establishment license: Prototype Fast Breeder Reactor Monju" [1]. The fuel pin shown in the left side of Fig. 7.16 is composed of core fuel pellets (MOX), axial blanket fuel pellets (depleted uranium dioxide or natural uranium dioxide), plenum, plenum spring, the tag gas capsule, and the upper and bottom end plugs. The function of plenum is to store fission product gas (FP gas). The plenum spring fixes the position of fuel pellets in the fuel pin. The tag gas capsule contains tag gases, such as krypton and xenon, which are used for fuel failure detection and location. After installation of the tag gas capsule in the fuel pin, both ends of cladding are sealed with end plugs by welding, the capsule is opened, and tag gases are diffused in the plenum. A wire (spacer) is wrapped along a fuel pin with a constant pitch in order to maintain the coolant flow area. A total of 169 fuel pins are bundled and they are installed into a hexagonal wrapper tube (duct). The entrance nozzle and handling head are welded at the bottom and top of the duct, respectively. Coolant flows into the entrance nozzle and flows out of the handling head. The handling head is also used when the fuel assembly is handled. The blanket fuel assembly has a structure similar to that of the core fuel assembly except for the pellet material and the number of fuel pins, whose number is 127.

3. Fuel specifications

The main items of fuel design and their characteristics are explained here. The plutonium content of MOX fuel is determined at a lower level in the inner core where fast neutron flux is higher and at a higher level in the outer core where fast neutron flux is lower. Radial power flattening is achieved by this arrangement of plutonium content. Fuel pellet density is determined at a low level in order to suppress pellet-cladding mechanical interaction (PCMI) due to fuel swelling (volume expansion) caused by accumulation of fission products accompanied with burnup. Pellet diameter is determined at a smaller level because fuel burnup becomes higher with pellets having a smaller diameter even if the LHR is the same. Fabrication of low density and small diameter pellets needs a high-level fabrication technique. Plenum volume is determined to be large enough to reduce the plenum gas pressure at high burnup. This causes less stress on cladding tubes. 316 SS equivalent stainless steel is used because this material has improved anti-swelling properties at high burnup. Thick cladding is used in order to reduce hoop stress exerted on cladding tubes. Wire spacer is tightly wrapped along the fuel pin in order to prevent coolant flow vibration, which may cause scar resulting from contact with fuel cladding. Porosity per ring (P/R) is used as the quantitative index of tightness.

7.6.2 Trend of Fast Reactor Fuel Design

Smaller-diameter, solid, and lower-density fuel pellets have been used for Joyo and Monju. Subsequently, a number of fuel irradiation experiments have been conducted to aim for high LHR and high burnup. Thus, the current fuel design tends to shift to larger-diameter, annular or hollow, and higher-density fuel pellets to achieve high performance. Annular pellets are effective for reducing the fuel temperatures near the centerline.

MA-containing fuel is being developed for reducing environment impact and nonproliferation. One MA element, ^{237}Np, has an extremely long half-life of 2.1×10^6 y. If ^{237}Np is mixed into high-level waste, high-level waste disposal becomes more troublesome due to its long-lasting radioactivity. ^{243}Am has also long half-life of 7370 y. Cm isotopes, whose mass number is smaller than 244, shift to Pu isotopes with rather long half-life via α-decay. Cm isotopes, whose mass number is larger than 244, have significant long half-lives. For example, the half-life of ^{245}Cm, ^{246}Cm, and ^{247}Cm is 8500 y, 4730 y, and 1.56×10^7 y, respectively. Therefore, if MA elements are mixed into high-level waste, environmental burden would increase. Np and Am are not defined as sensitive nuclear materials for safeguards by the International Atomic Energy Agency (IAEA) like U and Pu. However, Np and Am are fissile materials. Therefore, the recycling of MA elements using the FR is desired from the nonproliferation perspective. In this case MA elements are recovered together with Pu in the reprocessing process.

7.6.3 Fast Reactor Fuel Design

1. Design condition of fuel (use conditions)

Features of the use conditions of FR fuel include low-pressure and high-temperature sodium environment, hard neutron spectrum (high neutron energy), significant fast neutron fluence and burnup, high fuel pellet temperature, significant temperature difference across a fuel pellet, high structural material temperature, and significant spatial distribution in structural temperature. Accordingly, the use conditions of FR fuel are severer than those of LWR fuel especially from the perspective of temperature, LHR, burnup, fast neutron fluence, etc.

2. Fuel behavior under irradiation

Fuel restructuring occurs at an early stage of irradiation because the fuel pellet temperature is high enough to move as-fabricated pores toward the pellet center by repeating vaporization at higher-temperature position in pore and condensation at lower-temperature position. A central hole is formed as a result of this. Columnar grain is formed in the outer region of the central hole and surrounded by equiaxed grain, where grain growth occurs. A non-restructured region remains near the pellet surface. The formation of the central hole reduces the fuel centerline temperature, and the formation of columnar grain and the equiaxed grain increase the density and, hence, thermal conductivity of fuel pellets corresponding to these regions.

Plutonium and MA, such as americium and neptunium, are redistributed along the radial direction of fuel pellets at an early stage of irradiation. The concentrations of Pu and MA become higher in the center of fuel pellets due to thermal diffusion of these elements in solid phase and vaporization-condensation mechanism in pore caused by radial temperature gradient. Fuel melting temperature is a function of plutonium concentration, and it tends to decrease with increased Pu concentration. Thus, the margin for fuel melt is reduced by this phenomenon. In addition, oxygen is also redistributed along the radial direction, and the concentration tends to be lower in the center and higher in the outer region of a pellet. This causes a change in the oxygen to metal (O/M) ratio that occurs due to a steep temperature slope along the radial direction of a pellet. If O/M increases in the outer region of a pellet, oxygen potential also increases to enhance the formation of cesium and molybdenum oxides in the pellet-cladding gap with an increased burnup condition because precipitation of Cs and Mo on the pellet surface becomes remarkable at an increased burnup.

Fuel densification occurs at an early stage of irradiation resulting disappearance of as-fabricated pores, and it causes an increase in pellet density. However, fuel swelling becomes remarkable with increased burnup. Pellet volume is determined by the balance of densification and swelling. Consequently, pellet volume decreases at first and then increases.

A great amount of fission products (FP) is accumulated in fuel pellets, which causes swelling with increase in burnup. When swelling becomes significant, pellet-cladding mechanical interaction (PCMI) occurs, which causes high stress on cladding together with the plenum gas pressure. Stress due to PCMI tends to be relaxed because cladding swelling, which causes cladding diameter increase, becomes remarkable at high burnup.

The inner and outer cladding surfaces are corroded by chemical reaction with FP and sodium, respectively. Such corrosion causes reduction of cladding thickness. Since fuel cladding is always used at high temperatures and receives continuous stress, creep rupture of cladding may occur if cladding stress becomes high due to increasing stress by increasing plenum gas pressure and reduction of cladding thickness. Creep rupture is evaluated by the cumulative damage fraction (CDF) method. If the CDF value reaches unity, cladding is assumed to have failed due to creep rupture.

When cladding swelling becomes significant, the size of the fuel pin bundle increases and then the bundle comes into contact with the inside surface of the wrapper tube (duct), which is called BDI. Significant BDI causes bowing of fuel pins and binding force appears. BDI may also cause blockage of a coolant flow area, which would lead to an increase in cladding temperature. Therefore, the change in cladding outer diameter should be limited within an allowable limit.

Wrapper tubes (ducts) also swell at high burnup and duct-duct interaction (DDI) occurs. Significant DDI causes excessive stress on wrapper tube and induces deformation. Deformation of wrapper tubes exceeding the limit would cause difficulty in control rod insertion. This phenomenon is accelerated by flowering deformation of wrapper tubes. Flowering is caused by outward deformation of the upper section of wrapper tube. This is due to temperatures on the inner wrapper tube wall (the nearest wall to the core center) being higher than those on the outer wrapper tube wall (the farthest wall to the core center), and accordingly, thermal expansion is more significant on the inner wall than the outer wall of a wrapper tube. Then, wrapper tubes deform like a flower.

3. Items to confirm fuel integrity and design evaluation equations

Fundamental items to be confirmed in fuel integrity evaluation include the following: (1) no fuel melting should occur; (2) cladding should not fail due to excessive stress, which means that the stress on cladding should be less than the allowable design stress; (3) the change (increase) in cladding diameter should be less than the allowable level, which means that excessive cladding temperature increase should not occur; and (4) CDF should be less than 1.0.

The fuel centerline temperature is calculated by the procedure explained in Sect. 7.5.5 ("Temperature Distribution"). Design evaluation value of the fuel centerline temperature is obtained by multiplying nominal calculated temperature by the hot spot factor in order to take all uncertainties into account. However, it is very difficult to separately estimate the temperature difference across a fuel pellet and the temperature gap between pellet outer surface and cladding inner surface accurately. Thus, fuel melting test, which is called "power to melt" (PTM) test, is sometimes carried out in order to directly estimate the LHR to melt. The occurrence of fuel melting is judged by direct comparison between the power to melt observed by experiment and the transient over power in design.

FP gas is generated according to the increase in burnup, and it is released from fuel pellets and moves to plenum, and the released FP gas increases the plenum gas pressure. Increase in gas pressure proportionally increases the cladding stress, and if the stress level exceeds the allowable limit, cladding may fail. The relationship between gas pressure (P_{gas}) and cladding hoop stress (σ_θ) can be expressed by the following equation:

$$\sigma_\theta = P_{gas} \frac{2r_{c,m}}{t} \tag{7.8}$$

where $r_{c,m}$ and t are the mean cladding radius and cladding thickness, respectively.

The change in cladding outer diameter is evaluated as the sum of swelling strain and irradiation creep strain of cladding. These strains are given by equations unique to the cladding material. If the cladding outer diameter increases, the coolant flow area is reduced and the cladding temperature increases. The increase in cladding temperature reduces the strength of cladding material and leads to cladding failure. The increase in cladding outer diameter is limited to 7 % in the Monju design.

CDF is given by the following equation:

$$\text{CDF} = \sum_i \frac{\Delta t(i)}{t_r(i)} \tag{7.9}$$

where $\Delta t(i)$ is the duration of time interval i where temperature and stress state are considered to be constant and $t_r(i)$ is the creep rupture time corresponding to the time interval. It is defined that creep rupture of the cladding occurs when the CDF calculated by Eq. (7.9) reaches 1.0.

4. Integrity evaluation of fuel assembly

Integrity evaluation of core fuel assembly should show that all design items, such as the fuel centerline temperature, cladding stress, cladding outer diameter change, and CDF, satisfy the design criteria. Integrity evaluation results for blanket fuel assembly are expected to show that all design items satisfy the design criteria because design conditions, such as the LHR, burnup, and fast neutron fluence, are rather milder.

5. Technical standards for safety design and assessment of sodium-cooled fast reactor fuel

Fuel design is closely related to the fundamental requirements for reactor safety, such as control of reactivity (reactor shutdown), removal of heat from the reactor core (core cooling), and confinement of radioactive material (confinement). Therefore, it must be confirmed in fuel design that control rod insertion paths are kept available, the coolant flow area is maintained, and fuel integrity is ensured. It is also required to confirm fuel conditions during anticipated operational occurrences and postulated accidents. Fuel design shall verify that specified acceptable fuel design limits are satisfied in case of anticipated operational occurrences. Fuel failure is permitted when postulated accidents occur; however, the amount and the mode or scale of fuel failure should be limited. That is, the occurrence of large-scale fuel melting and failure and resultant severe core damage must be avoided even in case of postulated accident. It is very important to specify the requirements described

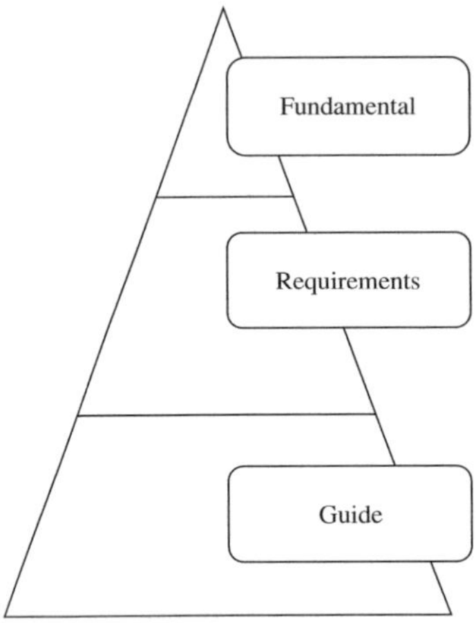

Fundamental of fuel safety design and safety assessment
Fundamental function of fuel should comply with reactor safety functions such as reactor shutdown, core cooling and confinement.

Requirements of fuel safety design and safety assessment	
Fuel safety design	Safety assessment
• To keep fuel integrity • To keep control rod insertion path • To keep coolable geometry	• To set specified acceptable fuel design limit • To set criteria for fuel failure

Guide of fuel safety design and safety assessment	
Fuel safety design	Safety assessment
• Evalution items • Criteria • Evaluation procedure	• Procedure for setting specified acceptable fuel design limit • Procedure for setting criteria for fuel failure

Fig. 7.17 Structure of technical standards for fast reactor fuel design

above as technical standards for safety design and assessment of sodium-cooled fast reactor fuel. In consideration of its importance, the IAEA established a consultant committee (meeting) in order to discuss and summarize the technical standards for fast reactor fuel. IAEA standards have a three-layer structure including fundamentals, requirements and guides. The structure of the technical standards for fast reactor fuel based on the IAEA concept is shown in Fig. 7.17.

Every possible fuel failure and deformation mode should be listed in fuel safety design. An example of fuel design items is introduced below. Deformation and failure of duct and cladding due to excessive stress and strain should be taken into account with respect to maintaining fuel integrity. Radial displacement of control rod guide tubes due to DDI should be taken into account with respect to maintaining control rod insertion paths. Fuel bundle deformation due to temperature change, irradiation, and load should be taken into account with respect to maintaining coolable geometry.

It is important to establish criteria and evaluation methods for each design item. Design stress intensity (Sm) is used as a criterion, and simplified methods such as those expressed by Eq. (7.8) and Eq. (7.9) are used as design methods for mechanical integrity evaluation items.

Acceptance criteria for the requirements for fuel should be determined in consideration of design margin based on the results of research and development (R&D), including irradiation experiments. Since the criteria certainly exert an influence on fuel specifications and use conditions via the process of fuel design, fuel R&D from the perspective of the development of acceptance criteria is important to develop high performance fuel. It is also important to promote R&D from the perspective of the improvement of the accuracy of physical properties and models used in fuel design, an accurate understanding of fuel behavior under irradiation, and the accumulation of knowledge required for fuel fabrication.

Table 7.4 Example of fast reactor fuel design

Parameter	Metal	Carbide/Nitride	Oxide	Monju
Pellet density (%TD)	100	80–82	92	85
Smear density (%TD)	75	78–79	80–85	80
Volume ratio of fuel to plenum	1.4	1	1	1.25
Fuel stack length (cm)	91	91	91	93
Pellet outer diameter (mm)	5	TBD	5.6	5.4
Pellet inner diameter (mm)	0	0	1.5	0
Bond material	Na	He	He	He
Cladding outer diameter (mm)	6.9	9.4	6.9	6.5
Cladding inner diameter (mm)	5.7	TBD	5.7	5.56
Maximum LHR (kW/m)	49–52	66–80	44–46	36
Maximum cladding temperature (°C)	620	620	620	675

Design by analysis is mainly used for fuel design. Two methods are applied in specific fuel design. These are simplified, including simple equations like Eq. (7.8), and rather complex, using a fuel performance analysis code. When using an analysis code, much attention should be paid to determining input conditions, analytical models, and physical properties in order to ensure conservativeness of the analytical results. In addition, the fuel performance analysis code needs to be validated and verified using fuel irradiation data.

6. Comparison of fast reactor fuel design examples

The result of fuel design for the Monju fuel (oxide) is compared with those for metal, carbide/nitride, and oxide fuels in Table 7.4. Although there was a lot of experience with oxide fuel, since Monju is the first prototype fast breeder reactor, the design was carried out very conservatively. The data described in the column titled "oxide" in this table are those updated by taking more recent R&D achievements on oxide fuel into account.

Since pellet swelling in carbide and nitride fuels is more significant than in oxide fuel, pellet density should be set at a lower level for the purpose of avoiding hard pellet cladding mechanical interaction. The density of carbide/nitride fuel almost corresponds to that of Monju pellets. This fact is easily explained by the conservative design of Monju. Smear density of metal should be set at a very low level to avoid fuel cladding mechanical interaction (FCMI) because the density of metal fuel is 100 %TD. Smear density is defined as the ratio of fuel mass to the volume of fuel and pellet-cladding gap region. The pellet-cladding gap volume works to reduce fuel swelling in the same manner as the pores in a fuel pellet. The ratio of the plenum volume to the fuel stack volume of metal fuel is set at a greater level because metal fuel has no buffer for swelling. Thus, the plenum length of metal fuel is set long. The plenum length of Monju fuel is set long to ensure conservatism against the increase in plenum pressure. Fuel stack length and pellet diameter are almost the same among all fuel types. Solid pellet design is applied to metal, carbide, and nitride fuels because their thermal conductivity is higher than that of oxide fuel. Annular pellet design is applied to oxide fuel because oxide fuel has very low thermal conductivity. Annular and high-density pellet design is selected for oxide fuel to prevent fuel melting. Helium gas is used as gap filling material (bond) in all fuel types except for metal fuel. Sodium, whose thermal conductivity is much higher than helium gas, is used as the filling material for metal fuel because eutectic reaction with metal fuel tends to occur at comparatively low cladding inner surface temperatures (around 650 °C). The cladding outer diameter is determined depending on the pellet outer diameter. Large-diameter pellets are selected for carbide/nitride fuel because their thermal conductivity is high. Thus, cladding outer diameter becomes large. When the cladding outer diameter is determined, the cladding inner diameter is unambiguously determined from cladding thickness. Cladding thickness is determined by hoop stress, and the hoop stress is dependent on cladding thickness and plenum gas pressure. The maximum LHR is determined by considering the thermal characteristics of fuel pellets and eutectic reaction. High LHR is possible for carbide and nitride fuels because their thermal conductivity is very high. High LHR is not practicable for metal fuel, however, because eutectic reaction, which occurs at rather low temperature, should be taken into account. Cladding midwall temperature is determined based on target burnup and cladding swelling characteristics. Swelling characteristics depend on cladding temperature.

7.7 Fuel Research and Development

7.7.1 Development of Cladding Material for Increased Burnup

The increase in fuel burnup contributes greatly to an increased natural uranium utilization factor and improved economy. In the case of fast reactor fuel, fast neutron fluence increases with increased burnup. With increased fast neutron fluence, cladding swelling and irradiation creep deformation become remarkable. In addition increased internal gas pressure may lead to creep rupture. Thus, development of core material should be an urgent issue for the mitigation of significant swelling and irradiation creep strains. Core material development should focus on characteristics such as void swelling, irradiation creep, high-temperature creep strength, corrosion by sodium, irradiation embrittlement, and mechanical strength. Void swelling is swelling due to void formation by irradiation damage. Irradiation creep deformation is accelerated by void swelling. High-temperature creep strength is called creep rupture strength, and it causes creep rupture when stress is continuously applied for a long period of time under high-temperature conditions. Corrosion of cladding by sodium reduces cladding thickness. Irradiation embrittlement, which is caused by void formation due to irradiation damage, reduces ductility (extension). An understanding of these characteristics is required for fuel design. Among these, void swelling is the most important as it controls the lifetime of fuel. A variety of experiments has been done for the improvement of

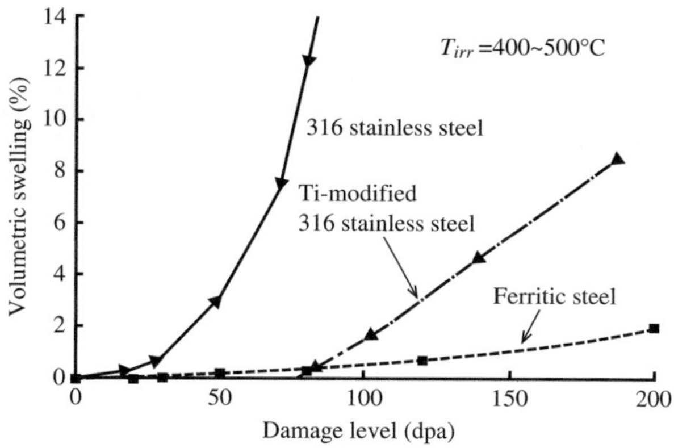

Fig. 7.18 Improvement of swelling characteristics [7]

void swelling. Improvement of void swelling characteristics is typically shown in Fig. 7.18. In this figure the horizontal axis is the displacement per atom (dpa), which roughly corresponded to fast neutron fluence, and the vertical axis is the percent volume swelling rate. Stainless steel containing a small amount of titanium, the so-called Ti-modified 316 stainless steel, has better void swelling characteristics than the usual 316 stainless steel. Ferritic steel has much better anti-swelling characteristics than austenitic steel. HT9, ferritic steel, which has been developed in the USA, shows good anti-swelling characteristics up to high neutron fluence. In Japan oxide dispersion-strengthened (ODS) ferritic steel has been developed as a promising anti-swelling material to be applied to the fast reactor cladding material.

7.7.2 *Basic Research and Physical Property Measurement of Fast Reactor Fuel*

A variety of research has been carried out in the field of fast reactor fuel. Crystal lattice structure examination and oxygen potential measurement are done as basic research. Melting point and thermal conductivity measurements are done as physical property measurements. Fuel swelling and FP gas release behaviors are investigated for modeling irradiation behavior. Fuel containing MA (MA-bearing fuel) has been aggressively studied, and it is found that the addition of up to 5 % MA to MOX has negligible effects on melting temperature and thermal conductivity.

The result of melting point measurement of MOX is shown in Fig. 7.19. The data shown by solid line, which were obtained more recently, are higher than those shown by dotted line, which were obtained previously. The difference increases with increased plutonium content. This difference is thought to be due to the difference of crucibles that were used for measurement. Crucibles made of rhenium were used in recent experiments, and tungsten crucibles were used in previous experiments. It is found that tungsten crucibles react with MOX of plutonium

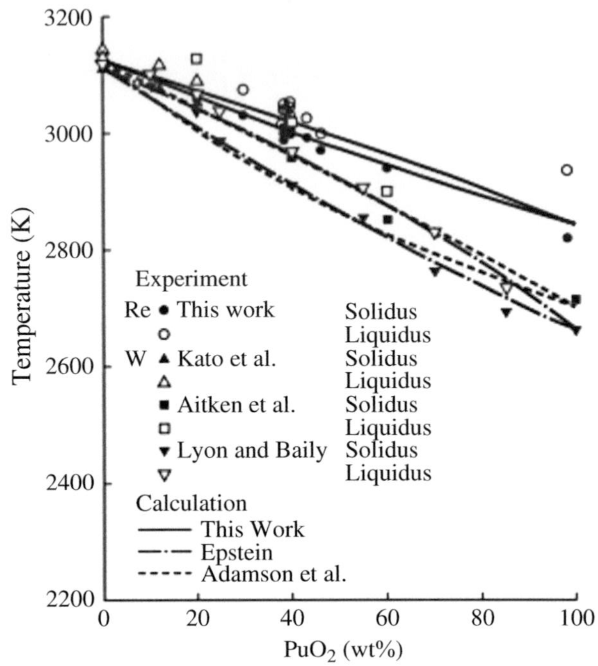

Fig. 7.19 Result of melting point measurement [8]

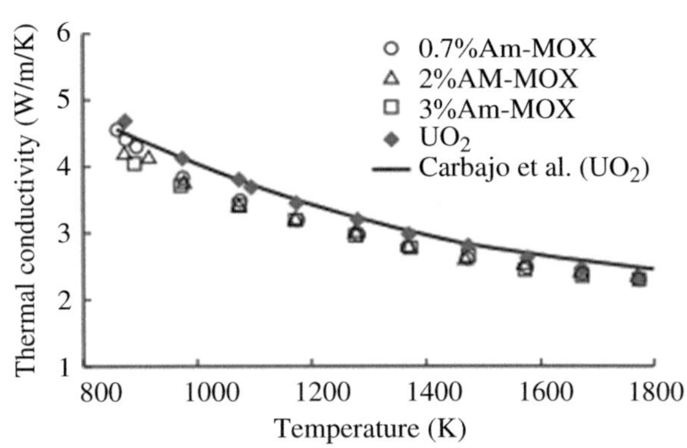

Fig. 7.20 Result of thermal conductivity measurement [9]

content more than 20 %, while no reaction is observed when rhenium crucibles are used.

Thermal conductivity of UO_2 and MOX having americium whose contents are 0.7, 2, and 3 % is shown in Fig. 7.20. The figure shows that thermal conductivity of MOX is lower than that of UO_2 and that no remarkable effect of americium addition on thermal conductivity is observed within a content of up to 3 %.

7.7.3 Fuel Irradiation Test

Since there are many unknown factors in the behavior of fuel under irradiation, fuel characteristics shall be confirmed by irradiation tests with the increase in burnup. Various types of irradiation tests should be performed with different objectives as follows: leading irradiation tests to check validity of new design concept and procedure, confirmation tests to verify analytical codes and fuel integrity, operational reliability tests to check and confirm the reliability of fuel during reactor operation, and tests to determine criteria in case of accidents. Fuel melting test, the so-called PTM test, is done to examine an allowable limit for the LHR to prevent fuel melting. The pellet microstructure is examined with the fuel that was irradiated at a high LHR estimated to cause fuel melting. Transient over power (TOP) test is carried out to examine the LHR which would cause fuel failure due to over power. Design margin for fuel failure by melting is confirmed by TOP test. Run to cladding breach (RTCB) test and run beyond cladding breach (RBCB) test are carried out to examine the limitation of achievable burnup and to examine failed fuel behavior after cladding breach and the possibility of fuel failure propagation. RTCB test results provide information, approximately the design limit of burnup, and RBCB test results provide information, approximately the effect of failed fuel on subsequent reactor operation. It is found by RBCB tests that reactor operation can be continued until the end of its operation cycle without failure propagation after fuel failure occurs. This finding is very useful from the perspective of reactor economy. However, in the case of fuel failure in Monju, it is quite probable that the reactor would immediately be shut down and failed fuel would be removed from the reactor core and examined. Detection methods of failed fuel and positioning (location) of failed fuel in reactor core are explained in Chap. 8.

A variety of irradiation tests have been performed using the experimental fast reactor, Joyo. One of the typical tests is the instrumented test assembly (INTA) test, which needs very high technical skill, and fuel centerline temperature can be measured in the core during irradiation. INTA test results provide valuable information for code validation and verification. PTM and RTCB tests have also been carried out in Joyo. Fuel irradiation tests have been conducted at various fast reactors abroad such as EBR-II and FFTF in the USA, Rapsodie and Phenix in France, DFR in the UK, and BOR-60 in Russia. Fuel safety research tests have been carried out in CABRI in France and TREAT in the USA. Fuel behavior under severe fuel damage is examined in these tests.

Results of a PTM test carried out in Joyo are shown in Fig. 7.21. The central hole is shown by the black circle indicated in the upper right-hand section of this fig. A dendrite structure is observed outside the central hole, and the structure was formed by cooling after melting. The LHR experienced in the test is estimated to be approximately 700 W/cm.

Irradiation tests would be useful only when three types of data, namely, fuel fabrication data, irradiation conditions data, and postirradiation examination (PIE) data, are available.

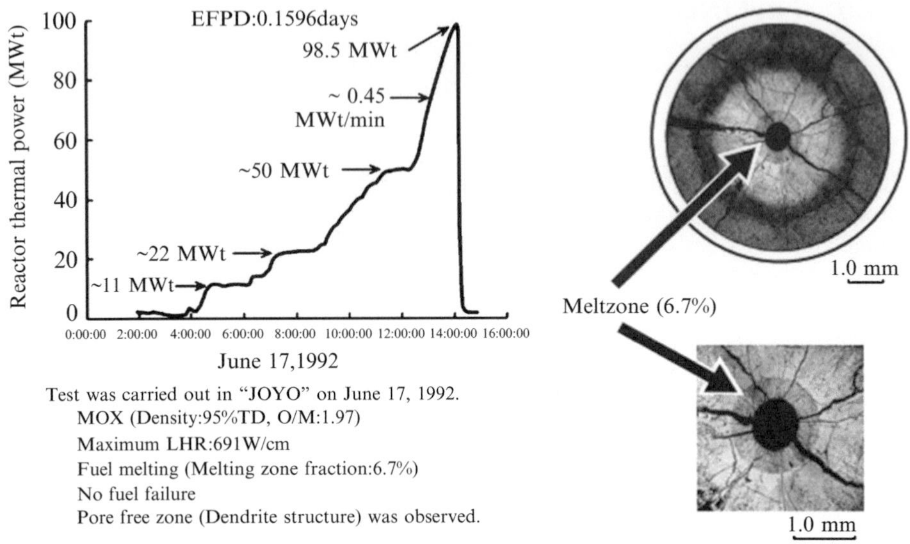

Fig. 7.21 Result of power to melt test performed in Joyo [10]

7.8 Integrity Evaluation of Control Rod Assembly

The structure of control rod assemblies is basically similar to that of fuel assemblies. Control rod elements are assembled in a wrapper tube (duct) instead of the fuel elements of a fuel assembly. Control rod elements have pellets containing B_4C, of which ^{10}B is enriched, and a plenum above the pellets. The role of ^{10}B is to absorb neutrons. If neutrons are absorbed by ^{10}B, alpha particles (helium nuclei) are formed in B_4C pellets by (n, α) reaction, and the alpha particles may cause absorber-cladding mechanical interaction (ACMI) due to swelling of the pellets. ACMI and the inner gas pressure due to helium gas release threaten cladding integrity of control rods. Thus, control rod assemblies must be replaced after a few cycle operations. In order to extend period of use of control rod assembly, sodium filling (bonding) and vent mechanism are studied. Sodium bonding has the advantage of high thermal conductivity, which increases the gap conductance and, thereby, allows for a broadening of the gap size, which reduces the ACMI effect. The vent mechanism has the advantage of reducing gas pressure because helium atom (gas) generated can be easily released from control rod elements to coolant sodium.

7.9 Control of Plutonium Content in Fuel Fabrication Process

The plutonium content of fast reactor fuel ranges from 20 to 30 %. Since the fraction of fissile plutonium (^{239}Pu and ^{241}Pu) to all plutonium isotopes recovered from reprocessing spent fuel is approximately 65 %, the fissile content of fast reactor fuel is much higher than that of LWR fuel, whose ^{235}U enrichment is approximately 4 %. The reason why high plutonium content is needed in the case of fast reactor fuel is the small fission cross section of plutonium in fast neutron energy region, as shown in Fig. 7.8. The fission cross section of plutonium is approximately 2 barns in a neutron energy region of a few hundred keV, which is the typical energy range of fast reactors. On the other hand, the fission cross section of ^{235}U ranges from 100 to 1000 barns in a neutron energy region from 0.01 to 0.1 eV, which is the energy range of LWR.

Plutonium has several isotopes. Since each plutonium isotope has different reactivity worth, which corresponds to the capability of fission, the amount of plutonium necessary for a certain reactor operation differs depending on the isotope composition of plutonium used. In addition, ^{241}Pu decays to ^{241}Am with a half-life of approximately 14 years, and the reactivity worth of ^{241}Am is much less than that of ^{241}Pu. ^{241}Am has negative reactivity worth. This means that the reactivity worth of plutonium decreases with an increased time elapsed from fuel fabrication to loading into a reactor core.

A physical accounting method for adjusting the fissile content (enrichment) has been introduced to solve these problems with the use of plutonium. The plutonium content should be controlled by this method, as explained in detail below.

Since plutonium content can be controlled only in a fuel fabrication stage, this method is applied in a fuel fabrication process. Firstly, the plutonium isotope composition at fabrication is modified to that estimated in loading with consideration of the decay of ^{241}Pu. The problem concerning the decay from ^{241}Pu to ^{241}Am is solved by this treatment. Next, reactivity worth, which is expressed by ($\nu\sigma_f$-σ_a), should be calculated by neutronic calculation. Here, ν is the number of neutrons produced per fission. Symbols σ_f and σ_a are microscopic fission and absorption cross sections, respectively. The equivalence factor (F_i) is given by the ratio of reactivity worth of each isotope to that of ^{239}Pu. Thus, the equivalence factor is defined as a factor equivalent to the reactivity worth of ^{239}Pu. The equivalence factor changes depending on the type of reactor and burnup conditions. The equivalence factor in the initial core (fresh core), where all fuel is fresh, is different from that in an equilibrium core.

The plutonium content required for reactor operation is determined by neutronic calculation for surveying the content. In this calculation the basic plutonium isotope composition, for example, $^{239}Pu : {}^{240}Pu : {}^{241}Pu : {}^{242}Pu = 58 : 24 : 14 : 4$, is assumed. The basic plutonium isotope composition and the plutonium content calculated are substituted into the following equation (Eq. (7.10)) to get the equivalent fissile content:

$$E_{239} = \varepsilon \sum_i \alpha_i F_i + (1 - \varepsilon) \sum_j \beta_j F_j \tag{7.10}$$

where E_{239} is the equivalent fissile content, ε is the plutonium content, α_i and β_j are the isotopic compositions of plutonium and uranium. If the isotopic compositions of plutonium and uranium at fuel fabrication are determined, the plutonium content can be easily calculated by the use of Eq. (7.10). MOX fuel pellets are fabricated by adjusting the plutonium content to a target value calculated by this method.

Further Readings

"Plutonium Fuel Technology" (in Japanese), Atomic Energy Society of Japan, (1998)
"Heat-transfer Engineering Data Document", Fourth Edition (in Japanese), Japan Society of Mechanical Engineers, (1986)
M. Akiyama (ed.), "Nuclear Thermal Engineering" (in Japanese), Nuclear Engineering Series, (University of Tokyo Press. Inc., 1978)
H. Bailly, D. Menessier, C. Prunier (eds.), "The nuclear fuel of pressurized water reactors and fast reactors design and behavior" (in Japanese), K. Konno, Translator, Maruzen Planet Co. Ltd
N.E. Todreas, M.S. Kazimi, Nuclear systems 1: Thermal hydraulic fundamentals, (Taylor & Francis Inc. 1990)
N.E. Todreas, M.S. Kazimi, N. Todreas, Nuclear systems 2: Elements of thermal hydraulic design, (Taylor & Francis Inc. 1990)
M. Hori (ed.), "Basic Fast Reactor Technology" (in Japanese), Nikkan Kogyo Shimbun, Ltd., (1993)
Edited by Sodium Educational Committee, "Sodium Technology Handbook" (in Japanese), Japan Nuclear Cycle Development Institute, JNC TN9410 2005–011 (2005)
Y. Oka, Writer and Editor, "Textbook of reactor design" (in Japanese), Ohmsha, Ltd
N. Nakae, et al., "Scope and content covered by LMFBR fuel design" (in Japanese), J. At. Energy Soc. Japan **53**(2), 40 (2011)
N. Nakae, et al., "Basic concept of fuel safety design and assessment for sodium-cooled fast reactor", JNES RE Report Series, JNES-RE-2012-0022, March, 2013
A.E. Waltar, A.B. Reynolds (eds.), Fast Breeder Reactors, (Pergamon Press, Oxford, 1981)
N. Nakae, "Management of plutonium content based on reactivity of each plutonium isotope", J. Nucl. Sci. Technol. **45**(4), 361–366 (2006)

References

1. Japan Atomic Energy Agency, "Application for reactor establishment license: Prototype fast breeder reactor Monju" (in Japanese)
2. Japan Nuclear Cycle Development Institute, JNC Technical Report, No. 18 (2003) (in Japanese)
3. JAEA Nuclear Data Center
4. T. Takeda et al., Study on detailed calculation and experiment methods of neutronics, fuel materials, and thermal hydraulics for a commercial type Japanese sodium-cooled fast reactor. Sci. Technol. Nuclear Install. **2012** (2012)
5. G. Chiba et al., JENDL4.0 benchmarking for fission reactor applications. J. Nucl. Sci. Technol. **48**(2), 172–187 (2011)
6. F. Yamada et al., Development of natural circulation analytical model in SUPER-COPD code and evaluation of core cooling capability in Monju during a station blackout. Nucl. Technol. **188**, 292 (2014)

7. S.J. Zinkle, "Advanced materials for future nuclear plants", fission energy workshop: Opportunities for fundamental research and breakthrough in fission, Global Climate & Energy Project, MIT, Cambridge, MA, November 29–30 (2007)
8. M. Kato et al., Solidus and liquidus temperatures in the UO_2-PuO_2 system. J. Nucl. Mater. **373**, 237–245 (2008)
9. K. Morimoto et al., Thermal Conductivity of (U, Pu, Am)O_2 Solid Solution. J. Alloys Compd. **452**, 54–60 (2008)
10. M. Inoue et al., Power-to-melts of uranium-plutonium oxide fuel pins at beginning-of-life condition in the experimental fast reactor JOYO. J. Nucl. Mater. **323**, 108–122 (2003)

Chapter 8
Plant Component Design (from Mechanism to Structure)

Kiyohiko Maeda and Naoto Kasahara

8.1 Thinking Process of Plant Component Design

Chapter 8 describes a method to determine the structure of major plant components based on the mechanisms determined by the FR plant system conceptual design described in Chap. 6 while satisfying constraints to prevent the occurrence of abnormal events in order to ensure safety and structural integrity, which are shown in Chaps. 4 and 5, respectively. As shown in Fig. 8.1, reactor structure comprises the most important components required to support the core and maintain coolant flow paths necessary for core cooling. Major systems include the primary and secondary cooling systems, consisting of heat exchangers, piping, and pumps to transfer heat from the reactor vessel by sodium. Other large-scale systems in the FR are the fuel handling and storage systems that exchange and transfer fuel and cool and store spent fuel. This chapter shows design examples of these major systems in which the mechanism and structure of each component are determined.

The reactor plant is an installation designed to remove heat from the core and transfer the driving force of generators to produce electric power. The general problem is to design high performance of plant components within the constraints of material, structure, etc. as described in Chap. 5 while ensuring safety and economic competitiveness.

K. Maeda (✉)
Japan Atomic Energy Agency, Ibaraki, Japan
e-mail: kiyohiko-maeda@ktj.biglobe.ne.jp

N. Kasahara
The University of Tokyo, Tokyo, Japan
e-mail: kasahara@n.t.u-tokyo.ac.jp

N. Kasahara (ed.), *Fast Reactor System Design*, An Advanced Course in Nuclear Engineering 8, DOI 10.1007/978-981-10-2821-2_8

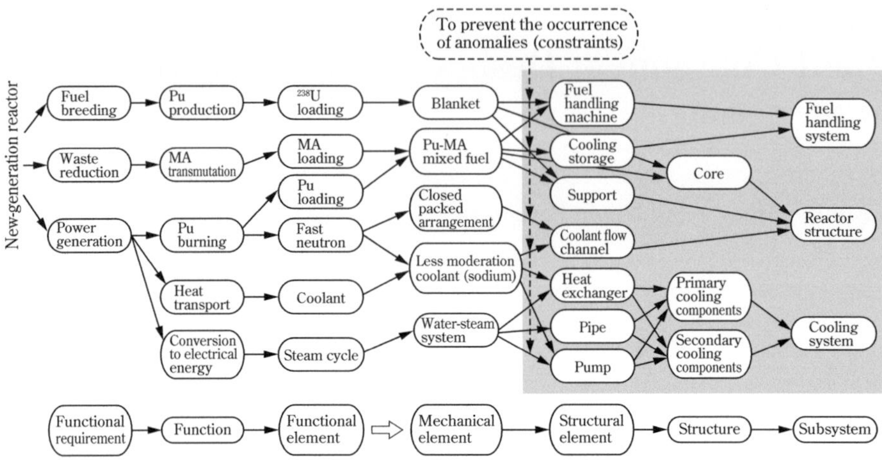

Fig. 8.1 Design flow of plant components from mechanism design to structural design

8.2 Reactor System Component Design

8.2.1 Reactor Vessel and Primary Tank

Reactor vessel and primary tank have similar functions; however, the former name is for loop-type and the latter one is for pool-type reactors. Figures 8.2 and 8.3 show the Monju reactor vessel and the Super Phenix primary tank (main vessel) as examples of the reactor vessel for loop- and pool-type fast reactors, respectively.

Coolant flow through fuel assemblies in a reactor vessel is generally upward in consideration of the buoyancy caused by temperature rise. Thus, the coolant entrance nozzle is usually provided in the lower part of the reactor vessel, and the coolant outlet nozzle is provided at the upper part. However, the level (height) of the coolant outlet nozzle must be adequately positioned in order to maintain coolant circulation in case of coolant leak accident, as described in 6.2.6.

Since fuel assemblies are withdrawn from the core and transferred under coolant surface during fuel exchange, the depth of the top of a fuel assembly installed in the core of the reactor vessel must be larger than the length of the fuel assembly.

The lower part of the reactor vessel contains the core support structure, which supports the fuel, control rod and shield assemblies, etc. The core support structure has the function of flattening the coolant temperature rise in the core by distributing the coolant flow to each fuel assembly according to power distribution. In an FR core, radial blanket fuel assemblies are installed in the periphery of the core to achieve a high breeding ratio. Power generation in the blanket fuel assemblies is smaller than that of the core fuel assemblies; however, the core fuel assemblies would be pushed up by pressure drop along the coolant flow if too much flow were supplied. To prevent this, a low-pressure plenum is installed in the lower part of the core support structure to supply less flow to the blanket fuel assemblies and to

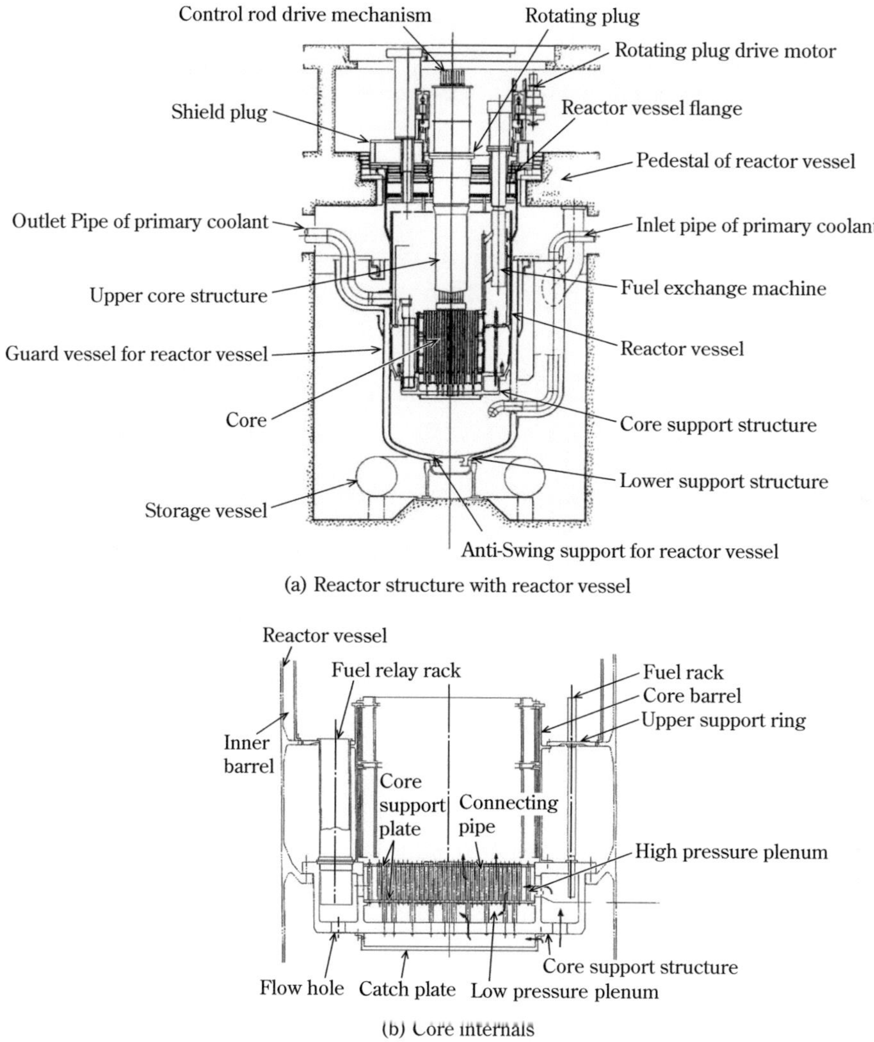

Fig. 8.2 Example of reactor vessel (Monju). (**a**) Reactor structure with reactor vessel. (**b**) Core internals [1]

supply greater flow to the core fuel assemblies from the upper part of the core support structure, which is called the "high-pressure plenum." In this configuration, the pressure difference between high- and low-pressure plenums produces downward force to the core fuel assemblies (i.e., hydraulic hold-down mechanism).

In addition to the core, intermediate heat exchangers (IHXs) and circulation pumps are installed in the primary tank of a pool-type reactor. Coolant heated in the core flows into the inner vessel installed in the primary tank and is cooled in the IHXs. Then, coolant flows out of the inner vessel and is pressurized by circulation

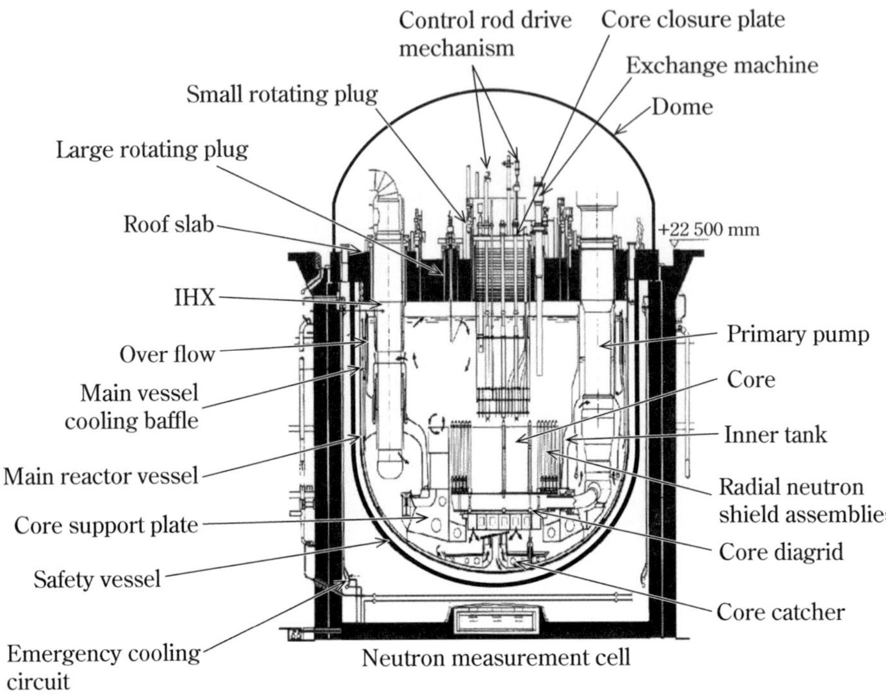

Fig. 8.3 Example of primary tank (main vessel) (Super Phenix) [2]

pumps to flow into the core support structure through a pipelike structure. This arrangement prevents excessive temperature rise and pressurization by primary tank circulation pumps. In addition, consideration must be given to thermal isolation and thermal stress characteristics of the heat partition wall between hot and cold coolant as well as the arrangement of neutron shields to prevent activation of the secondary coolant at the IHX [2, 6].

Merits and demerits of reactor vessel and primary tank from the viewpoint of component design are as follows [4]:

1. Merits of reactor vessel for loop-type reactor
   - Diameter can be small, and structural reliability of vessel including seismic strength can be high.
   - Fabrication and transportation of reactor vessel may be easy, and no fabrication work on site is required.
   - It is relatively easy to inspect after in-service condition.
2. Demerits of reactor vessel for loop-type reactor
   - Total size of nuclear steam supply system becomes larger, so construction cost may be larger.

- Inlet nozzles and outlet nozzles for primary cooling system are required, so structure of coolant boundary may be complex, and it may be weak point of integrity.

3. Merits of primary tank for tank-type reactor

- All components of primary system are in one tank, so total size of steam supply system can be compact, and construction cost may be smaller.
- No nozzle for primary cooling system is required, so primary tank can be simple, and there is no weak point on coolant boundary.

4. Demerits of primary tank-type reactor

- Diameter of tank must be large, so there is weak point for structural integrity of tank especially for seismic strength.
- Many coolant flow paths must be formed in tank, and both sides of those flow path formation parts are sodium, so it is difficult to detect defects of those parts.
- Tank size may be very large, so transportation of primary tank may limit site selection and way of fabrication of tank, and in many cases, welding work on site may be required.

As shown in the next section, fuel assemblies are supported by the core support structure installed in the lower part of the reactor vessel. Consequently, the elevation of the core support structure is changed depending on the temperature distribution in the reactor vessel wall, while the fuel assemblies themselves change their lengths depending on their temperature distribution in the core. The relative vertical positions of fuel assemblies in the core are determined by these two factors. On the other hand, the control rod drive mechanisms are supported at the top of the shield plug (or roof-deck) installed in the upper part of the reactor vessel and grip control rods at their bottom ends. Thus, the elevation of the gripping positions changes depending on the temperature distribution in the drive mechanism. Therefore, to smoothly move control rods to control the reactor accurately, or to firmly insert control rods in case of abnormal conditions, the relative position between the core and control rods should be maintained within an acceptable range. To this end, the accuracy of installed positions of all these components must be controlled strictly. Consideration must be given to the significant difference in the temperature distributions during the construction and operation stages [2].

Ductile materials that will not undergo rapid unstable fracture and with high heat resistance and less degradation by neutron irradiation should be selected for the reactor vessel and the primary tank. Furthermore, since a rapid change of heat generation and removal like reactor trip is anticipated, rapid changes in coolant temperatures should be considered. Therefore, the reactor vessel and its inner structure should be equipped with thermal shield on the surface wetted by sodium to mitigate thermal shock and should be arranged to prevent the occurrence of significant thermal stress in structural design [5].

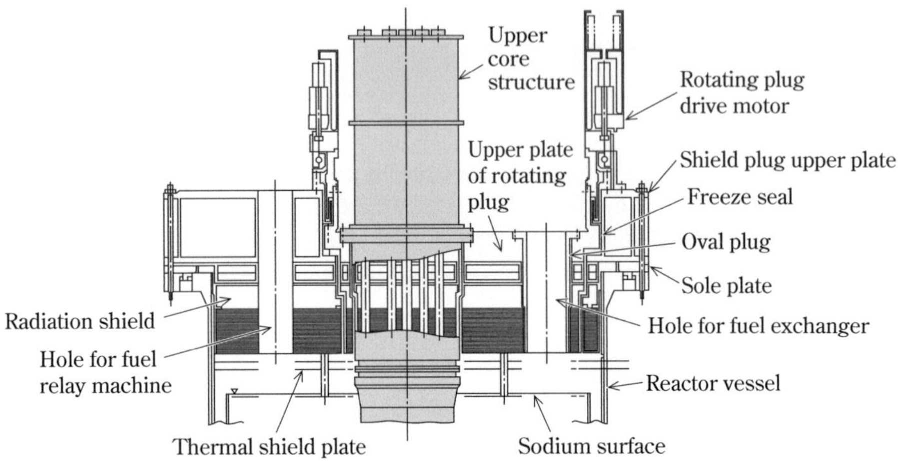

Fig. 8.4 Example of shield plug (Monju) [1]

8.2.2 Shield Plug and Roof-Deck

Since it is not necessary to pressurize coolant in the FR, the space above the coolant surface need not be included in the reactor coolant boundary. However, for a loop-type reactor vessel, the shield plug is installed as a lid to enclose the cover gas by taking the chemical activity of sodium into consideration. (See Fig. 8.4.) Of course, the shield plug has other functions, namely, to shield radiation from the core, to support the upper core structure, and to install components required for fuel exchange. For fuel exchange, a part of the shield plug is used as a rotating plug, for which an eccentric double rotating plug or a triple rotating plug may be used [1, 2, 3, 6].

For a pool-type reactor, the roof-deck (roof slab; see Fig. 8.3) having the function of supporting IHXs and circulation pumps as well as the function of the shield plug for a loop-type reactor is installed [1, 2, 3, 6].

8.2.3 Upper Core Structure and Control Rod Drive Mechanism

Above the core, the upper core structure is hung from the shield plug or roof-deck. The upper core structure has the function of supporting the control rod drive mechanisms, creating flow paths for coolant from fuel assemblies and providing instrumentation in the upper core part. In Fig. 8.5, an example of the upper core structure is shown.

The control rod drive mechanism has the function of controlling the reactor power by inserting and withdrawing the control rods and rapidly inserting the

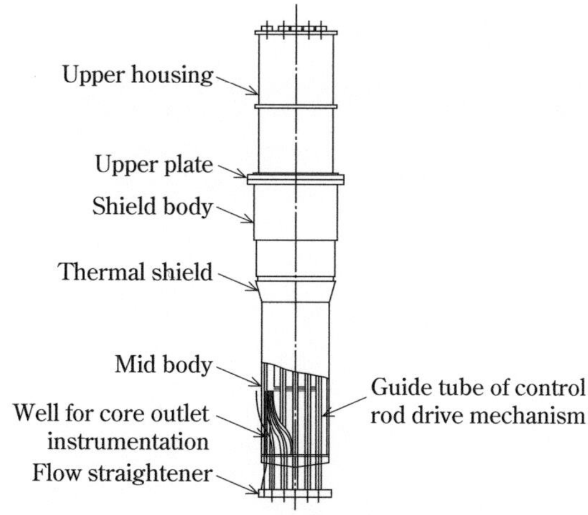

Fig. 8.5 Example of upper core structure (Monju) [1]

control rods into the core to shut down the reactor in case of emergency. Usually the control rods are driven by motors, but rapid insertion is performed by delatching the control rods from the drive shaft. Rapid insertion is achieved by gravitational drop and accelerated by spring force or gas pressure.

Under normal operating conditions, the control drive mechanism and all control rods are connected together, while in refueling, the control drive mechanism is separated from all control rods that are left in the core to maintain subcriticality and is pulled up sufficiently in the upper core structure to allow for turning of the rotating plug without interference with the core.

The control rod drive mechanism of Monju has three types: one for the fine control rods, one for the coarse control rods, and one for the backup control rods. (See Figs. 8.6, 8.7, and 8.8.) The main shutdown system consists of the fine control rods and coarse control rods. In emergency shutdown, these rods are delatched at the upper part of the drive shaft and accelerated by gas pressure to drop into the core. The backup shutdown system consists of the backup control rods that are delatched at the bottom end of the drive shaft and accelerated by spring force in emergency shutdown. These three types of control drive mechanisms are designed and fabricated by different corporations to maintain diversity. The control rods can be delatched by turning off the current for electric magnets for holding the control rods. Delatching is performed by redundant signals and at redundant points in the electric magnet circuits to ensure high reliability [1–4, 9].

There is a significant difference in outlet coolant temperature between control rods and the surrounding fuel assemblies, as well as between the outermost core fuel and innermost blanket fuel assemblies. Due consideration must be given to these significant temperature differences that may cause severe temperature fluctuation (thermal stripping) in the surrounding structure [5].

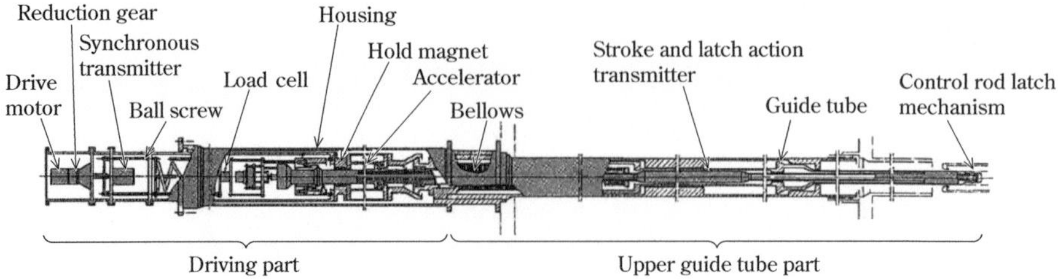

Fig. 8.6 Fine control rod drive mechanism [1]

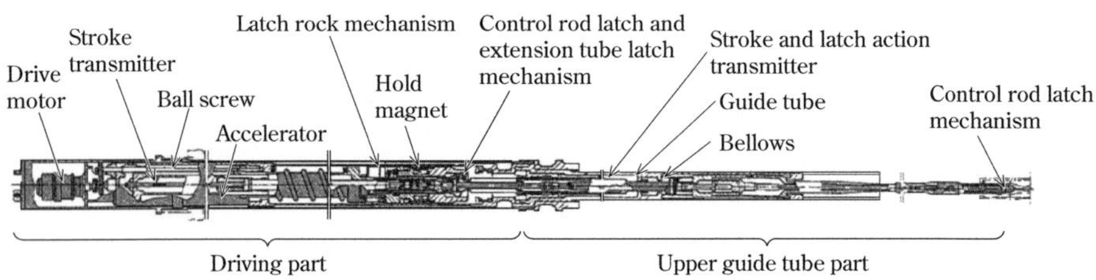

Fig. 8.7 Coarse control rod drive mechanism [1]

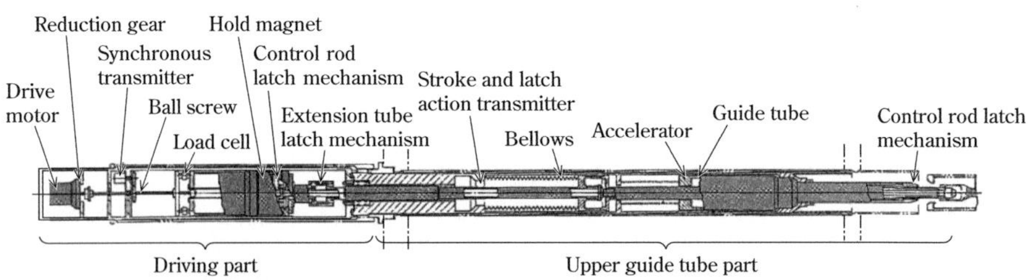

Fig. 8.8 Backup control rod drive mechanism [1]

8.3 Design of Main Cooling System Components

8.3.1 Main Circulation Pump

An example of the main circulation pump for a loop- and pool-type reactor is shown in Figs. 8.9 and 8.10, respectively. Generally, the main circulation pump used for the main cooling system is a mechanical pump with a vertical shaft. Since normal lubrication oil cannot be used in sodium, a hydrostatic bearing that uses the pump discharge pressure is used for the lower bearing. Some means is necessary for this type of pump to recover sodium delivered from the hydrostatic bearing (e.g., overflow column). Since the drive motor is installed in the air, the upper bearing also has the function of shaft seal.

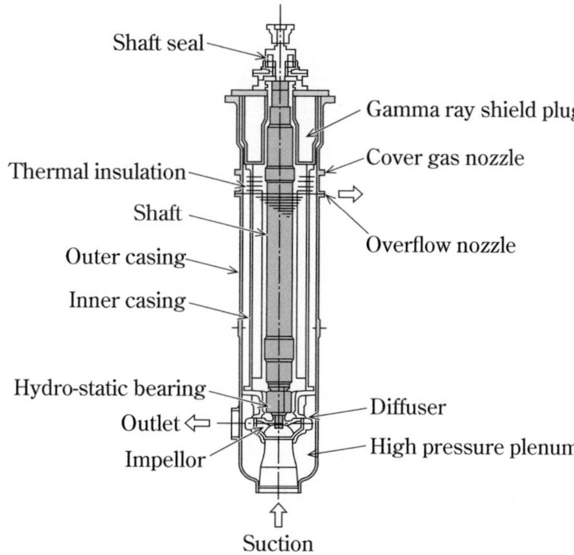

Fig. 8.9 Example of main circulation pump for loop-type reactor (Monju) [1]

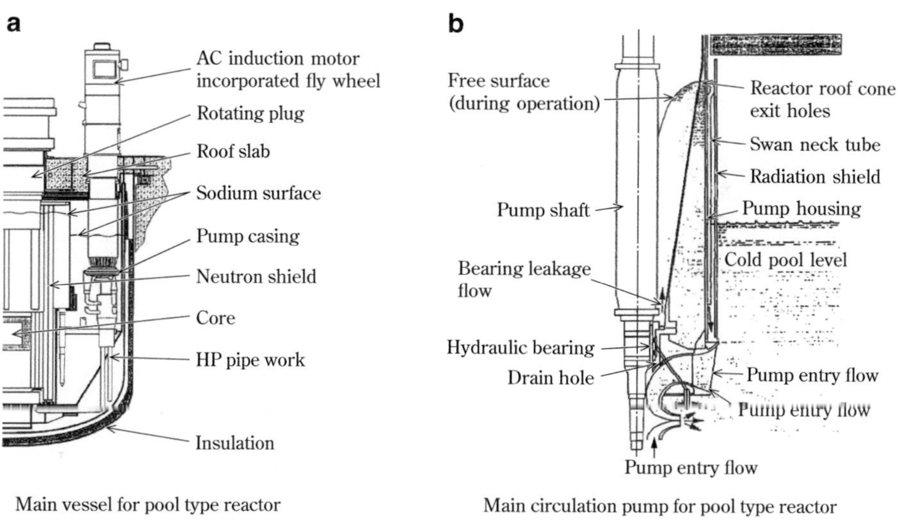

Fig. 8.10 Example of main circulation pump for pool-type reactor (PFR). (**a**) Main vessel for pool-type reactor. (**b**) Main circulation pump for pool-type reactor [2, 6]

In general, in addition to the main motor, a small motor (pony motor) to drive the pump with a slow speed is installed for use in case of reactor shutdown, including accidents. This concept is feasible thanks to the sodium's ability to circulate in the liquid state using almost the same circuits and procedures from normal operation to accident conditions.

To allow the change in coolant level and decrease the temperatures at the upper bearing and the drive motor, a vertical long pump must be adopted. As a result, such pumps tend to deform as a result of asymmetric temperature distribution. Therefore, measures need to be taken, for example, to prevent convection in gas spaces [1–4].

Next, a design example of the Monju rotational inertia of pump that determines the flow reduction rate in case of emergency shutdown is shown. The pump flow rate during normal shutdown is controlled to slowly decrease the system temperature to prevent excessive thermal stress on the relevant components. On the other hand, in case of emergency shutdown, since the system temperature should be decreased safely without any electrical control system, consideration should be given to the rotational inertia of the main cooling pump. The upper left section of Fig. 8.11 shows the result of prediction calculation for the coolant flow rate after emergency shutdown with a flow half-life parameter that depends on the rotational inertia. For example, in the case of the largest rotational inertia, the flow rate reduces to half at 106.5 sec. The lower left section of Fig. 8.11 shows the fuel cladding temperature predicted in the corresponding cases. It is suggested that long flow half-life provides an advantage to fuel safety, while when the flow half-life is short, decay heat is not fully removed from fuel, and the fuel cladding temperature tends to rise. On the other hand, the lower right section of Fig. 8.11 shows the predicted transient thermal stress on the inner surface of pipe. It shows that the longer the flow half-life, the larger the transient thermal stress, except in the case of an extreme short flow half-life which causes higher fuel temperature. This is because the larger the amount of heat removed for the same amount of decay heat, the more rapidly the coolant temperature decreases.

For this reason, it is necessary to select the optimum pump rotational inertia while maintaining fuel safety and mitigating transient thermal stress in case of reactor emergency shutdown.

Electromagnetic pumps are used to drive coolant at a relatively small flow rate.

Electromagnetic pumps utilize the merits of sodium, a liquid metal with excellent thermal conductivity. There are two types of electromagnetic pumps. One generates driving force on sodium by applying electric current in sodium according to the Fleming's left-hand rule, and the other generates drive force on sodium by electromagnetic induction, in the same manner as a linear motor. Neither type of electromagnetic pump has dynamic components. Both are advantageous in improving reliability, but attention should be paid to the cooling and heat resistance of magnetic coils. An example of electromagnetic pump is shown in Fig. 8.12, and the principle of electromagnetic pump is shown in Fig. 8.13. In general, the duct for driven liquid (sodium) is flat rectangular for improved efficiency.

To date, electromagnetic pumps have not been used for large-scale pumps because of their lower energy efficiency, caused by the loss of copper and iron in magnetic field generation, compared to mechanical pumps. To make electromagnetic pumps available for use, therefore, work has proceeded on the use of a circular duct with a central core to form an annular flow path and putting a magnetic coil in the center of a circular duct to recover heat lost while generating magnetic field by driven sodium for improved plant thermal efficiency.

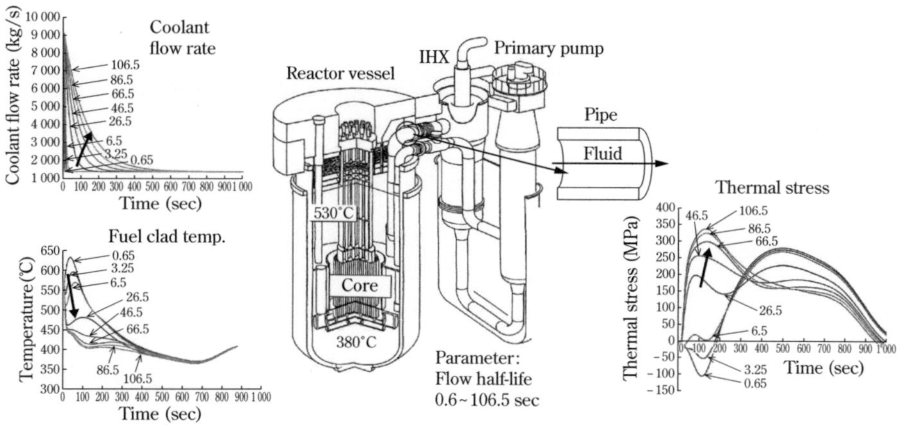

Fig. 8.11 Example of calculated fuel cladding temperature and thermal stress on pipe with a parameter of flow half-life in case of emergency shutdown

Fig. 8.12 Example of electromagnetic pump [3]

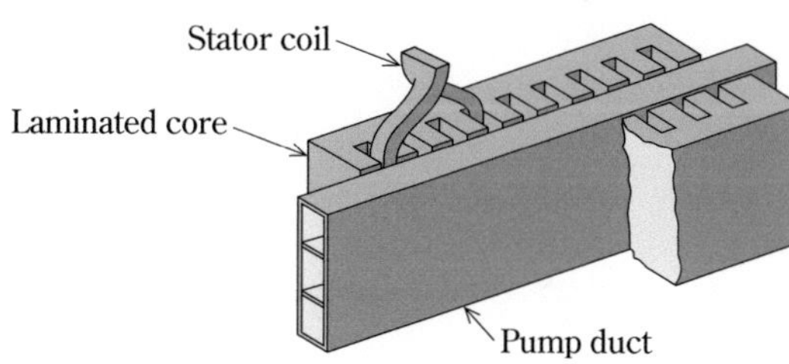

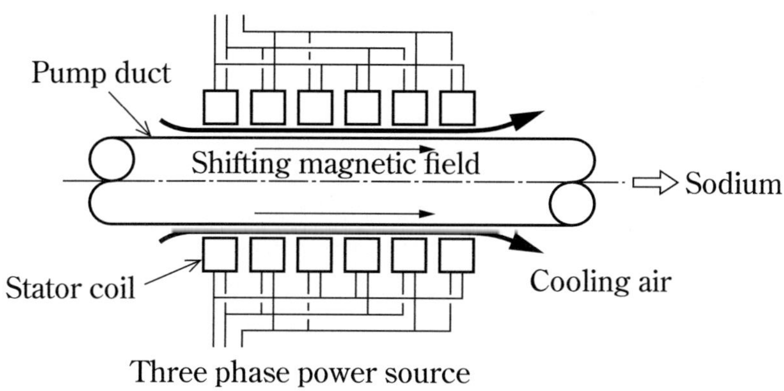

Fig. 8.13 Theoretical figure of electromagnetic pump

8.3.2 Intermediate Heat Exchanger

The intermediate heat exchanger (IHX) is a component designed to transfer heat generated in the core from the primary to the secondary cooling system. Since

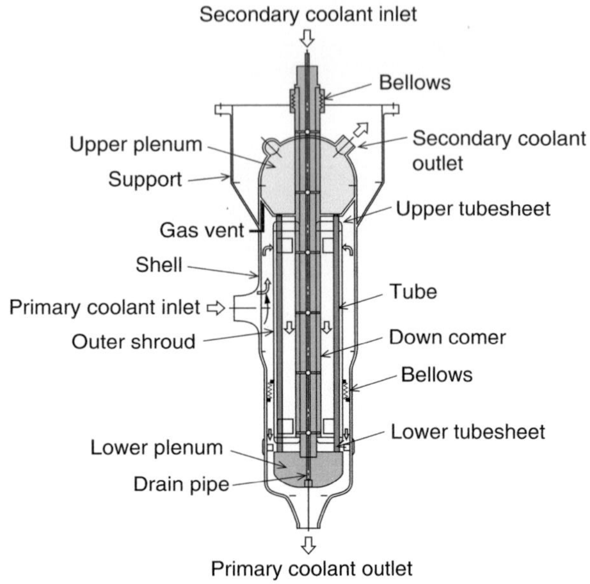

Fig. 8.14 IHX for loop-type reactor (Monju) [1]

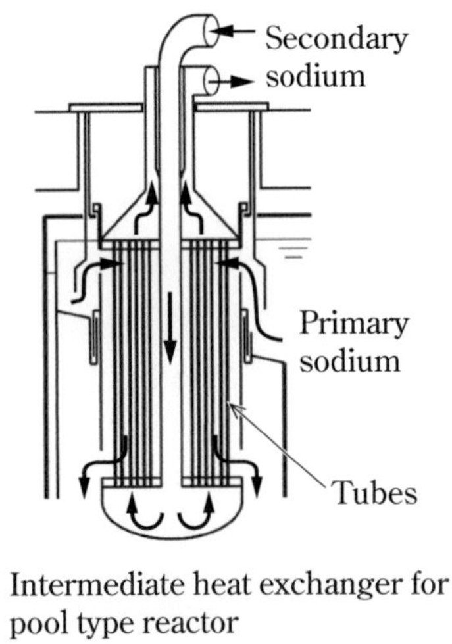

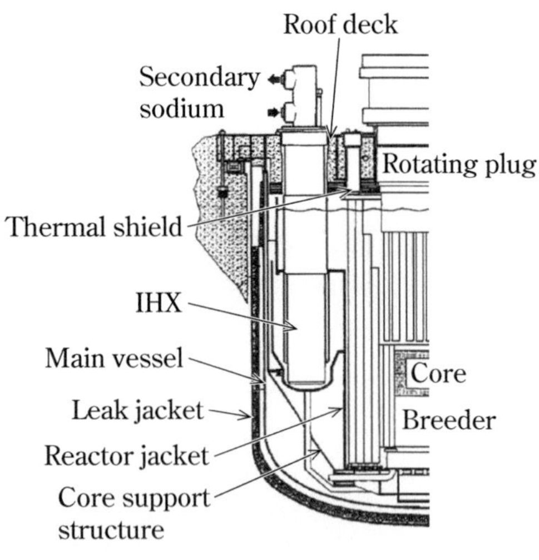

Fig. 8.15 IHX for pool-type reactor (PFR) [7]

sodium, which exhibits high thermal conductivity, is used as coolant, a relatively small-size IHX may be adopted. Figures 8.14 and 8.15 show examples of IHX for loop- and pool-type reactors, respectively. In the IHX, the primary and secondary coolants are separated only by heat transfer tubes. Consequently, the secondary coolant pressure is designed to be higher than the primary coolant pressure in order to prevent the leak of activated primary coolant into the secondary cooling system in case of tube rupture. Generally, this pressure difference is maintained by the difference in free surface level between the primary and secondary sides. To maintain the secondary coolant level higher than the primary coolant level is also advantageous in ensuring natural circulation capability. In the case of a hot leg pump system, however, since the pump discharge pressure is applied to the primary side of heat transfer tubes, it is necessary to increase the secondary cover gas pressure to maintain the necessary pressure difference.

Following the conventional IHX design method of setting the higher-pressure side to the inner side of heat transfer tubes, the secondary coolant flows through the inner side of the tubes, and the primary coolant flows through the shell side of IHX in many design examples. However, since the higher pressure does not require special pressure-proof design, the secondary coolant flows through the shell side of IHX in some design examples to reduce the size of the IHX. Figure 8.16 shows the structure of both types of IHXs [6].

From the viewpoint of structure, since heat transfer tubes are generally straight, it is necessary to absorb the thermal expansion of the tubes depending on the operating conditions and the difference in thermal expansion caused by the temperature difference between the primary and secondary coolant. For this reason, a lower tube plate for connecting tubes is arranged to move freely to prevent excessive thermal stress on the tubes, and a bellows is provided for sealed portions

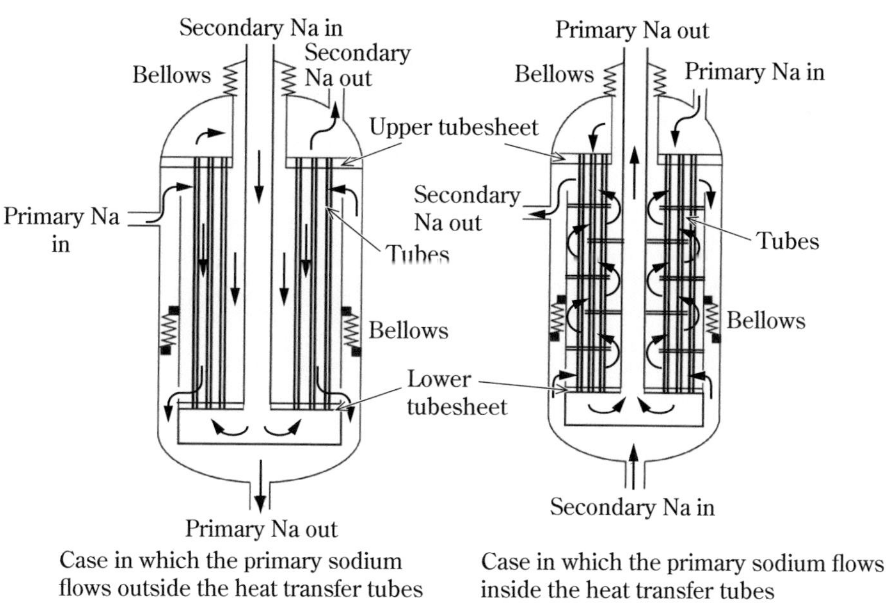

Fig. 8.16 Structure of IHX

to allow for movement of the tubes due to thermal expansion and to ensure seal performance (i.e., prevention of leak and shortcut flow) [2, 3, 5, 6].

8.3.3 Main Cooling System Piping

Since sodium is used as coolant, the inner pressure of the main cooling system piping can be set at a low level, and the piping thickness required for the inner pressure is very small. In many cases, therefore, determination of piping thickness is based mainly on the stiffness of piping to ensure seismic safety.

Preheaters are installed on the outside of the piping to heat up the piping to a specified temperature before charging sodium. The preheaters are fixed as close to the piping as possible and are in general covered with a thin stainless steel plate. Further, thermal insulator is installed over them. The gap between the piping outer surface and the stainless steel plate is used for the detection of sodium leak.

While plant design must satisfy both safety and structural constraints, it may occur that requirements from the respective design perspectives are inconsistent with one another and the inconsistency cannot be solved by existing technology. Monju's primary system piping design is a typical example. An explanation of the problems and how they were solved is provided below.

Important safety function requirements for the FR include the maintenance of primary sodium at a level higher than the top of fuel assemblies as well as coolant flow paths. As described in Chap. 6, the high boiling point of sodium allows for decay heat removal by natural circulation without electric power source when the above requirements are met. A representative scheme for safety design of the FR is the measure against sodium leak, as shown in Figs. 6.18 and 6.19. In case of pipe break, the sodium level can be maintained higher than the top level of the reactor vessel outlet nozzle, and the decay heat removal function can be maintained [1, 8, 9].

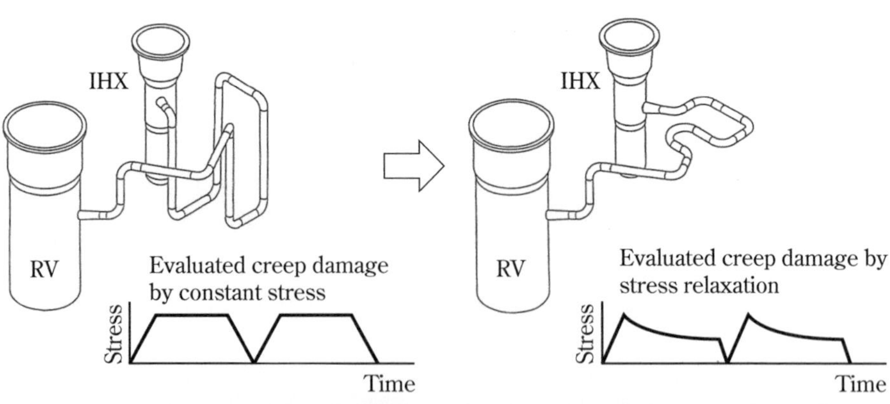

Fig. 8.17 Consistency with safety design based on development of structural design technology for Monju primary main cooling system piping

On the other hand, the structural design requirement is the prevention of creep fatigue failure caused by cyclic load from thermal expansion during normal start-up and shutdown. As shown in Fig. 8.17, both ends of the primary main cooling system piping of Monju are supported at the reactor vessel and IHX, respectively. Under such a condition, when the piping expands by reactor operation, deformation is restricted to generate thermal stress, and this condition continues during operation at high temperature. These cyclic loading and the continuation of stress at high temperature cause creep fatigue. To cope with these problems, elbows are generally applied. When many elbows are used, thermal stress by axial expansion is reduced; however, since pipes are bent upward and downward, the high-elevation horizontal piping layout cannot be achieved. The left figure in Fig. 8.17 shows the calculated stress history based on ASME CC1592 (the former code of the current ASME Sec. III, Subsection NH) and the pipe layout designed to satisfy allowable stress levels specified by the code. Since thermal expansion stress was assumed to be constant during operation, the stress limits were very severe, and many elbows were required. Because of this, the high-elevation horizontal piping layout could not be achieved. However, since a feature of thermal stress is displacement-controlled stress that would be reduced with increased displacement, as described in Fig. 5.4 of Chap. 5, thermal stress would reduce during high-temperature operation as shown in the right figure of Fig. 8.18. Then, a stress-strain evaluation method using the concept of elastic follow-up factor which expresses an intermediate state between the load-controlled and displacement-controlled state was developed, and a pipe layout that satisfies both safety and structural design requirements was achieved using this method, as shown in the right figure of Fig. 8.17. The means for solving the structural design problem is briefly explained in Fig. 8.18. Strictly speaking, thermal stress in actual components is not a perfect displacement-controlled load with a constant strain. It has an intermediate property between the displacement-controlled and load-controlled load with a constant stress. Previously, while a design method for perfect displacement-controlled load or for displacement-

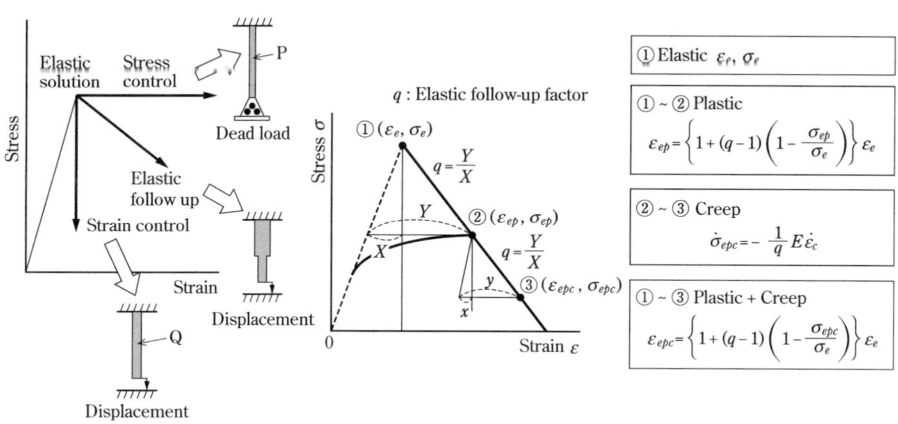

Fig. 8.18 Prediction method of stress strain based on elastic follow-up concept

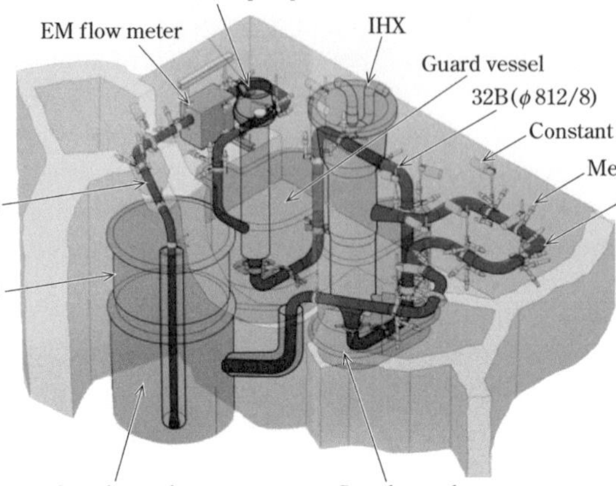

Fig. 8.19 Primary cooling system pipe for Monju [7]

controlled load was applicable, no method for the in-between state was developed. For this reason, thermal stress was assumed to be a perfect load-controlled stress to ensure conservative design. On the other hand, a new method was developed to predict the change in stress and strain caused by elastoplastic and creep strain, which pose problems at high temperatures, using a graphical solution on a stress-strain diagram by expressing the in-between state by the gradient of the downward-sloping line, q, as shown in Fig. 8.18 [1, 5].

Besides, when measures are taken to reduce thermal stress on piping, the earthquake resistance often deteriorates. Consequently, it is necessary to add hangers and snubbers that allow deformation by thermal expansion and constrain seismic vibration for supporting piping. The layout of the primary main cooling system piping of Monju is shown in Fig. 8.19 [3, 5].

A new design for the primary main cooling system piping layout for next-generation FRs has been proposed. In this design, the piping is not supported at the reactor vessel wall, which reduces and limits the axial expansion of piping. This design is intended to fundamentally solve thermal stress problems. In this design, the primary main cooling system piping is installed through the reactor vessel top flange to directly access the inside of the reactor vessel, as shown in Fig. 8.20 to moderate the constraint of piping on the reactor vessel side, and to release thermal stress.

8.3.4 Steam Generator

Since coolant sodium reacts vigorously with water and steam, the secondary cooling system is installed between the primary cooling system that removes heat

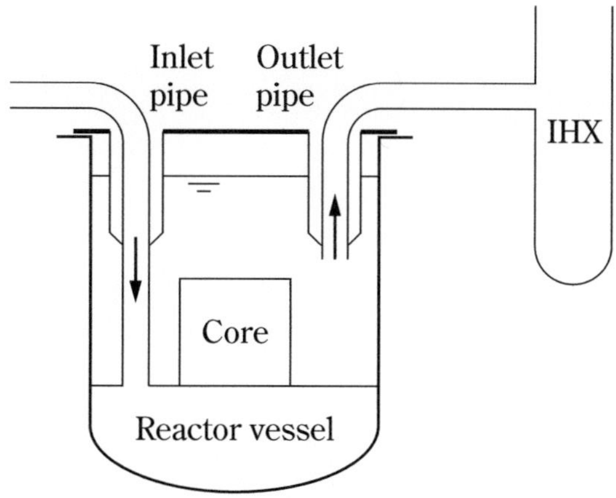

Fig. 8.20 Concept of overhead pipe layout

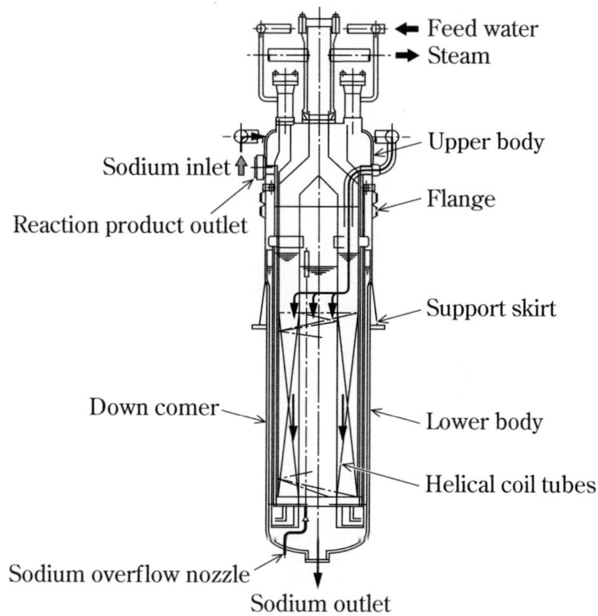

Fig. 8.21 Steam generator (Monju) [1]

from the core and the water-steam system that drives a turbine. While the primary cooling system is completely enclosed in the containment vessel, the secondary system penetrates the containment vessel, and the steam generators are installed outside the containment vessel. As the boiling point of sodium is very high, sodium can be used at high temperatures, and thereby, sufficiently superheated steam conditions can be applicable in FRs.

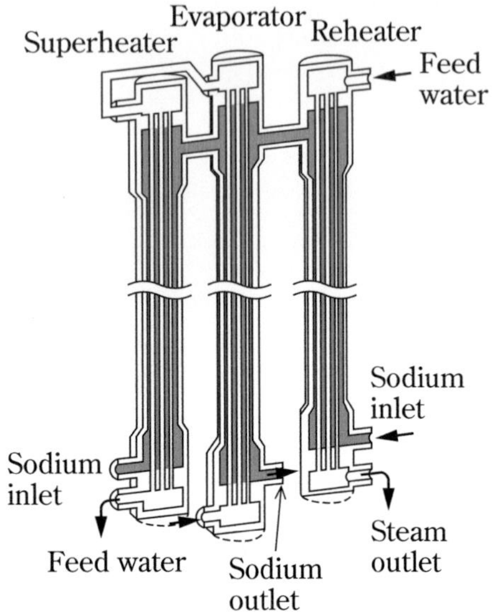

Fig. 8.22 Steam generator (BN-600) [2]

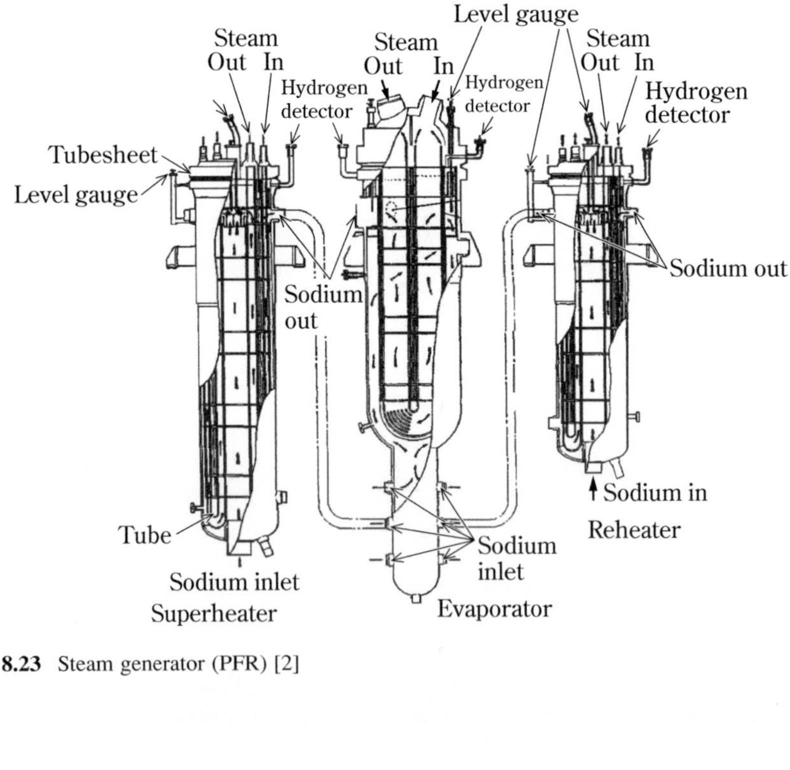

Fig. 8.23 Steam generator (PFR) [2]

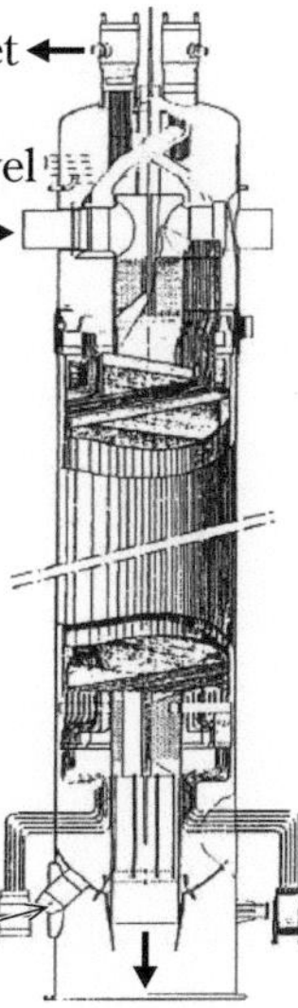

Fig. 8.24 Steam generator (Super Phenix) [2]

Many types of steam generators have been used for fast reactors. They are shown in Figs. 8.21, 8.22, 8.23, and 8.24. Depending on the anticorrosion characteristics of heat transfer tube material around the water boiling point and the heatproof characteristics of the tubes for superheaters, various types of steam generators, including those with a separated evaporator and superheater, all-in-one type, with a reheater, and with a recirculating evaporator, are used. The all-in-one-type steam generator is favorable for economy because of the reduced material mass and easier operation; however, material with both excellent heatproof and anticorrosion characteristics is required. For Monju, the evaporator and the superheater are separately installed. The evaporator was made of high anticorrosion material, and the superheater was made of high heatproof material.

In addition to straight tubes, there are many types of heat transfer tubes, such as the hairpin type, the U-tube type, and the helical coil type, that are designed to cope with the thermal expansion of tubes. Straight tubes are convenient for fabrication; however, the steam generator tends to be very tall to ensure the water-steam flow stability in the tubes, and it is difficult to repair damaged tubes and mitigate thermal stress. The helical coil-type tube has excellent heat transfer characteristics by cross flow and is easy to repair by drawing out the tube bundle; however, this type is difficult to fabricate. Here, the tube plate to which the tubes are gathered is where the severest thermal stress occurs in a plant. The tube plate can be placed both in the cover gas and in sodium. The former type is easy to test and repair while it is subject to severe thermal stress.

As the steam generator is a component to separate sodium from water, it is necessary to pay attention in design to prevent tube failure, as well as to mitigate the effects of sodium-water reaction in case of tube failure.

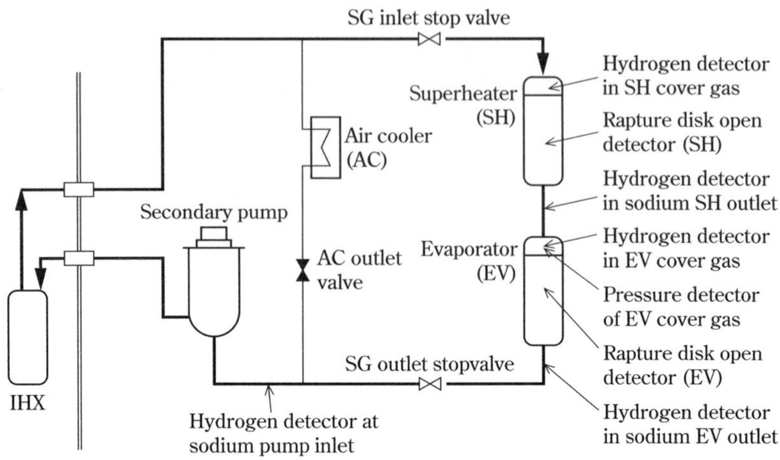

Fig. 8.25 Water leak detector (Monju)

In case of tube failure, high-pressure steam blows into sodium, and water reacts with sodium to produce heat and hydrogen. In case of a very small leak rate, the hydrogen concentration in sodium would increase slowly; therefore, this leak could be detected by hydrogen concentration meter in sodium or in the cover gas. In case of an intermediate leak rate, the sodium-water reaction zone would reach the adjacent tube and damage it (wastage). If this condition is left as it is, the adjacent tube would break (propagation of failure). Therefore, measures, such as early detection of leak (pressure meter) and early steam blow, need to be taken. An example of the arrangement of these water leak detectors is shown in Fig. 8.25.

In case of a large leak rate, since sodium reacts vigorously with water, a momentary high pressure (spike pressure) is produced and propagates to the surroundings as a pressure wave. Furthermore, the accumulation of produced hydrogen increases the pressure in the steam generator in a quasi-static manner. If this is left as it proceeds, the secondary cooling system would be overpressurized. Therefore, measures need to be taken, such as early detection of large leak (detection of the opening of rapture disk), early blowout of steam and reaction products (including hydrogen gas) through rapture disk, receiving reaction products in the container, and the separation and burning of hydrogen gas. An example of the arrangement of sodium-water reaction containers is shown in Fig. 8.26 [2, 3, 6].

Steam generators for the FR should be considered under development, and sodium-water reactions occurred many times for various reasons. Troubles occurred especially at welds between the tube and tube plate and between the tubes. For those portions, due consideration needs to be given to material properties and welding methods. In addition, welded parts must be strictly inspected, and it is desirable to adopt a structure that allows for sufficient inspection.

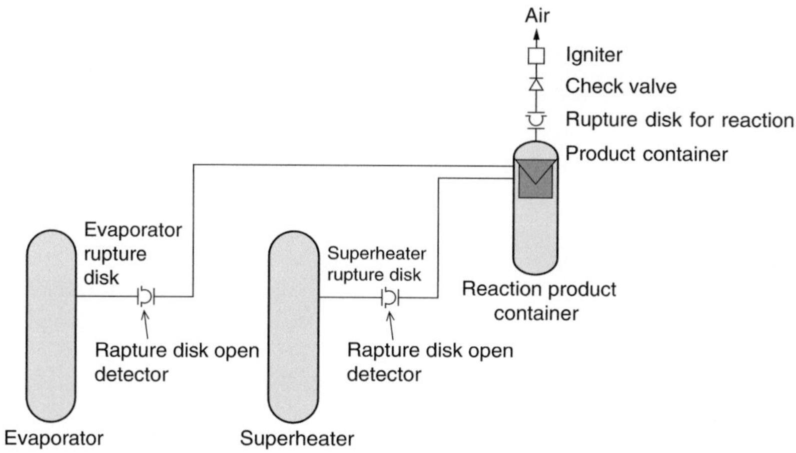

Fig. 8.26 Reaction product containers (Monju) [1]

8.4 Fuel Handling System Design

As shown in 6.3.4, for a small reactor that does not require many fuel assembly changes during one refueling period, fuel racks for cooling spent fuel may be installed in the reactor vessel. For a large-scale reactor, however, that requires many fuel assembly changes during one refueling period, an ex-vessel storage tank is installed outside the reactor vessel to store and cool spent fuel, and core refueling only is done in a refueling period. Cooldown and cleaning of the spent fuel assemblies are frequently performed after reactor start-up. In general, therefore, sodium is used as the coolant for the fuel cool-down system (ex-vessel storage tank), and water is used as the coolant for the system for waiting time to transfer fuel for reprocessing (fuel storage pool).

For fuel exchange, in the case of a single rotating plug system, the arm of the fuel exchange machine and the rotating plug are employed to move the fuel gripper to the required fuel assembly position. In the case of a double rotating plug system, the large and small rotating plugs, which are positioned eccentrically, move the fuel export hole to the required fuel assembly position while removing the upper core structure from above the core. There are two types of fuel exchange machine arms. One is an expandable pantograph type, and the other is a fixed-length type. In general, since the diameter of the upper core structure is nearly same as the core diameter, the reactor vessel diameter must be larger than twice the core diameter to completely remove the upper core structure from above the core. However, in the case of a double rotating plug-type or a pantograph-type fuel exchange system, which has good accessibility to above the core, since the space required for removing the upper core structure is smaller, the diameter of reactor vessel can be reduced (at least 1.5 times the core diameter), though their driving mechanisms are rather complex (See Fig. 8.27).

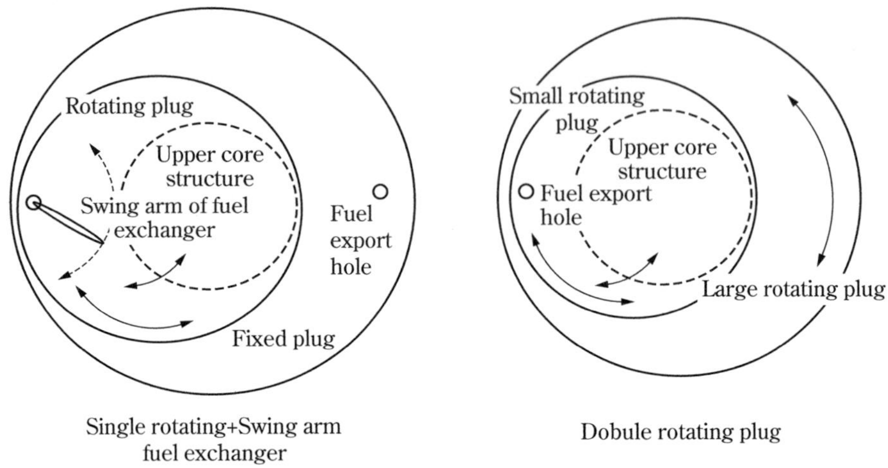

Fig. 8.27 Example of fuel exchange system

Fig. 8.28 Concept of door valve

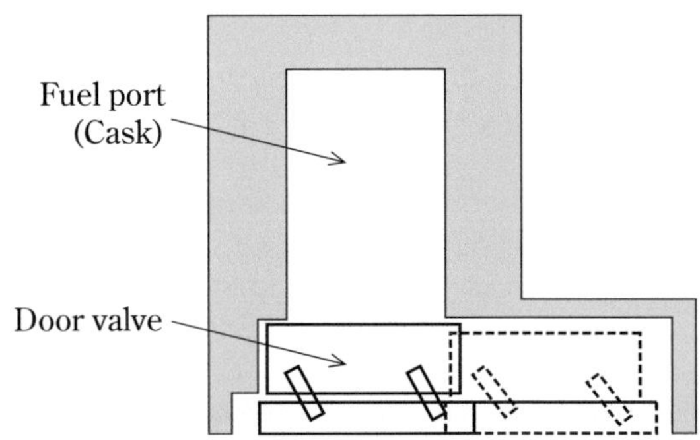

The fuel transfer machine must contain an extremely large amount of radioactive materials and transfer spent fuel producing a large amount of decay heat from the reactor vessel to the ex-vessel storage tank. Consequently, the area for containing fuel (cask) must have high airtightness with a door valve (Fig. 8.28) that enables containment of radioactive materials in case of a fuel failure accident, and a highly reliable cooling system must be installed to maintain the integrity of fuel through the transfer processes [1–4, 6].

Spent fuel and new fuel must be transferred through the containment vessel wall during refueling. Applicable transfer methods are listed below [2–5, 7]. Consideration must be given to the containment function during refueling (reactor shutdown period), quickness of refueling, complexity of components (reliability and cost), etc. to select a suitable method [1–4, 6].

1. Install an airtight vessel with rotating racks (transfer rotor) at the boundary of containment vessel, and transfer fuels through the vessel with airtight tools (door valves) (Fig. 8.29) [7].

Fig. 8.29 Transfer rotor system (Joyo) [7]

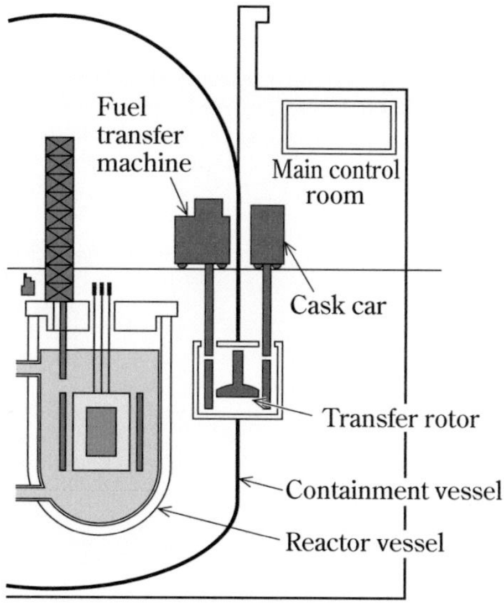

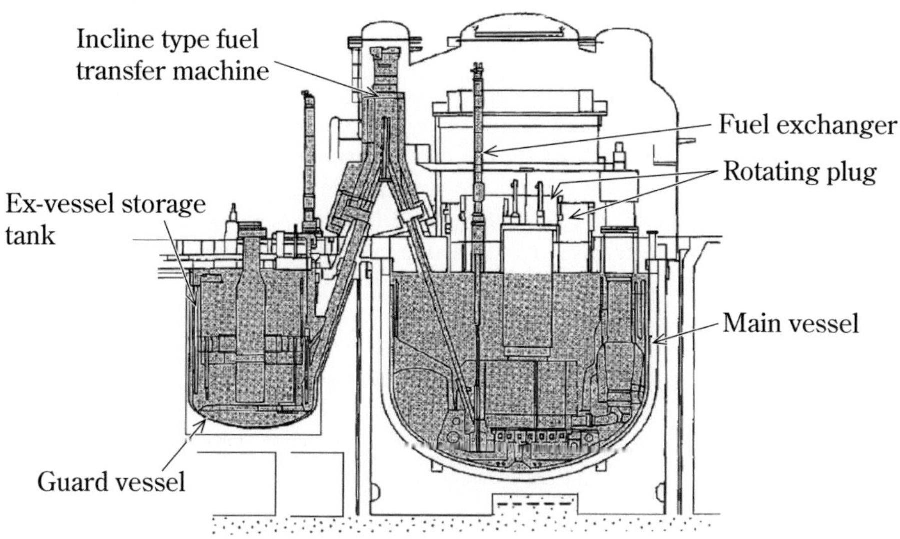

Fig. 8.30 Inclined type fuel transfer system (SPX) [3]

2. Install the inclined component through the containment vessel boundary, and transfer fuels through an airtight tool attached to the inclined component (Fig. 8.30) [3, 6].
3. Set a specified area outside the containment vessel to prevent the release of radioactive materials, open a hatch of the containment vessel during refueling, and run the fuel transfer machine loaded with fuel through the hatch to the specified area (Fig. 8.31) [1, 7].

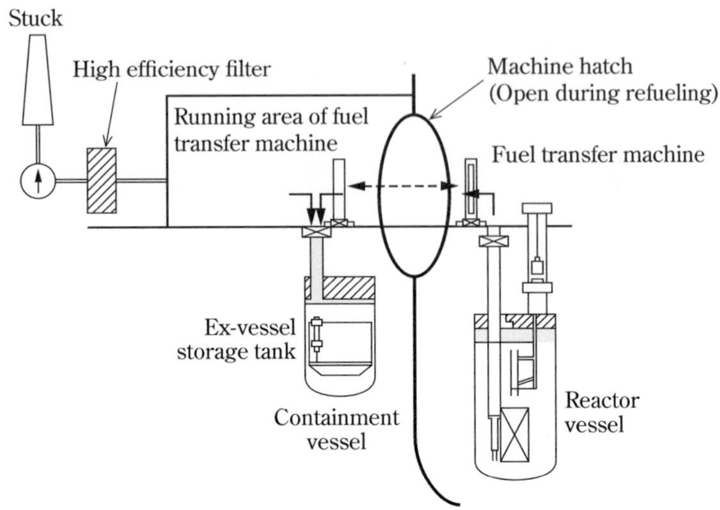

Fig. 8.31 Running type fuel transfer system (Monju)

Since use of the fuel transfer machine requires the handling of fuels with a large amount of radioactive material just after removal from the reactor vessel outside the containment vessel, the fuel container of the fuel transfer machine must have good airtightness, and the running area of the fuel transfer machine must be kept at a negative pressure and be equipped with an exhaust system with high efficiency filters to prepare for the release of radioactive materials from fuels (Monju) [1].

8.5 Instrumentation and Control System Design

Since FRs use sodium as coolant, important measurement items and methods are quite different from those of LWRs. The importance of coolant pressure is relatively low because coolant is not pressurized. The main items to be measured for coolant are the flow rate, temperature, level, and purity, as described in 6.3.7.

The flow rate is generally measured by electromagnetic flowmeter, which measures the voltage generated by liquid metal flowing through a magnetic field based on the reverse principle of an electromagnetic pump. There are two methods of magnetic field generation, use of a magnetic coil and use of a permanent magnet.

Temperature is generally measured by thermocouples. In some cases, however, resistance-temperature detectors are used, for example, to calibrate thermocouples.

There are three types of level gauges: electromagnetic, resistance, and point of contact.

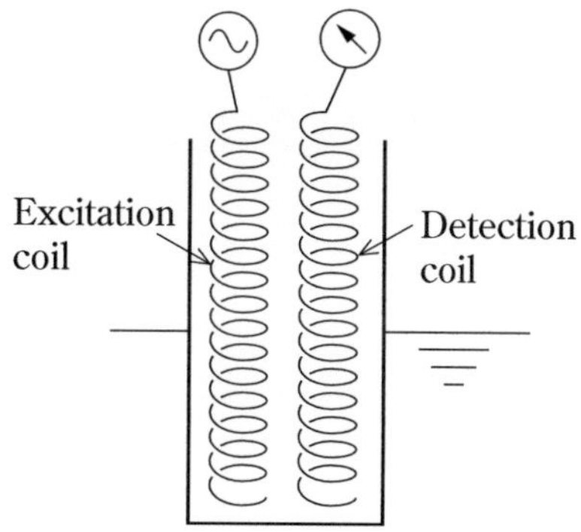

Fig. 8.32 Continuous-type level gauge

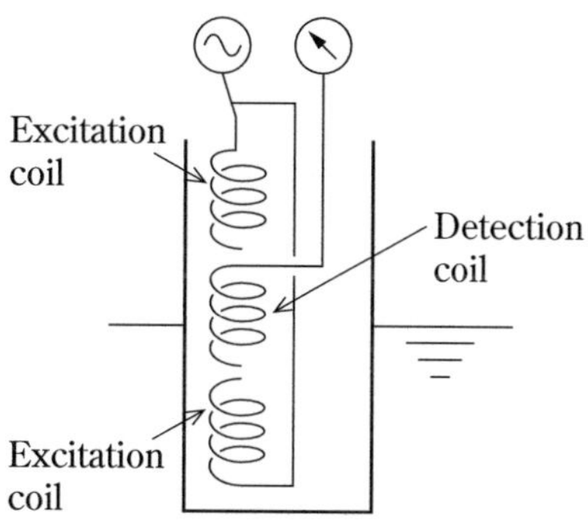

Fig. 8.33 Contact point-type level gauge

Electromagnetic-type level gauges include continuous and contact types. Continuous-type level measurement is based on the change in mutual inductance between the double coils according to the depth to which the gauge is inserted into the sodium (Fig. 8.32). Contact-type level gauge measurement is based on the generation of electromotive force at the central coil due to the change in mutual inductance between the upper and lower coils when only the lower coil is dipped in sodium (Fig. 8.33).

The resistance-type level gauge measures the resistance of a guide tube with relatively high electric resistance. When the guide tube is dipped into sodium, since electric current flows also through sodium, the resistance of the guide tube changes with the depth to which the gauge is inserted into the sodium (Fig. 8.34).

The contact-type level gauge utilizes the electric conductivity of sodium, that is, when the electric contact touches the sodium surface, an electric connection is made (Fig. 8.35).

For purity measurement, the plugging meter is generally used. The plugging meter is based on the same principle as a cold trap. That is, when sodium is cooled to below the saturation temperature, impurities are precipitated at an orifice to decrease the sodium flow rate through the orifice. The purity of sodium can be obtained from the plugging temperature and the saturation curve. The

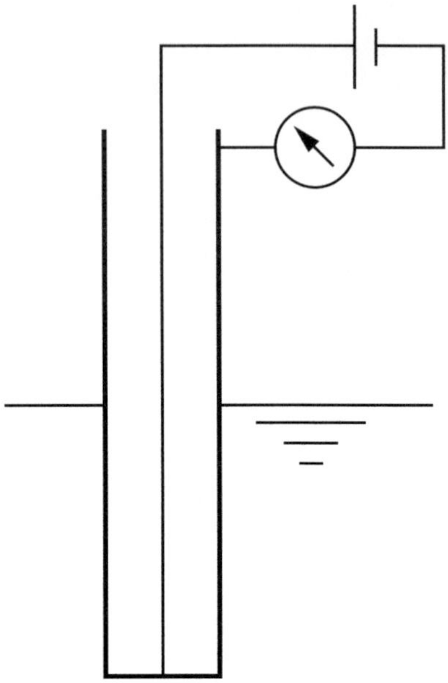

Fig. 8.34 Resistance-type level gauge

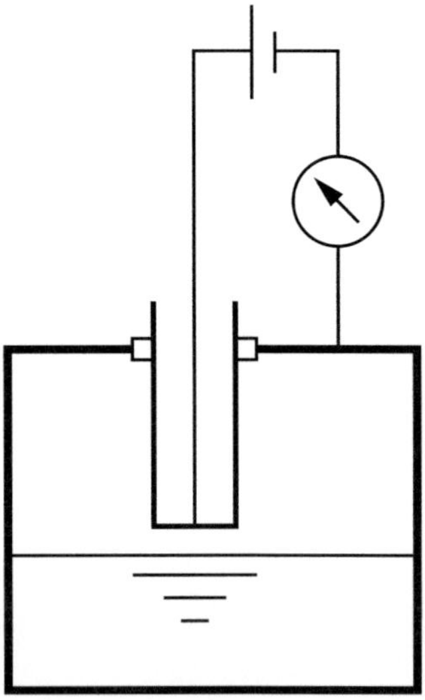

Fig. 8.35 Contact-type level gauge

concentrations of multiple impurities (e.g., hydrogen and oxygen) can be measured individually by sophisticated techniques (Fig. 8.36).

An instrumentation system unique to FRs is the fuel failure detection (and localization) (FFD(L)) system. When a fuel failure occurs, it is necessary to take appropriate measures to detect the failure and shut down the reactor. In case of a fuel failure, neutron precursors and the fission product (FP) gases (Xe, Kr, etc.)

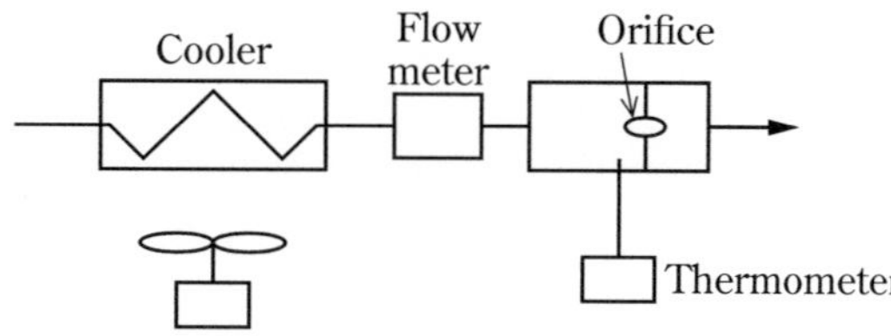

Fig. 8.36 Principle of plugging meter

confined in fuel pins leak out through the failed parts. Consequently, fuel failure can be detected by detecting neutrons in sodium or the FP gases in the cover gas. The delayed neutron method detects neutrons which are released from neutron precursors in sodium. Prompt neutrons cannot be used for fuel failure detection because they are generated in the core. On the other hand, neutron detectors arranged around the primary coolant pipe away from the core can detect neutrons released from neutron precursors in the primary coolant in case of a fuel failure. The cover gas method (using a precipitator) detects the beta decay noble gases in the cover gas by selectively precipitating the beta decay gas ions using a wire. The device that controls the wire is called the precipitator.

To detect the position of failed fuels (failed fuel detection and location), the tagging gas and the selector valve methods are used. For the tagging gas method, tagging gases consisting of Xe and Kr isotopes are filled in fuel pins at fabrication with specified mix ratios of Xe/Kr for respective fuel assemblies. Failed fuel assemblies can be identified by comparing the detected mix ratio and the filled gas mix ratio in case of a fuel failure. On the other hand, for the selector valve method (SV-FFDL: selector valve mechanism for a failed fuel detection and location), sampling tubes are set up near the coolant outlet of every fuel assembly to detect radioactive FPs contained in sodium samples. In this method, failed fuel assemblies are identified by exchanging the sampling tube with a selector valve.

The tagging gas method is mechanically simple; however, management of the mix ratios of tagging gas is complex, especially in the case of an increased number of fuel assemblies (i.e., large core), leading to difficulty in the exact identification of failed fuel assemblies. On the other hand, the selector valve method is mechanically complicated because it requires that many sampling tubes be set up; however, it is able to exactly identify failed fuel assemblies even in a large core.

References

1. Power Reactor and Nuclear Fuel Development Co., in *"Safety Evaluation Document of Prototype Fast Breeder Reactor 'Monju'* (in Japanese)
2. IAEA, in *TECDOC-1531 Fast Reactor Database 2006 Update*
3. Japan Atomic Energy Relations Organization, in *Collection of Public Relations Material on FBR*, 2nd edn. (in Japanese)
4. Power Reactor and Nuclear Fuel Development Co., ed by M. Hori, in *Basic Fast Reactor Engineering*, (Nikkan Kogyo Shimbun, Ltd., 1993) (in Japanese)

5. ASME, in *ASME Boiler and Pressure Vessel Code, Sec. III, Subsection NH*, (New York, 2010)
6. IAEA, in *TECDOC-1691, Status of Fast Reactor Research and Technology Development*
7. Japan Atomic Energy Agency, Pamphlets
8. Nuclear Safety Commission, in *Philosophy in Safety Evaluation of Fast Breeder Reactors* (1980) (in Japanese)
9. Nuclear Regulation Authority, in *Regulation on the Standards for the Location, Structure and Equipment of a Commercial Nuclear Power Reactor and its Associated Facilities* (2013) (in Japanese)

Chapter 9
Maintenance

Satoru Nakai

A nuclear power plant (NPP) has to maintain safety, reliability, and structural integrity throughout plant life. Therefore, careful operation, such as the avoidance of rapid temperature change that may cause significant thermal stress due to deviation from a design condition, is necessary. In addition, it is difficult to ensure safety and reliability throughout plant life by design alone because the prediction of the behavior of a large, complicated system, such as an NPP, during lifetime in the design phase is accompanied by great uncertainties. Therefore, appropriate maintenance activities are necessary, and the consideration of maintenance in the design stage becomes relatively important. For a new type reactor in particular, the importance of maintenance becomes more significant because of the larger uncertainties described in Chap. 2. Furthermore, it is necessary to improve the maintenance technologies through experiences of plant operations and to feedback to design of maintainable plants. Therefore, it is hoped that the reader understand the basis for the maintenance of NPPs and further learn unique ways of thinking about the maintenance of the fast reactor (FR).

9.1 Relationship Between Design and Maintenance

The main philosophy underlying the prevention of accidents and to mitigate the influence of an accident in a nuclear power plant is to apply the defense in depth concept that is specified in the IAEA's safe design requirements (SSR-2/1) [8]. The defense in depth concept is applied to all activities related to the safety. In other words, the first protection level to prevent deviation from normal operating

S. Nakai (✉)
Japan Atomic Energy Agency, Ibaraki, Japan
e-mail: nakai.satoru@jaea.go.jp

N. Kasahara (ed.), *Fast Reactor System Design*, An Advanced Course in Nuclear Engineering 8, DOI 10.1007/978-981-10-2821-2_9

conditions and system failure is applied to design, manufacturing, construction, in-service inspection, maintenance, and examination.

In addition, since NPP operation is long term, appropriate margins are provided in design to manage aging degradation during plant operation, and the evaluation of aging mechanisms assumed in design and detection of unanticipated deterioration are performed by monitoring and inspection.

Therefore, the components important to safety in an NPP must be designed to allow for easy maintenance for the purpose of maintaining the functions required in the design conditions, preventing the loss of functions by detecting unanticipated deterioration, and restoring failed functions by repair.

In addition, when the design of components important to safety is such that the performance of maintenance is difficult, it is necessary to show that maintenance is unnecessary due to sufficient safety margin.

The objective of maintenance is "to ensure that a plant is operated as intended within the design assumptions and operating limits and conditions." To achieve the objective, confirmation of the deterioration of components, evaluation of the integrity, confirmation of the functions, and necessary care and repair are performed by means of overhaul, opening check, nondestructive testing, condition monitoring, periodical operation examinations, etc.

As mentioned previously, maintenance capability for components important to safety must be taken into account in design. It is thought that the optimization of maintenance is to minimize maintenance cost while maximizing safety and reliability. Therefore, the securing of access routes and spaces for maintenance and the setting of sufficient radiation shield walls should be considered in design for easier maintenance. However, when all these requirements are met in design, the amount of materials for systems, structures, and components (SSCs) tends to increase, and construction costs may rise. Therefore, it may be effective to adopt maintenance-free components or reduce the frequency of maintenance during plant life by examining the degradation phenomenon and failure modes based on the characteristics of systems, such as coolant, operating conditions, material, and component structure, and by adopting a sufficiently conservative design. In other words, due consideration must be given to maintenance in design to achieve the required safety, reliability, and economy.

9.2 NPP Maintenance [1]

9.2.1 What Is Maintenance?

According to the Japanese Industrial Standards (JIS Z 8143), equipment maintenance is defined as "all the procedures and activities carried out to maintain the conditions under which equipment can perform its original functions, is operable and usable, and is allowed to recover from failure and defect." The JIS provides for action such as the checking, servicing, and lubricating of equipment to prevent deterioration, the inspection of equipment to examine structure and functionality before use, the assessment of equipment deterioration during use, the repair of equipment to allow recovery from deterioration, and the evaluation of the state of recovery after repair as specific maintenance activities. In addition, *The Maintenance Handbook* (edited by the Japan Institute of Plant Maintenance) states that the purpose of maintenance is to help increase the productivity of a company by ensuring that production, quality, delivery, and other requirements are met, by improving the work environment for workers, and by creating a workplace that can enhance employee morale. *The Maintenance Handbook* specifies "production," "quality," "cost," "delivery date," "safety," and "motivation" as the outputs of production activities in a company and defines maintenance as an important part of production activities for maximizing these.

Industries conduct a wide variety of equipment maintenance activities and have chosen the most suitable maintenance method for each piece of equipment based on experience. However, there has been a common understanding across different industries over time that the goals of maintenance are minimizing cost and maximizing reliability. These two goals can, in a sense, be regarded as constraints in performing maintenance activities. To date, engineers have created maintenance programs on the basis of their experience or precedents for the purpose of overcoming these constraints.

9.2.2 Major Features of the NPP from the Perspective of Maintenance

Difficult maintenance conditions have to be overcome in cases where direct human access for inspection is practically impossible or where replacement work is extremely difficult. This is exactly the case in an NPP, which has to be maintained in a radioactive environment once operation starts. A concept more commonly adopted in such cases is, for example, maintenance to satisfy the required functionality rather than to restore the original functions, by taking advantage of the tolerance in structural strength incorporated in the design phase.

To optimize maintenance for a plant system, attention to individual components constituting the system is not enough. Consideration must be given to the interaction between system components; and because the NPP is a large and complicated system, and the influence at an accident is significant, public concern is high, and strict regulation is applied.

It is necessary to clarify the essential maintenance elements to optimize maintenance. Although maintenance activities in industries vary widely, an easy way to gain a full picture of maintenance is to consider maintenance as a single system that consists of the following three subsystems:

1. Equipment and Components: equipment, components, and other hardware to be maintained
2. Technology and Information: technology and information needed for performing maintenance (including supporting devices and systems)
3. Organizations and Personnel: entities that perform maintenance

The Equipment and Components subsystem is the direct subject of maintenance, while the Organizations and Personnel subsystem conducts maintenance using various elements of the Technology and Information subsystem. As a result of maintenance, there is feedback to the maintenance-related technology and information, the maintenance structure, and necessary capacities and skills.

To resolve challenges for maintenance, it is essential to consider maintenance optimization for the entire plant equipment maintenance system in light of the constraints imposed by the social and business environments as requirements for that system. In this consideration, temporal factors of each of the three subsystems, that is, individual phases including design, construction, operation, and decommissioning for the Equipment and Components subsystem, the progress of innovation in terms of the Technology and Information subsystem, and generational changes for the Organizations and Personnel subsystem, should be taken into account.

Key points in these considerations are the development, application, and management of high-endurance equipment, components, and materials for the Equipment and Components subsystem, the sophistication of maintenance methods and techniques (for diagnosis, evaluation, management, and work assistance) for the Technology and Information subsystem, and measures to prevent errors, the education and training to preserve knowledge and skills, and the establishment of organizational management structure for the Organizations and Personnel subsystem.

Figure 9.1 shows the concept of maintenance optimization in the Technology and Information subsystem. The figure indicates that optimal maintenance in all plant equipment is performed by developing and implementing a long-term maintenance program for the entire plant life, while meeting the required performance criteria for integrity derived from the maintenance requirements (constraints) of minimizing cost and maximizing reliability, as previously described.

1. Maintenance program for the entire plant life cycle

A long-term maintenance program for the entire service life of a plant, including the design, construction, operation, and decommissioning phases, should be developed, and the plant should be maintained in accordance with this program. For this purpose, the maintenance program should be developed in accordance with changes in constraints, such as the sophistication of operational methods and the enhancement of regulations.

2. Requirements for minimizing cost and maximizing reliability

Requirements (criteria) for reliability (safety) should be met, while maintenance costs should be reduced to the minimum.

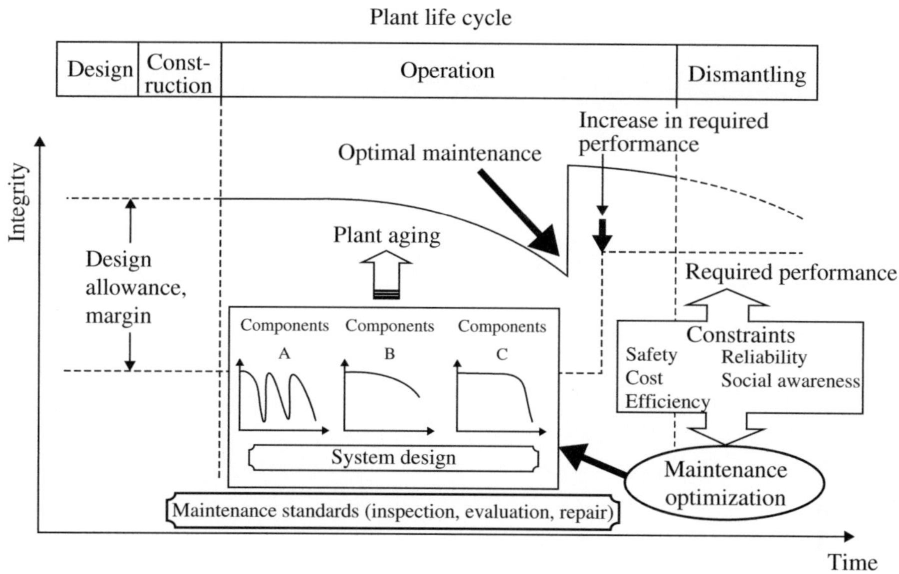

Note: In design, integrity is secured by providing design margins from the perspective of the prevention of abnormality. Failures that have significant effects on safety and reliability should be prevented by maintaining equipment functions and the integrity of facilities throughout the plant life cycle. Since the progression of deterioration varies according to equipment materials, operating conditions, and installation environments, the security of functions and the integrity of facilities can be ensured by maintenance activities with appropriate frequencies and methods corresponding to the speed of progression.

Fig. 9.1 Plant life management and the concept of maintenance optimization

3. Sophistication of maintenance methods and technologies

Maintenance methods and technologies should be advanced to achieve above objectives (1) and (2).

9.2.3 Deterioration and Maintenance

Equipment and components of NPP systems are inevitably subject to degradation over time. Such degradation includes thermal aging at high temperatures, embrittlement, fatigue in vibration environments, fretting in impact environments, corrosion and erosion in corrosive environments, and stress corrosion cracking (SCC). NPP operators must develop maintenance programs to prevent failures or troubles caused by degradation. Operational management, patrols, and periodic

Table 9.1 Major deteriorations requiring maintenance in NPPs

Aging deterioration in NPPs		Feature
Neutron irradiation embrittlement of reactor pressure vessel		Ductile-brittle transition temperature increases with neutron irradiation
Stress corrosion cracking (SCC)	Intergranular SCC (IGSCC)	Independent of aging
	Primary water SCC (PWSCC)	Possibility increases with elevated temperatures and aging
	Irradiation-assisted SCC (IASCC)	Occurred beyond a certain neutron fluence
Fatigue	Low-cycle fatigue	No cracking has occurred to dateFatigue accumulates with the number of reactor start-ups and shutdowns
	High-cycle fatigue	Greatly dependent on local conditions (vibration, temperature change)
Cable insulation degradation		Slow degradation except for those under high-temperature and irradiation environments
Pipe thinning (corrosion, erosion)		Greatly dependent on local conditions (structure, fluid conditions, etc.)

examinations are performed during plant operation. During plant shutdown, periodical inspections are performed, and the replacement, repair, and improvement of equipment and components are performed as needed. Troubles, damage, and operational and maintenance experience in domestic and foreign plants should be properly reflected in the maintenance activities. In addition, continuous improvement efforts, including the development of new inspection methods, are required.

The major deterioration requiring maintenance, except deterioration of dynamic components such as abrasion in sliding parts of rotating machinery including pumps, is shown in Table 9.1. Since these deterioration data are accumulated using the operating experience in domestic and foreign NPPs and knowledge about the deterioration mechanisms is deepened by research and development, it is necessary to continually obtain the latest information.

In addition to the physical phenomena described above, deterioration includes standards that have become outdated due to new knowledge and technical innovation. In this case, back-checking or backfitting based on new standards is carried out, and modification is carried out when standards are not met. A typical example is the update of design basis earthquake ground motion.

A classification of maintenance methods is shown in Fig. 9.2.

9.2.4 PDCA Cycle of Maintenance Program

To ensure that the functions of equipment in a nuclear plant system expected in the design phase are maintained, it is essential to perform maintenance suitable for the use conditions, deterioration, and changes of equipment in a timely manner. These

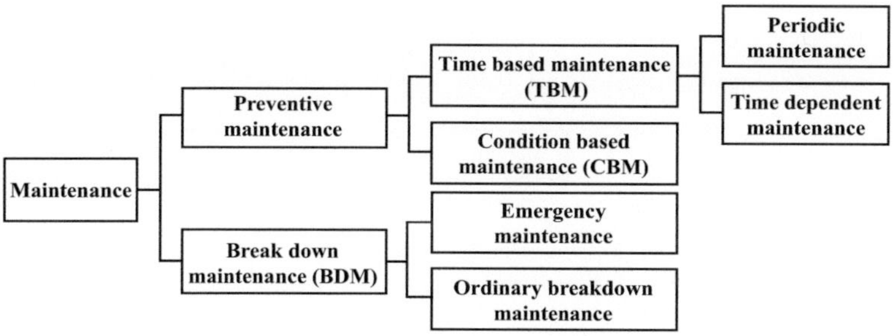

Preventive maintenace: Maintenance performed to reduce the probability of function deterioration or trouble of an item (component, equipment, etc.), and thereby, to prevent failure during operation of the item.
Time based maintenance (TBM): The preventive maintenance performed according to a predetermined time schedule.
Condition based maintenance (CBM): Action to monitor process values and trends at certain time points to confirm the specifications and the opering conditions of items, detect deterioration tendencies, identify troubles and faults, record the progress of failue, etc. Such monitoring is performed by means of devices and methods, such as checking, examination, measurementl, and alarms, continuously, indirectly or periodically.
Periodical maintenance: Preventive maintenance performed at a predetermined time interval.
Age-based maintenance (Time depend maintenance): Preventive maintenance performed when an item has been used for a planned accumulated operating time.
Break down maintenance (BDM): Maintenance performed to restore the required functions of an item after the discovery of an item failure.
Emergency maintenance: Repair performed as a result of the failure of a component to be covered by preventive maintenance.
Normal BDM: Originally BDM

Concepts specified in the "Code for Maintenance at Nuclear Power Plants" (JEAC 4209-2007), Japan Electric Association Code (JEAC) [3]
TBM
• To be performed when required by laws and ordinances.
• To be performed for the periodical replacement of consumables, etc.
CBM
• To be applied to cases in which a sign of failure can be detected by the appropriate collection and evaluation of condition data so that appropriate measures, including inspection and repair, can be carried out in a timely manner.
BDM
• To be applied to cases in which there is minor influence on the safety of NPP, environmental burden or continued operation in case of failure, etc.

Fig. 9.2 Maintenance method classification

maintenance activities consist of four elements: (1) development of a maintenance program (Plan), (2) implementation and monitoring of the maintenance program (Do), (3) confirmation of the results and effectiveness evaluation (Check), and (4) review of the maintenance program (Act) (PDCA cycle). Maintenance activities are performed for the checking, maintenance, or improvement of the functionality of equipment related to plant operation. On the premise of securing the safety of nuclear reactor facilities, maintenance activities are conducted to maintain the reliability of electricity supply according to the significance of the equipment while taking into account efficiency and economy. A maintenance implementation flow is shown in Fig. 9.3.

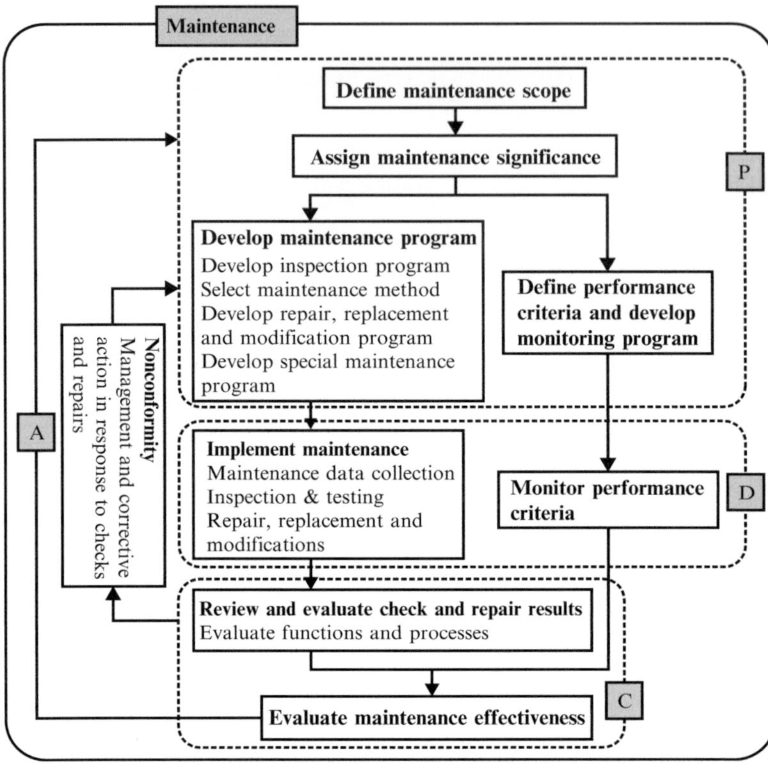

Fig. 9.3 Maintenance implementation flow [4]

Since basic knowledge for maintenance activities is updated and maintenance technology is innovated over time, maintenance activities need to be continuously improved in light of the results of effectiveness evaluation. For example, nuclear plant system operators should review domestic and foreign trouble information and conduct maintenance that incorporates new findings. Therefore, it is important to continue maintenance activities as a PDCA cycle, as shown in Fig. 9.3.

After the Fukushima Daiichi NPP Accident, continuous improvement activities based on the responsibility of electric utilities are further required.

9.2.5 Maintenance Technology

The objectives of maintenance are regarded as the detection and mitigation of the deterioration of components, the restoration of functions of failed SSC, and the improvement of component reliability. In-service inspection (ISI), a maintenance subject, is to confirm whether SSCs are acceptable for continued safe operation, or remedial measures should be taken by mainly nondestructive examination for possible deterioration. The maintenance methods are shown in Fig. 9.2.

Maintenance activities should be accomplished on a routine basis and may include any combination of external inspections, alignment or calibration, internal inspections, overhaul, and replacement of both active and passive components and equipment. Maintenance should involve continuous or periodic monitoring and diagnosis in order to predict equipment failure. On the other hand, in ISI, emphasis should be placed on the examination of static components, including the pressure boundaries of the primary and secondary coolant systems because of their importance to safety and the potentially severe consequences of their failure. ISI involves nondestructive tests, such as the visual test, surface test, volumetric test, and pressure leakage tests.

9.3 Maintenance of Fast Reactors

9.3.1 Features of Fast Reactors from the Perspective of Maintenance

Major features of a sodium-cooled fast reactor (FR), specifications of which are shown in Table 9.2, from the perspective of maintenance are summarized below with a focus placed on the maintenance of important functions, including the coolant boundary function, the heat transfer function from the core, and the reactor control functions. These features are quite different from light water reactors (LWRs) and require particular maintenance technologies to FRs.

1. Since the coolant is sodium with high boiling point, the coolant is operated as liquid phase with a sufficient degree of subcooling. That enables low-pressure operative conditions and avoids the rapid coolant loss from decompression boiling. Therefore, leak before break or leak detection before break (LBB) concept can be applied to sodium-containing components and piping, as described in Chap. 5. Sodium leakage from these components and piping

Table 9.2 Features of FR

Plant parameter	LWR (PWR)	FR (Monju)	Feature of FR
Coolant	Water	Sodium	Large thermal stress
RV outlet temperature	320 °C	529 °C	Elevated temperatures
Temperature difference between RV inlet and outlet	30 °C	132 °C	Large temperature difference and large thermal stress
Operating pressure	16 MPa	1 MPa	Low pressure
RV inner diameter	4 m	7 m	Large diameter
RV shell thickness	200 mm	50 mm	Thinner shell thickness
Inner diameter/shell thickness (RV)	20	140	Thinner shell thickness and larger inner diameter

Note: All numbers are approximate values
Ref: Committee on Power Generation Facility Codes of the Japan Society of Mechanical Engineers (JSME)

can rapidly and surely be detected using different methods, such as the detection of sodium aerosol in gases sampled from an annular section between the component and thermal insulator and the decrease in sodium level in the reactor vessel.

Therefore, in sodium-cooled fast reactors (FRs), even if a crack penetrates the sodium boundary, it is possible to prevent a large-scale break by detecting sodium leak caused by the defect and taking necessary measures, such as reactor shutdown.

2. Components containing sodium are unlikely to fail due to corrosion if the components consist of materials with good compatibility with sodium, such as stainless steel and low alloy steel, and sodium impurity is adequately controlled. In addition, flow-accelerated corrosion (FAC) and SCC are also unlikely to occur since the sodium components are under a reductive environment to their materials. Therefore, overhaul and disassembling inspection of components are not required in principle for an area wetted by sodium. However, the opening of sodium boundary is necessary during periodical replacement of a steam generator rupture disk and a hydrogen meter that is designed to detect a sodium-water reaction.
3. FRs have special systems, including the cover gas system and various instrumentation systems, such as the sodium leak detector, the hydrogen meter in sodium to detect the sodium-water reaction in case of a steam generator heat transfer tube failure, and the induction-type sodium level meter using the electromagnetic characteristics of sodium. In addition, FRs have unique equipment, such as the fuel handling equipment having specifications and structure different from those of light water reactors (LWRs).

9.3.2 FR Maintenance and Operation [2]

The features of the maintenance program for fast reactors are described below taking the program for Monju as an example. Monju established a maintenance program for the construction stage until the end of the core confirmation examination on the basis of the preventive maintenance concept in January 2009. The program is based on in-house standards and the fabricators' standards in consideration of operation and maintenance results obtained in Monju, as well as operation and maintenance experiences of the fast experimental reactor Joyo, and the failure events of foreign FRs and domestic LWRs.

The Monju maintenance program is designed for continuous improvement of maintenance activities. The specific elements include the definition of maintenance scope, the determination of performance criteria (PC), the development of PC monitoring programs and maintenance programs, the implementation of maintenance, the review and evaluation of check and repair results, nonconformity management and corrective actions, and the evaluation of maintenance effectiveness.

In addition, in order to establish a maintenance scheme suitable for FRs, a new type reactor, the preventive maintenance concept was selected as the basis to determine the maintenance methods of components unless the loss of component's function gives minor influence to the safety of NPP facilities.

In addition, since there is limited experience in operation and maintenance of the facilities peculiar to Monju such as sodium systems, the cover gas system, and the fuel handling equipment, TBM-based preventive maintenance was selected, and the maintenance interval was determined based on fabricators' recommendations, except those components whose maintenance intervals are based on out-of-pile testing, such as pump mechanical seal parts. It will be necessary in the future to optimize the maintenance scheme in consideration of the functions, importance, and influence in case of trouble, deterioration phenomena, and inspection methods for the equipment peculiar to Monju by accumulating maintenance experience and establishing maintenance management methods.

For those facilities common to LWRs, such as the seawater system, the water-steam system, the containment vessel, the ventilation and air-conditioning system, emergency generators, and the electric equipment, the maintenance experience that has been accumulated at LWRs for more than 30 years can be utilized because they have the same specifications and are used under similar environments.

For those facilities peculiar to FRs, it is necessary to perform appropriate maintenance activities through identification of deterioration phenomena peculiar to FRs and a good understanding of deterioration progress and influences on the NPP.

A feature of FR design is that fuel is loaded in the core, and the operation of the cooling systems is necessary to remove decay heat even during reactor shutdown. Therefore, additional operation, such as the switching of the cooling systems (i.e., filling sodium to a system that needs decay heat removal and draining sodium from a system to be checked), is necessary before performing maintenance, and it is one of the factors of a long maintenance period. One possible solution is the installation of an independent cooling system that allows for easy and short-term maintenance without depending on the other cooling systems.

From the viewpoint of operation, including start-up and shutdown, one major feature of FR design is that strict system pressure control is not required because of low operating pressures in FRs. Another feature of FR design is that the most significant load next to mechanical load caused by earthquake is the thermal stress caused during start-up and shutdown. Consequently, the power increase rate during start-up is limited to prevent an excessive temperature change rate of components including the reactor vessel (RV) and the intermediate heat exchanger (IHX). During reactor shutdown, the ratio between power and flow rate is controlled within a certain limit to prevent an excessive component temperature change rate except under abnormal conditions.

9.3.3 *Activities Necessary for FR Maintenance [7]*

There is no sufficient operation and maintenance experience for FRs, and there is only one FR with power generation capability in Japan, the prototype FR Monju.

Therefore, it is necessary to establish a rational maintenance scheme with high reliability in consideration of FR characteristics concerning the performance and deterioration assumed in design. The required activities are listed below.

1. Organization of maintenance data based on FR operation and maintenance experience

For FRs, there are limited data of operation and maintenance experience (particularly for maintenance). Therefore, systematic organization of the operation and maintenance experience in the experimental fast reactor Joyo and the maintenance data of Monju with reference to foreign FR operation experience and trouble data is necessary to identify the deterioration phenomena and necessary measures (e.g., inspection). The results are organized into a list of FR deterioration mechanisms as described below.

2. Organization of deterioration mechanisms peculiar to FRs

The knowledge of LWRs summarized in the Atomic Energy Society of Japan standards [5] about the deterioration phenomena is not directly applicable to the equipment peculiar to FRs, such as the components of sodium systems, the argon gas system, and the fuel handling system. Therefore, the degradation phenomena occurring at the component parts are identified in consideration of the material and environmental conditions, and the appropriate evaluation methods are organized. An example of such organization regarding the functions and deterioration phenomena for the steam generator is shown in Fig. 9.4.

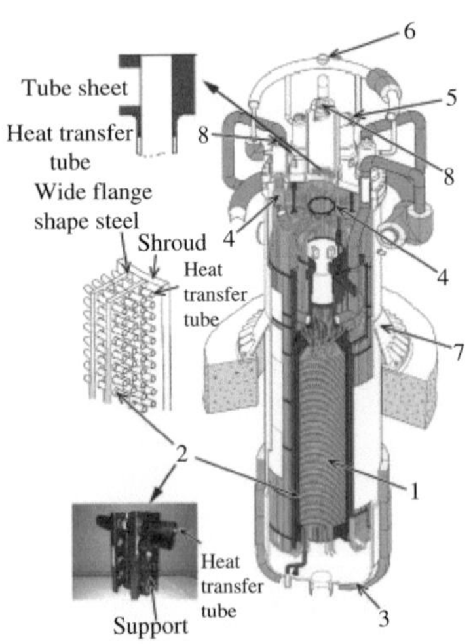

Function	Failure mode	Major parts		Degradation
Pressure boundary	Internal* leakage	1	Heat transfer tube	Fatigue (crack), wear, corrosion, creep
			Low alloy steel	
		4	Tube sheet (water,steam outlet)	Fatigue (crack), corrosion
			Low alloy steel	
	External leakage	5	Steam outlet nozzle	Fatigue (crack), corrosion
			Low alloy steel	
		6	Water inlet nozzle	Fatigue (crack), corrosion
			Low alloy steel	
		7	Shell	Fatigue crack), creep
			Low alloy steel	
		8	Maintenance hole	Fatigue (crack), corrosion
			Low alloy steel	
		3	Lower end plate	Fatigue (crack)
			Low alloy steel	
Heat transfer	Performance deterioration	1	Heat transfer tube	Scale adhesion
			Low alloy steel	
		2	Heat transfer tube support	Fatigue (crack)
			Stainless steel	

Fig. 9.4 Steam generator functions and deterioration mechanisms

3. Determination of maintenance significance based on probabilistic risk assessment (PRA)

Risk significance evaluation is based on the characteristics of FRs, and maintenance significance is based on the evaluation result. For the components of high safety significance, the maintenance significance is based on the redundancy and risk significance. For the equipment needed for accident management, the maintenance significance is appropriately determined as well.

4. Determination of maintenance methods based on the deterioration and functions of each component part

For the equipment peculiar to FRs and that common to LWRs, the functions and structure of components, examination results of deterioration mechanisms of each component part, the past maintenance experiences, and the condition monitoring parameters during plant operation are organized. Based on these results, inspection items necessary for each component are identified to evaluate the sufficiency of the present maintenance methods, and, thereby, to determine the most suitable inspection items, methods, frequencies, etc.

9.3.4 In-Service Inspection (ISI)

The objective of ISI is to examine SSCs for possible deterioration so as to determine whether a plant is acceptable for continued safe operation over the plant's operating lifetime. Accordingly, ISI is one maintenance measure to ensure that a plant is operated safely in accordance with the design assumptions and requirements and within the operational limits and conditions. Activities necessary to establish ISI for FRs are described below.

1. Establishment of the prototype FR ISI principle

In LWRs, ISI is performed in accordance with the Rules on Fitness-for-Services for Nuclear Power Plants [6] of the Japan Society of Mechanical Engineers (JSME). In LWRs, an evaluation of crack progress and the possibility of break is required when a crack is detected by nondestructive testing because the crack may grow and lead to break due to high internal pressure. On the other hand, in Monju, where LBB is applicable to components containing sodium as described in Sect. 9.3.1, the safety of nuclear facilities can be secured by continuous leak monitoring to detect sodium leakage before a crack leads to a break. Therefore, inspections and evaluation required for ISI differ between FRs and LWRs. In addition, since the inspection environments in FRs differ greatly from those in LWRs, devices and examination techniques applicable to ISI also differ from each other. For this reason, it is important to establish an FR ISI principle on the basis of the characteristics of Monju while following the concept of the Rules on Fitness-for-Services for Nuclear Power Plants [6]. A flow for the establishment of FR ISI principle is shown in Fig. 9.5.

Features of FR

LBB is applicable because of the use of ductile material, low pressures, and early detection of sodium leakage.

1. Leak detection is possible before break (LBB)
2. Sufficient time margin for safety operation after leak

⇩

FR ISI requirements

For the components that have a significant effect on nuclear reactor safety at the time of leak, continuous leak monitoring is effective as a substitute for non-destructive examinations, such as surface test and volumetric test.

⇩

Establishment of FR ISI principle on the basis of the characteristics of Monju while following the concept of LWR maintenance standards

Fig. 9.5 Flow for establishment of FR ISI principle

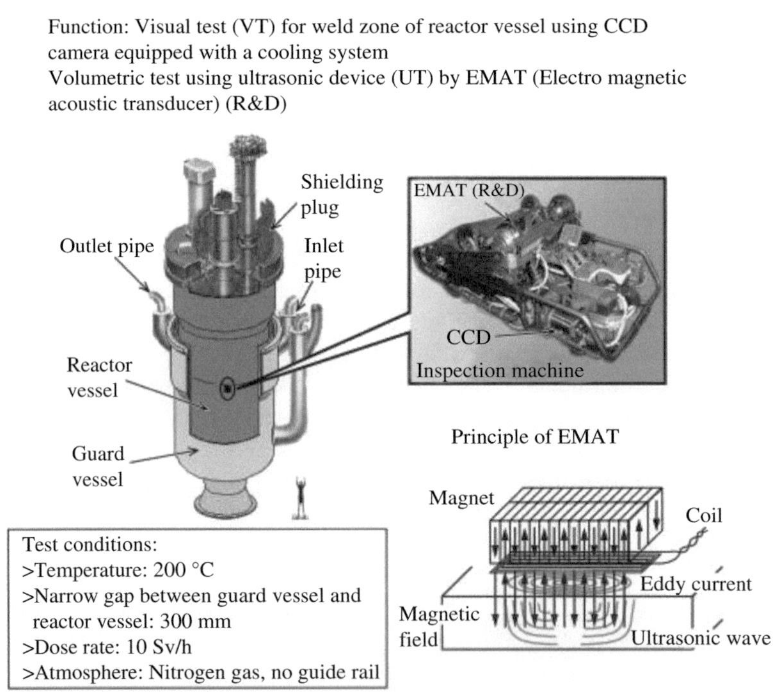

Fig. 9.6 Reactor vessel visual test (VT) and ultrasonic test (UT) device

2. Development of ISI technology

In addition to the continuous leak monitoring system, a major element of FR ISI, efforts have been made to develop inspection devices for components peculiar to FRs with high safety significance by incorporating the latest development results. Such devices include (1) the reactor vessel inspection device (Fig. 9.6),

(2) the primary cooling system inspection device (Fig. 9.7), and (3) the steam generator heat transfer tube inspection device (Fig. 9.8) that have been developed since the 1970s.

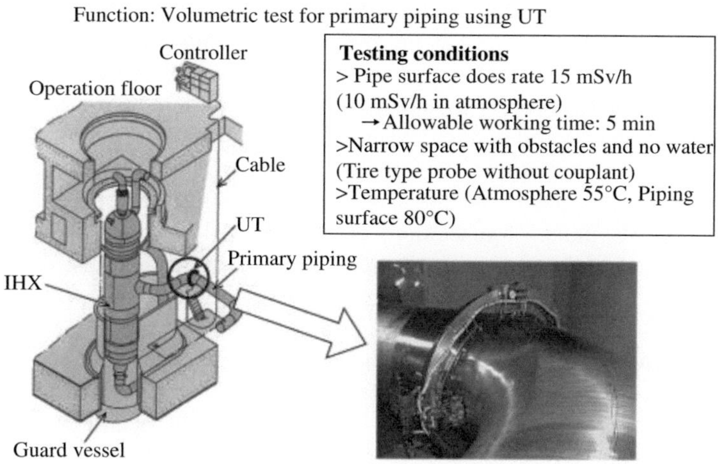

Fig. 9.7 Primary piping inspection device

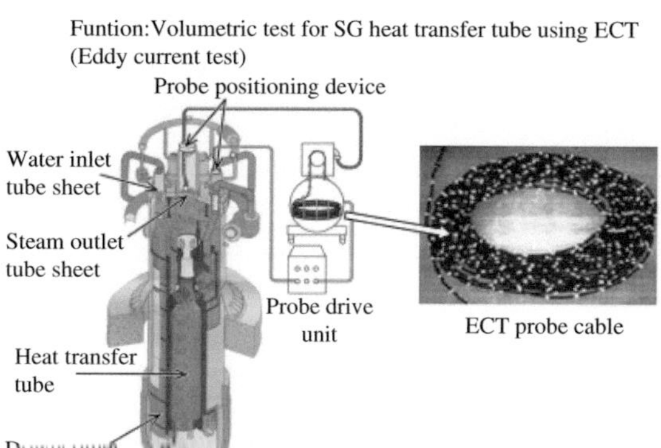

Feature of tube: Ferromagnetism, a long and intricate shape, thin wall thickness etc.

Items	Evaporator	Super heater	PWR
Tube number / unit	140	147	3400
Tube length (m)	85	46	23
Tube material	STBA24	SUS321	Inconel
Tube outer diameter (mm)	31.8	31.8	22.2
Tube thickness (mm)	3.8	3.5	1.3
Helical coil diameter (m)	1.2-2.5	1.4-2.7	-
Bending radius(mm)	160	160	120

Advancement (optimization of the sensor, evaluation of the noise signal) of the eddy current test technology by the three-dimensional simulation

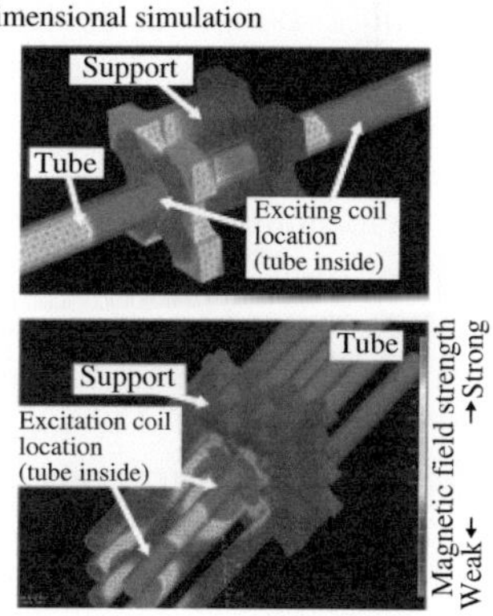

The simulation of the magnetic field strength near a support plate at eddy current test in a single pipe model and multi pipes model. An actual structure (multi pipes) is predicted precisely by comparison between similation of the single tube and single tube test result which is relatively easy test.

Fig. 9.8 Steam generator heat transfer tube inspection device

9.3.5 FR Repair Technology

1. Principles of Repair

When a crack is found in equipment during in-service inspection, the growth of the crack should be predicted to assess the structural integrity of the equipment. If the assessment is not satisfactory, the equipment should be repaired or replaced. Equipment may be repaired or replaced to correct a problem found during a check or operation of other equipment or for preventive maintenance purposes. Repair is conducted to restore or improve the structural integrity of a component required for its operation by thermal or mechanical means, without the replacement of the component or its parts. Replacement is conducted to restore or improve the intended function of the component by replacing the entire component or its parts.

2. Elemental technologies used in repair

In addition to the technologies used in the construction phase of a nuclear power plant, newly developed or improved technologies that have been proven are used to repair a component or structure after the start of plant operation. The repair technologies used in LWRs include electro-discharge machining (EDM) and grinding for defect removal; plasma cutting, laser cutting, and abrasive water jet cutting used to remove the existing structure for partial replacement; and various types of welding, relief of weld residual stress, and plugging for heat transfer tubes of heat exchangers.

3. FR repair technologies

Since overhaul and disassembly inspections are not required for components containing sodium, as described in 9.3.1, repair for the sodium-containing components has been carried out to fix trouble or failure, such as the Monju secondary loop sodium leakage, or improve the performance of a component, such as the replacement of the Joyo IHX.

The techniques used for defect removal, cutting, and welding in FRs are similar to those used in LWRs. However, for components containing sodium, attention needs to be paid to the existence of residual sodium on the internal surface of the components even after sodium draining.

The features of FR repair technologies are described below based on the experience in the experimental fast reactor Joyo, at which the IHX was replaced by cutting the primary and secondary piping under the conditions where fuel was loaded in the core and sodium at a temperature of 200 °C was charged in the reactor vessel.

Since the coolant boundary was opened when cutting sodium pipes, removing the remaining sodium, and performing welding, it was necessary to take appropriate measures to prevent the ingress of oxygen in air to sodium system; and oxygen concentration control, argon gas pressure control, cleanliness control, and welding control were performed to prevent weld defects resulting from the residual sodium.

To control sodium purity and suppress the inflow of oxygen from the air into the coolant boundary during modification, such as pipe cutting, sodium removal,

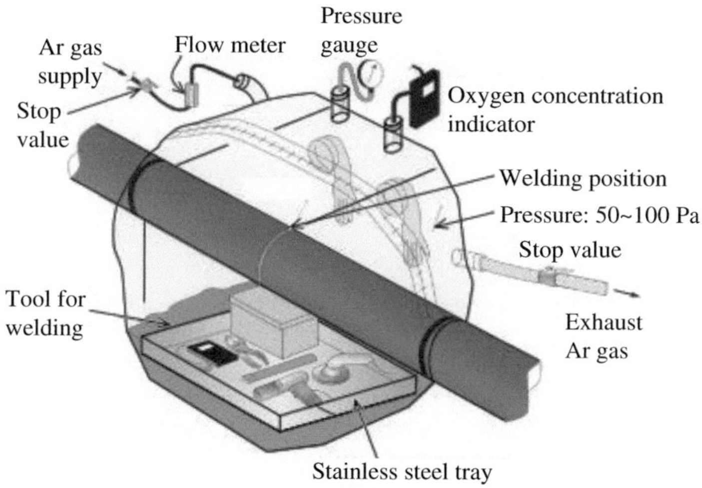

Fig. 9.9 Prevention of air inflow to sodium system

groove alignment, and welding operations, cutting and welding locations were covered with a seal bag (Fig. 9.9), which could be used in a narrow space, to limit the amount of oxygen flowing into the pipe to as little as possible, and then the air in the pipe was replaced with argon to reduce the oxygen concentration.

Points to remember for the cover gas control were the low-pressure control of the cover gas in the sodium system and the sealing performance and pressure resistance of the plug used to temporarily close the cut end of the pipe. When performing operations to open the coolant boundary, such as pipe cutting, sodium removal, groove alignment, and welding, the pressure of the cover gas in the system was maintained below 300 Pa to prevent any damage to the seal bag and improve the ease of operation.

Points to remember for cleanliness control were to prevent cut debris and tools from being dropped into the system, and to select a suitable method to remove and clean sodium remaining on the inner surface of the pipe, which could cause weld defects. When a large amount of sodium was deposited on the pipe inner surface, it was removed with a scraper or drill. A small amount of sodium was washed off with a denatured alcohol solution. The cloths used for final cleaning were analyzed for the amount of sodium deposited on the inner surface of the pipe.

Points to remember for weld control were to control the pressure of back seal gas and to monitor the temperature of the pipe area where sodium was deposited. For the modification of piping, TIG welding was applied to all areas. During TIG welding, it is necessary to supply a small amount of back seal gas to prevent impurities, such as oxygen, nitrogen, and hydrogen, from being mixed into the molten area and causing weld defects. It was necessary to maintain the pressure of the argon cover gas in the cooling system that was used as the back seal gas at an appropriate level. Residual sodium on the pipe inner surface over a length of about 300 mm from the weld position was removed and washed off to prevent it from melting during welding and affecting the weld. During welding, the temperature of the pipe where sodium was deposited was monitored to maintain it below 70 °C, sufficiently lower than the melting point.

9.3.6 Reflection to Design

For LWRs, there is approximately 40 years of operation and maintenance experience since the start-up of Japan's first commercial unit (Tsuruga NPP unit 1) in 1970. LWRs have experienced various troubles caused by thermal fatigue and flow-induced vibration due to poor design, the lack of assumptions in design, or inappropriate maintenance, such as poor adjustment after overhaul. The improvement of maintenance activities is based on these experiences and these experiences are reflected in the design of new plants.

As for FRs, there is limited operation and maintenance experience, and the design performance of components and the change with time have not been sufficiently demonstrated. FR maintenance methodology will be established by developing technologies necessary for the maintenance of FRs and performing maintenance activities using these technologies to evaluate the effectiveness, validity, and ease of maintenance, including the response to troubles. Therefore, sufficient operation experiences and feedback to maintenance methods are necessary to establish maintenance technologies. In addition, obtained knowledge can be reflected to improve the effectiveness and ease of maintenance in future plant for further safety, reliability, and economy.

Further Readings

K. Miya, G. Yagawa, in *Reactor Structural Engineering*. Series in Nuclear Engineering, (University of Tokyo Press, 1977) (in Japanese)

Y. Ando, K. Okabayashi, in *Genshiryoku Puranto no Kozo Sekkei (Structural Engineering in Nuclear Plants)*. Series in Nuclear Engineering Vol. 3, (University of Tokyo Press, 1977) (in Japanese)

S. Ishino, in *Shosha Sonsho (Irradiation Damage)*. Series in Nuclear Engineering Vol. 8, (University of Tokyo Press, 1979) (in Japanese)

F. Masuyama, K. Maruyama (translation), in *Evaluation of Damage Mechanism and Lifetime of Elevated Temperature Components*, (Nikkan Kogyo Shimbun, Ltd., 1993) (in Japanese)

Penny, Marriott, in *Design for Creep*, 2nd edn. (Chapman & Hall 1995)

References

1. Kazuyuki Demachi, the Author and Editor, Nuclear Power Text Book, in *Nuclear Maintenance Engineering*, (Ohmsha, Ltd., 2010) (in Japanese)

2. S. Nakai, Y. Kaneko, K. Mukai, Restart of prototype FR Monju after long-term shut-down. J. Jpn. Soc. Maint **9**(4) (2011) in Japanese
3. Japan Electric Association, Code for maintenance at nuclear power plants. JEAC 4209, (2007) (in Japanese)
4. Japan Electric Association, Guide for maintenance at nuclear power plants. JEAG 4210, (2007) (in Japanese)
5. Atomic Energy Society of Japan (AESJ) standards, Code on implementation and review of nuclear power plant ageing management programs 2008 (AESJ-SC- P005: 2008) (in Japanese)
6. Japan Society of Mechanical Engineering (JSME), Codes for power generation facilities rules on fitness-for-services for nuclear power plants. (JSME S NA1) (in Japanese)
7. S. Nakai, M. Uchihashi, Y. Kaneko, Maintenance of prototype FBR Monju. Jpn. Soc. Maint. 8th Annual Conference, October, (2011) (in Japanese)
8. IAEA, in *"Safety of Nuclear Power Plants": Design Specific Safety Requirements、 SSR-2/1*, (International Atomic Energy Agency, Vienna, 2012)

Chapter 10
Actual Monju Design

Tadao Takahashi

The latest technologies for fast reactor (FR) plant system design are described in the remaining chapters of this book. In this chapter, however, the history and principal technologies used in the design of Monju, aiming to have the breeding characteristic for the major object, are introduced with reference to actual design experience. At the time of Monju design stage, since there was little information on FR plant design, the Monju design was performed step by step from a comprehensive perspective. For this reason, Monju design was not conducted according to the usual design process, namely, from function to mechanism and next to structure. The design process had many repetitions by performing calculation processes and investigations on important technical problems for FR design from the initial design stage. Describing these design processes and important problems would help readers to understand the overall technical view of FR design and is in line with the purpose of this book.

10.1 Policy for the Development of Monju

10.1.1 Background and Technical Level at the Time of Monju Design

Although Japan has few energy resources, it has great demand for future energy. For this reason, Japan should have independent energy resources and own energy production technologies to secure its future energy. The Japanese government decided to promote FBR development as a national project. Monju is the prototype reactor, an important step for developing the FR project. The Monju project was

T. Takahashi (✉)
FBR Senior Research Laboratory, Yokohama, Japan
e-mail: tadao_takahashi2002@yahoo.co.jp

N. Kasahara (ed.), *Fast Reactor System Design*, An Advanced Course in Nuclear Engineering 8, DOI 10.1007/978-981-10-2821-2_10

started in 1967, when there was limited information available. Such information included some books on reactor physics, a few research reports on fast reactors, and limited technical information on FR plant design. Of course, there was no computer program for FR plant design.

At that time, the design of the experimental reactor Joyo, which preceded Monju, was underway, and its construction was scheduled to start in 1970. In addition, construction for the advanced heavy water reactor Fugen, which was developed in parallel with Monju, was scheduled to start in 1970. Consequently, technical information, such as confirmed technical design data or results from safety evaluation by the authorities, of these reactors could not be used as a complete set for Monju plant design. In each technical field, the experience of fundamental design, safety evaluation, and detailed design of Joyo were reflected separately to those of Monju. Fabrication technologies were also developed under long-term plan to scale up FR particular components without technological gap, through testing and fabrication of Joyo components and partial mock-up components of Monju.

At the same time, thermal reactors were being introduced for construction through technical cooperation with overseas countries. That is to say, a gas-cooled reactor was imported from the UK and achieved first criticality in 1965, and the construction of light water reactors (LWRs) imported from the USA started. Since these introduced technologies were peculiar to thermal reactors and contained confidential information, they had difficulties to be utilized for FR plant design.

10.1.2 Monju Design Policy and Procedures

In the policy of FR development, Monju was defined as the prototype reactor following the experimental reactor Joyo and was expected to contribute to the establishment of FR technologies aimed at future commercialization. Since development should be performed using domestic technologies, the Power Reactor and Nuclear Fuel Development Corporation (PNC) was established in October 1967 as the main development organization.

The main specifications of Monju were as follows: the electric power was 250 to 300 MW, approximately 1/3–1/5 of that of a commercialized reactor. The fuel material selected was mixed uranium-plutonium oxide, following the uranium oxide fuel utilized in light water reactors. The coolant was liquid sodium having suitable nuclear and thermal characteristics for FR.

Based on the above policy, the fundamental principles for design were established in the then design report, as shown in Fig. 10.1.

Firstly, for the Monju technology, very difficult but important requirements were given: (1) development should be performed using domestic technologies with full consideration of FR characteristics, and (2) performance suitable for a prototype reactor should be achieved. At the same time, harmonization between design and R&D was required. In addition, available technical information, such as the experience of preceding reactors, related information on foreign FRs, and the progress of LWR technologies, should be taken into consideration.

(Goal)	(Development)	(Technology)	(Time)	(Information)
• Develop using domestic technologies				
	• Fully consider FR characteristics			
	• Establish FR technologies			
	• Harmonize between design and R&D			
		• Exploit characteristics suitable for a prototype reactor		
		• Design useful for future commercialization		
		• Emphasize both safety and reliability		
		• Improve economic aspect based on ensured safety		
		• Consider conformity with licensing requirements		
			• Meet the specified construction term	
			• Put top priority on implementation	
				• Utilize experiences of preceding reactors
				• Refer to information from foreign FRs
				• Take latest LWR technologies into consideration

Fig. 10.1 Fundamental principles for FR development

In consideration of the role and future use of Monju, the establishment of FR technologies to exploit future commercialization was also required. Safety was the most important requirement, but consideration of the economic aspect was also a factor.

Since the policy stipulation that "construction should start within five years" was the most severe restriction for the actual design, the top priority was placed on the substantiation of design and conformity with the licensing requirements. This meant that Monju design should use achievable technologies and be based on R&D results. Actually, these restrictions caused technical limitations for the Monju design.

To promote the development, Monju was designated a national project, and a large part of the development budget came from the national budget. Large-scale R&D test facilities were built, and many research organizations joined in the Monju project with their superior technologies. In addition, the five big nuclear manufacturing groups capable of constructing a nuclear power plant were asked to join in the project so that technical information obtained through R&D could be shared among the participants. From the perspective of the development of independent domestic technologies, the materials used were limited to domestic products, and as a matter of course, neither foreign technology nor technical cooperation with foreign organizations was involved.

To build an effective development system, a new main development organization was established with the participation of many excellent engineers from related agencies and test facilities to promote the Monju project with the total power of Japan. However, the main contractor, the sponsor usually selected for a big project, was not designated; therefore, the main development organization had full responsibility for the project. The procedure used was as follows: the main organization held a design meeting with attendance of all project members, led discussions based on the results designed by the members, made design decisions after discussion, and determined the next design subjects and the responsible members. At the initial stage, since the discussion points were related to the whole plant, more than one member was designated as responsible for design. According to design progress, however, a single member was selected for a particular design work and had detailed discussions with all members. In addition, a main contractor was not appointed after construction started. Instead, an organization, named the FBR Engineering Company, was established by some major members for coordinating the schedule, technology adjustment, and communication with the main development organization. It made great efforts from the design to construction stages.

R&D activities were conducted in good harmonization with design work, large sodium test facilities were built, and many engineers from the relevant organizations were asked to join in the tests. The responsibility of determining the fundamental policy on FR plant design, rules and standards on structural feasibility and materials, the policy of safety, and the methods for safety evaluation was given to the main organization for collective examination and effective advancement. When determining the policy, rules, and standards, they were named, respectively, using either "... of FBRs" for all FBR plants or "... of the prototype reactor" exclusively for Monju. To study a commercialized fast reactor, a large-sized power plant was designed on a trial basis using the design technology for Monju to discuss the applicability of the design technology and identify the problems to be solved.

10.1.3 History of Monju Design

The history of Monju design is shown in Fig. 10.2. Design steps in the initial stage include the determination by development policy, the decision of basic concept, the selection of fundamental specifications for prototype reactor, and the investigation of concepts and specifications of reactor and main cooling systems. Following these steps, the survey study of core nuclear and thermal characteristics, the adjustment study of whole plant system, the conceptual study of a large power plant, and the integration of conceptual study results and licensable design for getting permission by regulatory authority were performed [1].

The safety examination for application for reactor establishment license was started by the regulatory authority in December 1980, and design adjustment was made with the progress of the examination.

Step 1	Determination by development policy
Step 2	Decision of basic concept
Step 3	Selection of fundamental specifications for prototype reactor
Step 4	Investigation of concepts and specifications of reactor and main cooling systems
Step 5	Surveillance study of nuclear core and thermal characteristics
Step 6	Adjustment study of whole plant system
Step 7	Conceptual study of a large power plant
Step 8	Intergration of conceptual study results
Step 9	Licensable design for getting permission by regulatory authority

Fig. 10.2 History of Monju design

However, the start of construction activities was delayed due to time-consuming preparation activities, such as obtaining permission from local authorities. As a result, the policy stipulation that "construction should start within five years" was not satisfied.

The abovementioned design steps were reviewed more than twice per year for intensive discussions and cautious investigations.

10.1.4 Relation Between the Flow to Decide Main Specifications and Design Thinking Process

The relation between the flow to decide main specifications and design thinking process is shown in Fig. 10.3.

The figure shows that, in the process from step 1 to step 4, design was performed following the design thinking process, namely, the function, mechanism, and structure, while detailed structural design was performed in parallel in each step. It is also shown that the design process from step 5 to step 8 also followed the design thinking process as a whole.

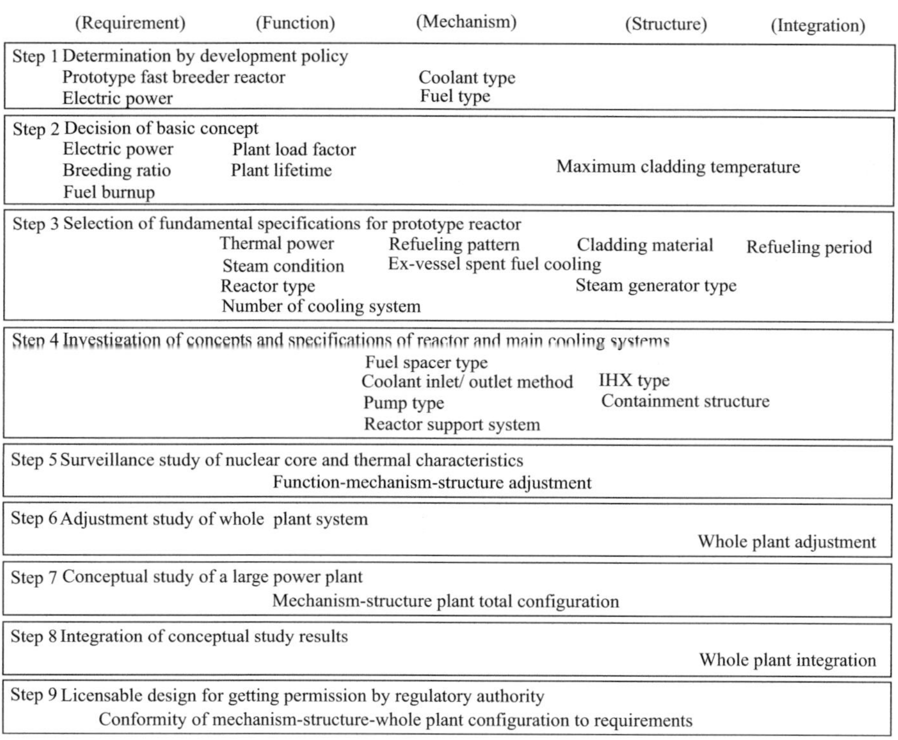

Fig. 10.3 Relation between the flow to decide main specifications and design thinking process

But, the actual design procedure was not as simple as shown in the figure. Some of the determined specifications were often revised according to the design progress, and feedback and feedforward methods were often used for assuming and correcting the specifications, respectively.

The next section describes why the main design specifications were determined in the respective steps, how the specifications were determined, and what kind of problems were faced in these steps. It also describes what changes were made in specifications in the design process.

Moreover, the determined specifications are compared with those of other FR plants.

The author hopes that readers can better understand how to perform a system design process by learning the Monju design process described in this chapter.

10.2 Flow of Monju Conceptual Design

10.2.1 Design Step 1: Philosophy

At first, in step 1, the specifications determined by development policy are shown in Table 10.1. These are the philosophy for FR development and the functions required in design.

There is no intention to discuss this philosophy in this chapter. For reference, the power of prototype FRs is shown in Table 10.2, and the fuel and coolant types of all FR plants are shown in Figs. 10.4 and 10.5, respectively. In the figures, Monju data are shown by large circles. The electric power of Monju plotted in the figure is 280 MW, which represents a change from the initial value. Uranium and plutonium mixed oxide fuel is used in many plants, and liquid metal sodium is used for coolant in almost all plants.

10.2.2 Design Step 2: Function

Next, the basic plant concepts determined in step 2 are shown in Table 10.3. The basic concepts represent the power of the plant and are aimed at high thermal efficiency, fuel burnup, and breeding ratio, namely, the major characteristics of FR. In addition, the plant load factor was decided, and the maximum fuel cladding temperature was set up from the perspective of the balance between heat generation and removal functions.

Table 10.1 Step 1 specifications determined by development policy

Step 1 determination by development policy	
Reactor type	Prototype fast breeder reactor
Electric power	250–300 MW
Fuel	Pu-U mixed oxide
Coolant	Liquid metal sodium

Table 10.2 Power of prototype-class fast reactors

Plant	Nation	Electric power (MWe)	Thermal power (MWt)
Phenix	France	255	563
SNR-300	Germany	327	762
PBFR	India	500	1250
Monju	Japan	280	714
PFR	UK	250	670
CRBRP	USA	380	975
BN-350	Russia	130	750
BN-600	Russia	600	1470
ALMR (PRISM)	USA	303	840
KALIMER-105	Korea	162	392
SVBR-75/100	Russia	80	265
BREST-OD-300	Russia	300	700

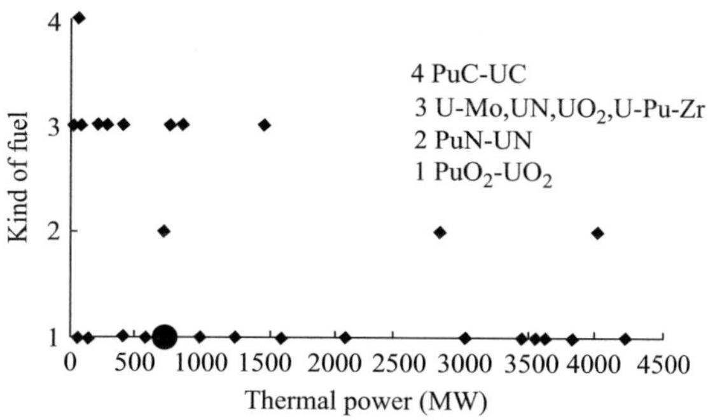

Fig. 10.4 Fuel type

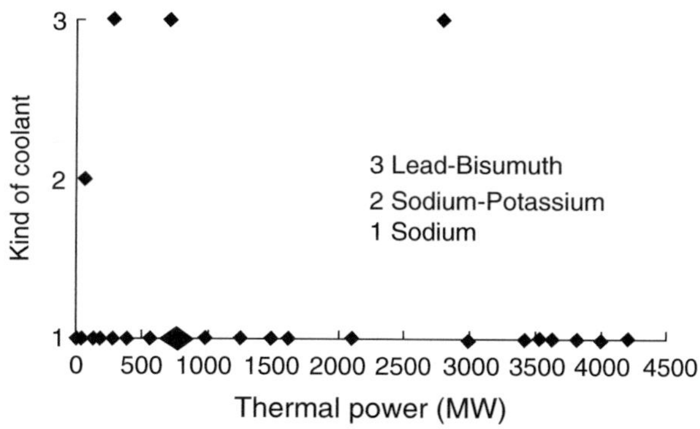

Fig. 10.5 Coolant material

Table 10.3 Step 2 decided basic concepts

Step 2 decision of basic concept	
Electric power	Approx. 300 MW
Fuel burnup	100 GWd/t
Maximum cladding temperature	Below 700 °C
Breeding ratio	Over 1.2
Plant load factor	80%
Plant lifetime	30 years

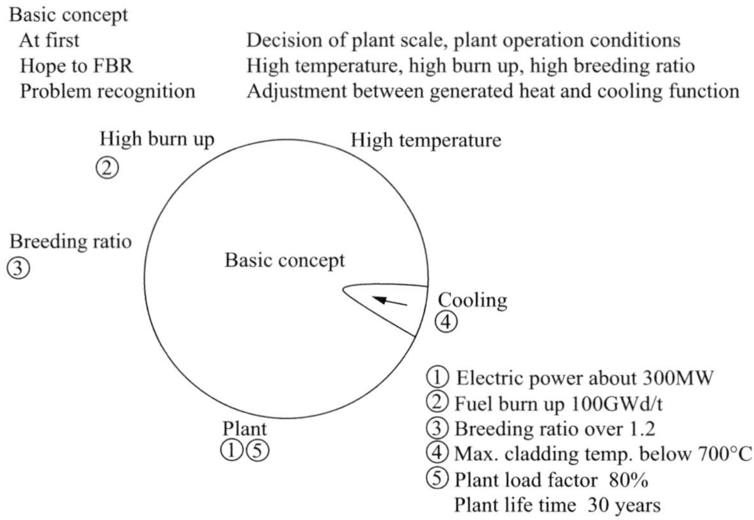

Fig. 10.6 Design step 2

The basic concepts are represented by a circle, on which the decided major concepts, namely, high system temperature, high burnup, high breeding ratio, and major plant functions are written using circled numbers. This is called a design progress model and shown in Fig. 10.6. Using the model, design progress was shown by reducing the radius in the direction of the respective concepts.

The relationships between the fuel burnup, breeding ratio, and maximum cladding temperature between the electric power, which were referred to from the [2], are shown in Figs. 10.7, 10.8 and 10.9, respectively. In these figures, Monju data are shown with large circles.

The burnup of Monju was changed to 80 GWd/t from the initial value, 100 GWd/t, which is an ambitious target even now. The breeding ratio of 1.2 was very high compared with that of other fast reactors. The maximum cladding temperature of 700 degrees Celsius was very high; however, it was changed by detailed design to 675 degrees Celsius of the maximum cladding tube wall center temperature.

Anyway, it is obvious that the basic concepts decided in step 2 were very high targets even from the present viewpoint, suggesting a strong attitude toward development with ambitious targets.

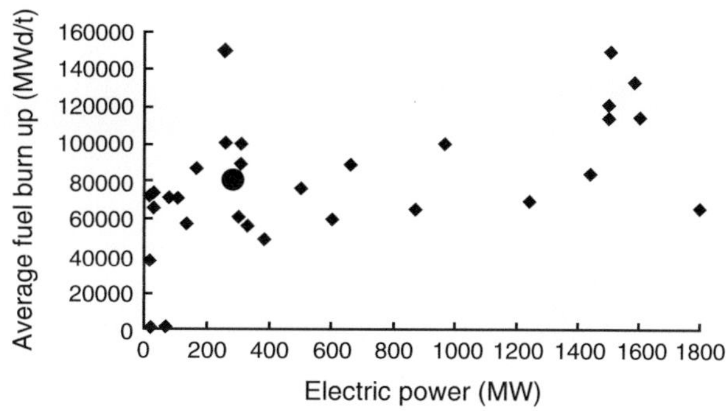

Fig. 10.7 Relationship between fuel burnup and electric power

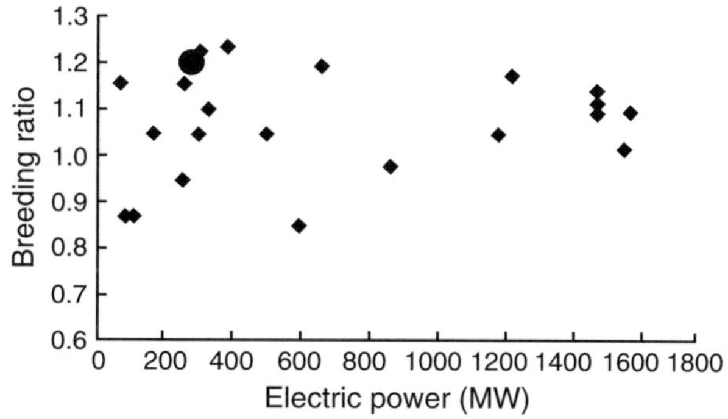

Fig. 10.8 Relationship between breeding ratio and electric power

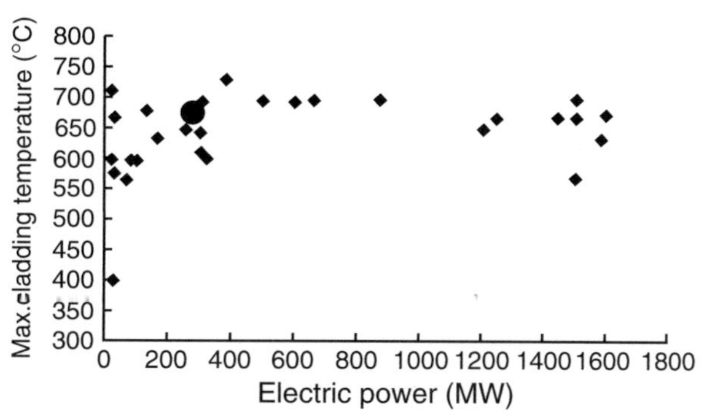

Fig. 10.9 Relationship between maximum cladding temperature and electric power

10.2.3 Design Step 3: From Function to Mechanism

In step 3, suitable characteristics for prototype reactor were investigated, and the selected specifications are shown in Table 10.4.

Step 3 produced a great advance in design from function to mechanism and partly to structure. The design progress model of this step is shown in Fig. 10.10, showing the reduction in radii of the respective concept directions and the concentration from circumference to the circle center.

Table 10.4 Step 3 selected specifications

Step 3 selection of fundamental specifications for prototype reactor	
Thermal power	750 MW
Cladding material	316SS
Steam condition	169 kg/cm^2, 510/510 °C reheating cycle
Reactor type	Loop type
Number of cooling systems	Three loops
Location of primary pumps	Cold leg
Steam generator tube type	Helically coiled, one through, non-separated
Refueling method	Single rotating plug + fixed arm pantograph
Cooling of spent fuel	Ex-vessel storage and cooling
Refueling period	6 months

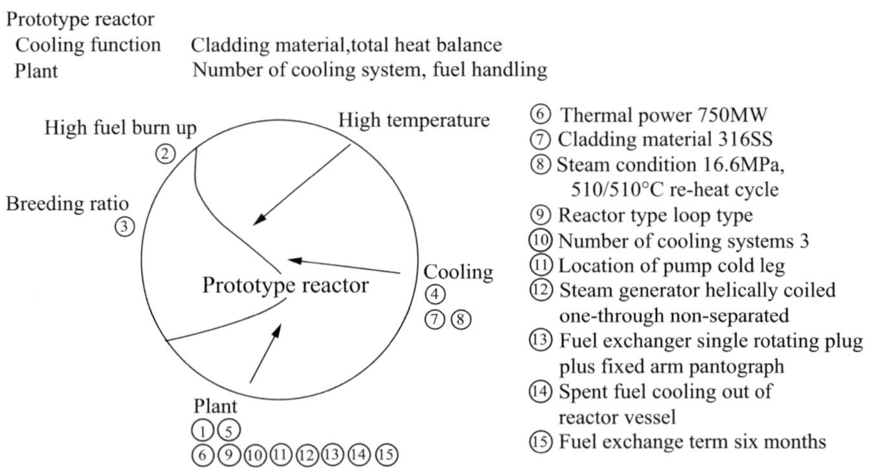

Fig. 10.10 Design step 3

It is evaluated that step 3 was of great importance and was valuable for advancing the design.

In this step, discussion was made on what characteristics would be suitable for the prototype reactor because the selection of characteristics was needed following the basic concepts decided in the previous step to determine the specifications of the prototype reactor. In order to improve the balance between heat generation and removal functions and clarify the features of the prototype reactor, the reactor type; the number of cooling loops; the position of primary circulation pumps; the type of steam generator, as the most important component; the refueling method that has an important relationship with the reactor; and the philosophy of plant operation were determined.

These items were selected as important for both prototype reactor development and future commercialization. The then design report, however, showed that the

Fig. 10.11 Relationship between thermal power and electric power

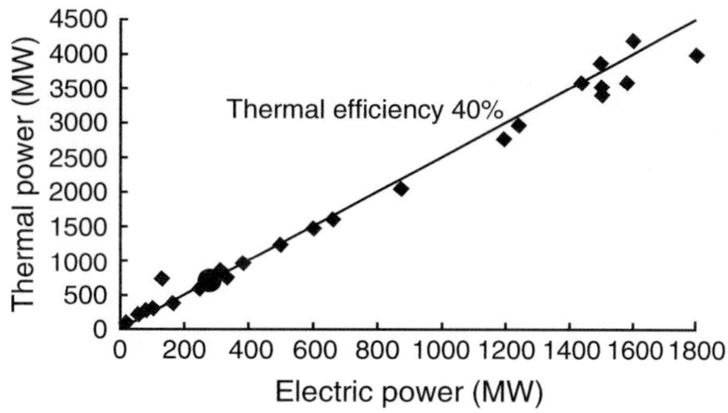

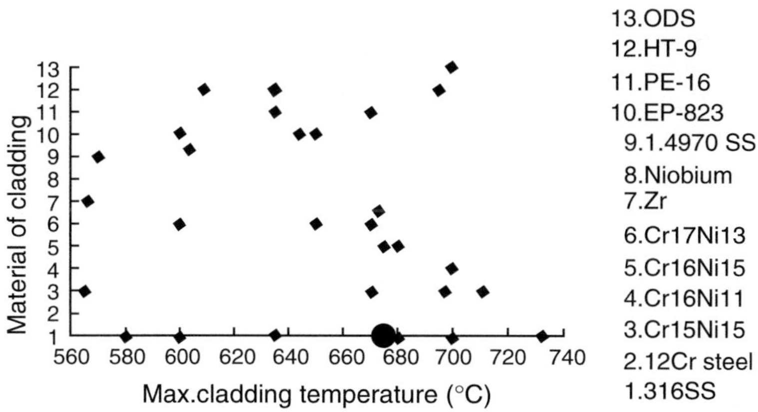

Fig. 10.12 Relationship between cladding material and maximum cladding temperature

refueling period of 6 months remained a problem for designers from the perspective of long-term operation.

The design progress model of this step shows the reduction of the corresponding radii and the approach to the circle center. It also shows that both the breeding ratio and fuel burnup were determined as a result of design progress.

The relationships between the thermal power, cladding material, turbine inlet steam conditions, the reactor type, the number of primary cooling loops, the position of the primary coolant pumps, type of steam generator heat transfer tubes, refueling method and period, and the electric power, which were referred to from the IAEA data book [2], are shown in Figs. 10.11, 10.12, 10.13, 10.14, 10.15, 10.16, 10.17, 10.18 and 10.19, respectively. In these figures, Monju data are shown with large circles.

The ratio between electric power and thermal power is the thermal efficiency. A straight line shown in Fig. 10.11 represents the thermal efficiency of 40%. The thermal efficiency of Monju is approximately 40%, a level similar to that of other plants.

Fig. 10.13 Relationship between turbine inlet steam pressure and temperature

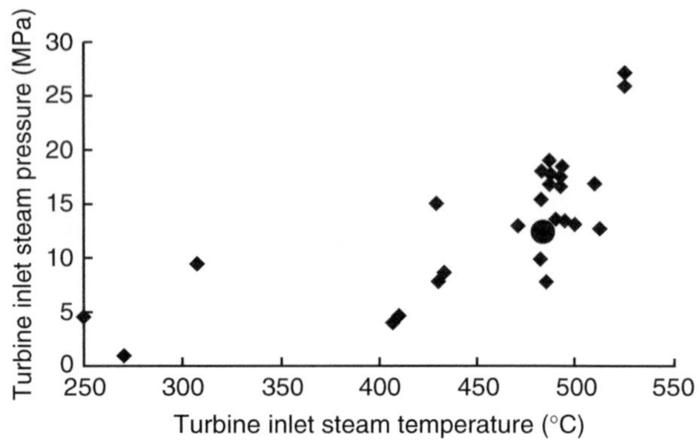

Fig. 10.14 Relationship between reactor type and electric power

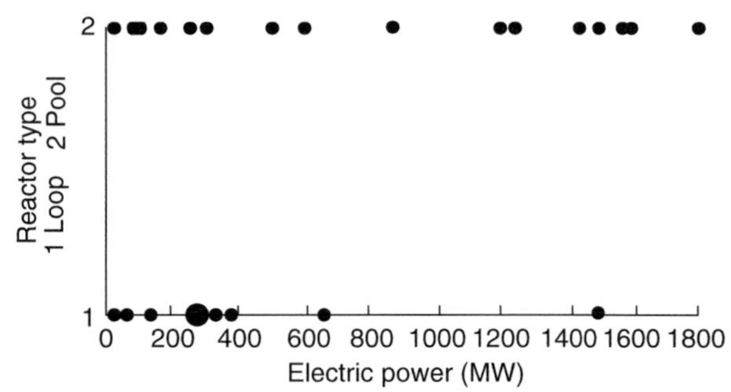

Fig. 10.15 Relationship between the number of main cooling loops and electric power

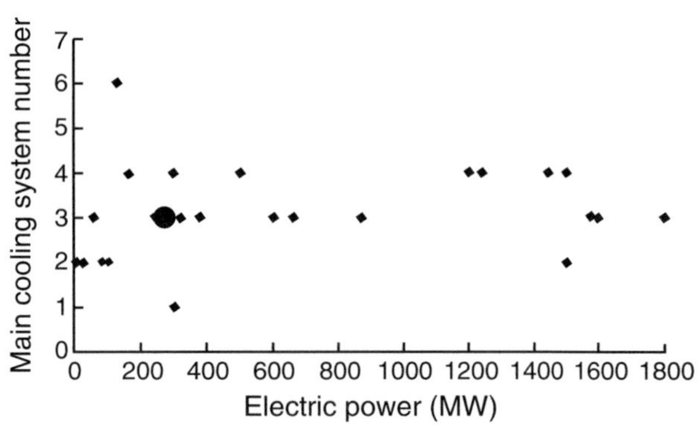

Fig. 10.16 Relationship between the location of primary pump and electric power

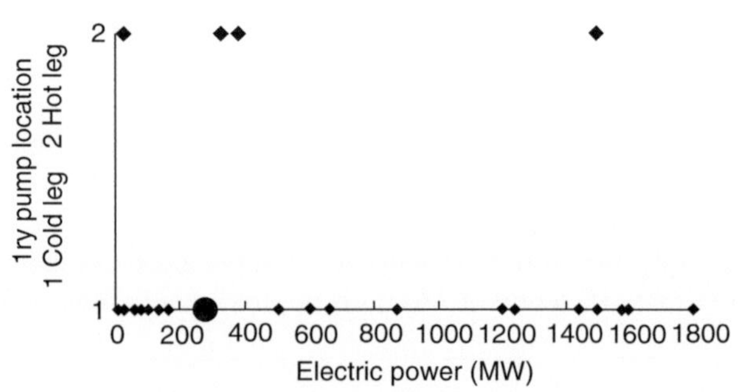

Fig. 10.17 Relationship between steam generator tube type and electric power

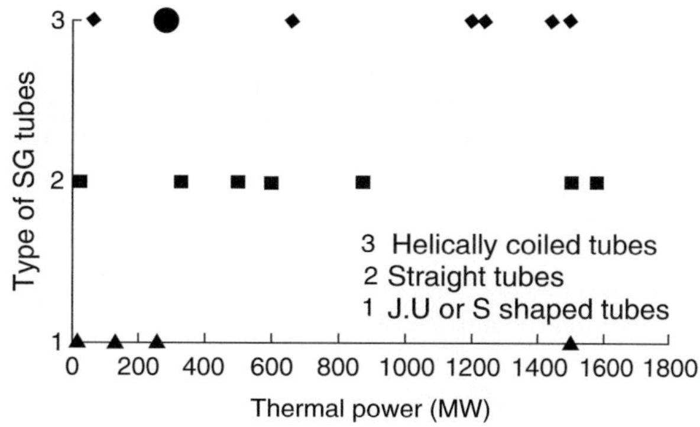

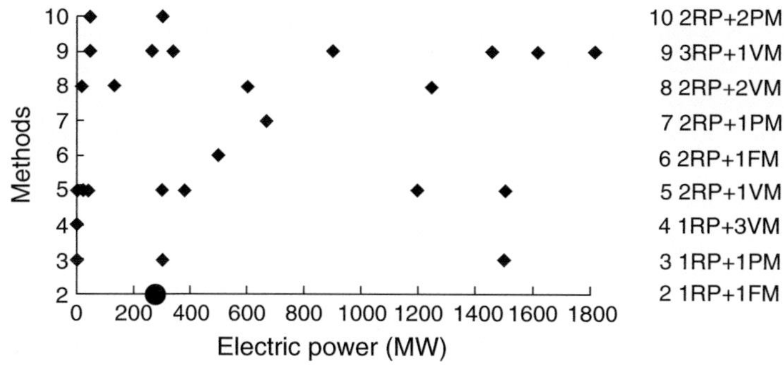

Fig. 10.18 Relationship between refueling method and electric power

Fig. 10.19 Relationship between refueling period and electric power

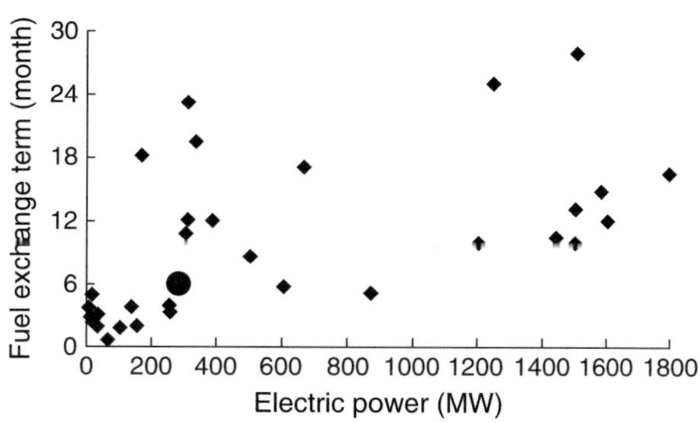

The relationship between maximum cladding temperature and cladding material is shown in Fig. 10.12. The cladding material used for Monju is 316SS, and the figure shows this material can be used at high temperatures. Although 12Cr steel was used, no maximum cladding temperature is found in the IAEA data book [2].

The steam turbine condition, i.e., the relationship between turbine inlet temperature and pressure, is shown in Fig. 10.13. The turbine for Monju was selected from a series of existing turbines for the commercial thermal power plants because the development of a new turbine would have required an extended period of time. As a result, the inlet steam pressure is somewhat lower, while the temperature is higher than the average values.

The relationship between electric power and reactor type is shown in Fig. 10.14. In the Monju design, a loop-type reactor was selected, because of the ease of incorporating the R&D results for individual components that are separately installed in the loop-type reactor. The loop-type reactor is often adopted for prototype fast reactors having a low or middle power level.

The relationship between electric power and the number of main cooling systems is shown in Fig. 10.15. Since Monju uses the primary main cooling system also for decay heat removal, it was designed to have three main cooling loops to ensure heat removal function at all times. However, a four-loop system was also investigated for future commercialization.

The relationship between electric power and the location of primary cooling pump is shown in Fig. 10.16. It was determined to install the Monju primary pumps at the cold leg because of insufficient time for testing hot shock on the pumps. It is believed that the Monju design is reasonable in consideration of the design of other fast reactors.

The relationship between electric power and the type of steam generator heat transfer tubes is shown in Fig. 10.17. The decision to adopt helically coiled heat transfer tubes was an ambitious idea from the viewpoint of compact structure and high heat transfer characteristics. However, this decision was considered suitable because R&D results were incorporated and manufacturing technologies were innovated. Considering that some plants had afterward adopted this type, it is evaluated that the Monju decision played a leading role in this technical field.

The relationship between electric power and refueling mechanisms is shown in Fig. 10.18. In this figure, RP stands for rotating plug, FM pantograph method with fixed arm, PM pantograph with variable arm, and VM vertical movable mechanism, respectively. Monju adopted the single RP plus fixed arm pantograph method. The figure shows that various types of this combination (RP+FM) are used.

Monju adopted an ex-vessel system for cooling spent fuel different from that installed for Joyo to reduce the reactor vessel diameter.

The relationship between electric power and refueling period is shown in Fig. 10.19. The design book states that the Monju refueling period, 6 months, was a reluctant decision in consideration of the relationship between fuel burnup and refueling period of such a small-sized reactor.

10.2.4 Design Step 4 : From Mechanism to Structure

In step 4, the design of core and major components of plant main circuits was investigated, and the determined specifications are shown in Table 10.5. In this step, design progressed from mechanism to structure.

Table 10.5 Step 4 determined specifications

Step 4 investigation of concepts and specifications of reactor and main cooling systems	
Fuel spacer type	Wire-wrapped fuel spacer
Reactor support system	Core side support
Coolant inlet/outlet method	Lower inlet/upper outlet flow
Pump type	Mechanical vertical type/overflow
Intermediate heat exchanger type	Parallel straight tubes/counter flow
Containment structure	Vertical self-standing pressure-resistant steel container

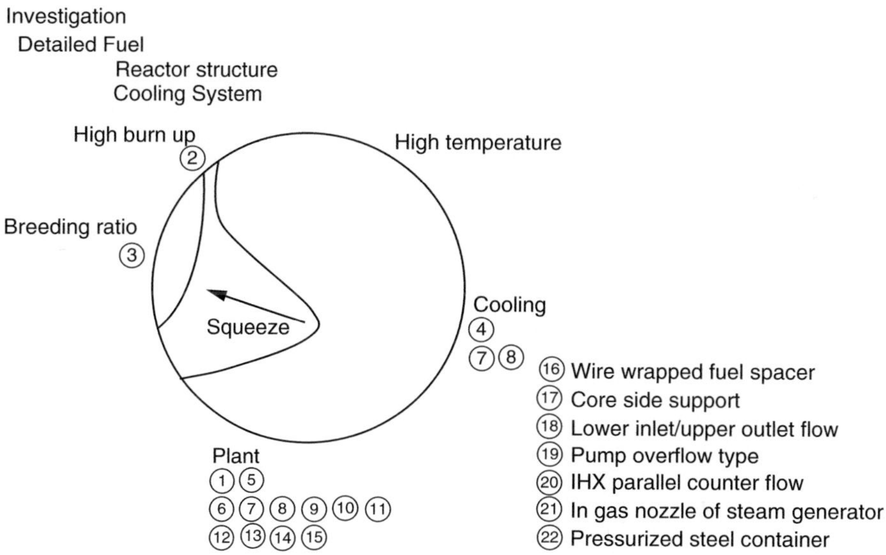

Fig. 10.20 Design step 4

The design progress model of this step is shown in Fig. 10.20, and it shows the concentration from circumference to the circle center.

In this step, many specifications of mechanisms and structure were determined, and it was expected that the two most representative specifications, high burnup and high breeding ratio, were close to the final results.

10.3 Research and Development (R&D) Activities for Monju

For development of an advanced reactor, R&D activities are planned and performed to surely reflect the results in the reactor design. From the perspective of this relationship, R&D can be considered a part of design. The Monju R&D activities are described below from the design perspective.

10.3.1 Design and R&D

It is difficult to harmonize design and R&D. Design proceeds assuming the results of R&D, while R&D produces results after design progress. Accordingly, design should be modified using the results of R&D. For this reason, design should be made assuming R&D results in advance. In the Monju project, the slogan "design and R&D are two halves of the same whole" was used to enhance harmonization between design and R&D. In fact, the Monju design was confirmed by the R&D results, while many modifications were made to by harmonizing the design and R&D results.

10.3.2 Important R&D Items

Since the FR development project was the first experience, great hope was placed on the R&D results. However, due to limited budget and time, the R&D items were significantly limited.

Accordingly, focus was placed on fundamentally important R&D items, including the experimental reactor for developing the fundamental technologies and testing the reactor characteristics, major sodium components testing for dynamic characteristics, steam generator testing for thermal characteristics and safety behavior, plant safety tests including nuclear safety, and material and structure tests to obtain data for establishing standards. R&D activities for these items are described below.

First, the scale of R&D tests is described. It is practically difficult to perform tests using the same scale model as the plant components from technical and economic aspects. In addition, it is not always necessary to reflect the full-scale test results to design. Therefore, various test scales were applied as follows: for coolant circulating pumps, tests using a full-scale model, modified impellor model, and scale-down model were available. For heat exchangers, tests using a scaled-down model, partial structure model, and a model having the minimum heat capacity of automatic controlled operation range were available. For theoretical extension methods, tests using a partially scaled-up model and a model considering nondimensional numbers were available. In any case, since the possibility of reflecting the test results to design was the most important, sufficient consideration is required before planning R&D tests and performing works for obtaining practical data.

1. Experimental reactor

In the experimental fast reactor Joyo, the preceding reactor in Japan's FR development program, development was made on fundamental fast reactor technologies, such as nuclear characteristics including fuel behaviors, basic sodium technology, core heat removal characteristics, and major components, like refueling machines, intermediate heat exchangers, and sodium circulating pumps. In addition, design methods and safety evaluation technologies specific to FRs were developed.

Joyo construction started in 1970 after completion of the design. The experiences obtained in Joyo were reflected step by step to the Monju design following the progress of Joyo activities. Moreover, since Monju design took longer than expected, the operating experiences of Joyo were also reflected to it. In addition, troubles experienced in Joyo were rapidly reflected to the Monju design.

It is needless to say that the reflection of Joyo design and experiences to the Monju design and safety examination was required. However, since there were some differences in the design standards and design evaluation technology between Joyo and Monju due to their different features, the comparison between them was an important consideration in the Monju design.

Considering these aspects, it is thought that there was a steady succession of technology from the experimental to prototype reactors.

2. Main components

R&D activities focused on main components. Sodium circulation pumps, which are dynamic components used under sodium, were tested following the testing of Joyo pumps. Control rods were tested from the viewpoint of the reliability of the drive mechanism. Steam generators, from a small-sized model to one fifth scale model, were tested. Since the hydraulic properties of water (i.e., specific gravity and viscosity) are similar to sodium, hydrodynamic tests for sodium systems were performed using water. While there were test results on heat transfer characteristics obtained at a small system model designed for academic research, sodium thermal characteristics tests were conducted with sodium because the Prandtl number differs between sodium and water.

When a number of ideas for candidate structures were proposed, a new scheme for selecting the component structure was introduced to overcome the shortage of time and budget. In this scheme, a reference model was specified as the design target, the alternative was examined to provide equivalent characteristics and functions, and a backup model was prepared to fabricate the component using conventional technologies when all candidate structures could not be adopted. The actual structure that was determined by this scheme included the type of steam generator heat transfer tubes, the primary cooling system piping, the spacer of fuel elements, the coolant flow scheme in the intermediate heat exchanger, the support structure of the reactor vessel, the fuel transfer mechanism from reactor to storage, and the location for cooling spent fuel after withdrawal from the core.

3. Safety

The evaluation of FR plant safety requires a variety of information, including the conditions for jeopardizing integrity, initiating events leading to an unusual state or accident, the scale of rupture, detection function for accidental events, the propagation of an accident, and accident propagation prevention systems.

Such information can be obtained from the R&D results, by selecting similar data from past information, or it can be assumed based on reliable technologies and highly certain consideration. However, since information concerning nuclear safety depends strongly on test conditions and facilities, the relevant tests were performed using foreign test facilities. In addition, since trouble data, such as the component failure rate, were collected worldwide and shared, it was possible to utilize these data in design.

Using the above information, a series of accidents were selected for safety evaluation in consideration of the initiating events and their degrees as well as the methods of preventing propagation of accidents.

On the other hand, it was very difficult to comply with the regulation titled "Philosophy in Safety Evaluation of Fast Breeder Reactors" (in Japanese) that was provided by the Nuclear Safety Commission and published just before submission of the application for Monju construction permission. Article 5 of the regulation states "Events for which the probability of occurrence is lower than the 'Accident' that is defined in (2.2) while the consequence is more significant should be sufficiently investigated with reference to the initiating events and the protection measures against the successive events progress, taking into account that the operating experience of LMFBR is scarce, and thereby, appropriate limitation of radioactivity release should be ensured." To satisfy this article, various phenomena, such as abnormal states of the core, guillotine rupture of large-diameter piping, and the loss of reactivity control functions, were selected and evaluated in a reasonable manner.

4. Material and structure

In the Monju design, since new materials and operating conditions that differ from those of previous reactors were required, the current standards and rules were not applicable. To solve these problems, it was necessary to clarify the principle of design and establish new standards before submission of the application for Monju construction permission.

For this purpose, R&D activities were performed with the participation of many professionals from many organizations, namely, test organizations, universities, and private companies, and many universal standards were established based on the data obtained by the activities. These standards were discussed, evaluated, and approved at meetings by many specialists of universities, etc. Significant efforts were made for these processes.

5. Fuel

Development of FR fuel fabrication technology was conducted as one of the major R&D items for the main development organization (i.e., PNC) in addition to the development of spent fuel reprocessing technology.

Development of fuel requires handling techniques and the characterization of uranium-plutonium oxide, pellet fabrication technique using a pressing sintering process, the development of cladding material, fuel element assembly method, fabrication of wrapper tubes, and fuel subassembly assembly method. In addition, it was necessary to perform testing of fuel elements and subassemblies under similar nuclear and thermal conditions in the core. Furthermore, irradiation tests and safety confirmation tests, including those related to fuel rupture, were very important; however, it was very difficult to find a reactor in which such tests could be appropriately performed.

Normally, it takes approximately 20 years to develop new fuel, but the development of Monju fuel was based on the technical information accumulated over a long period by the Nuclear Fuel Corporation, the forerunner of the main development organization. Since the number of Monju fuel pellets is approximately four million, they are products of mass production. From this viewpoint, the adjustment of accuracy of the products was needed to meet both requirements of detailed reactor physics design and industrial design of the pellet fabrication process.

10.4 Monju Design Process Flow and Adjustment of Specifications

10.4.1 Design Process Flow of Each System

A Monju design process flow is summarized below.

The fuel linear power density was analyzed from the geometrical configuration of fuel and coolant and their volume ratios as a specification to be used for reactor physics analysis, heat generation, transportation analysis, and plant design.

In the Monju design, some specifications were not changed from the values selected in the initial stage, while other specifications were changed through the design progress, due to newly established standards, or by reflecting the results of R&D. These processes can be seen in the design flow described below:

1. Reactor design flow

The reactor design flow for Monju is shown in Fig. 10.21.

Using the type of fuel and coolant specified by the policy, analyses of criticality and neutron flux distribution were performed after calculation of the volume ratios of fuel and coolant in the core, and analysis of fuel linear power density was performed by calculating heat generation in the fuel elements and heat conduction through the fuel pellets and cladding material. Adjustment between the fuel linear power density and the coolant flow rate distribution was made for balanced heat removal from fuel under design limitations of fuel, cladding, and coolant.

In this figure, a hollow circle, shown in the left of each specification, means specifications that were not changed from the initial values, while a solid circle shows specifications changed according to the design progress. For example, there are many solid circles after the heat generation process in Fig. 10.21. This shows

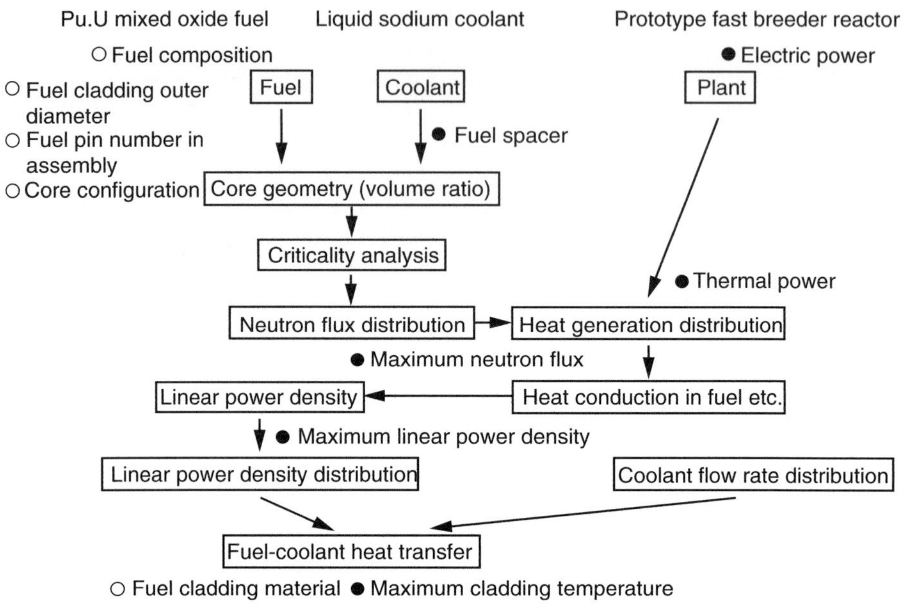

Fig. 10.21 Flow of reactor design

that modification and adjustment were performed iteratively in the heat removal and transport analysis processes, namely, the most important part of an FR design.

As shown in this and the subsequent figures, modifications were made for many parameters according to the design progress and feedback processes. This suggests that the design flow was neither simple nor straightforward.

2. Main cooling system design flow

The main cooling system design flow for Monju is shown in Fig. 10.22. Hollow and solid circles are used in the same way as in Fig. 10.21.

As is understood from the figure, analysis of cooling characteristics was performed using the fuel linear power density estimated in the reactor design flow; however, the design was adjusted in consideration of the R&D results on heat exchangers. In addition, considering the total plant characteristics, the water-steam cycle properties were changed to optimize the plant heat balance.

3. Fuel handling system design flow

In the Monju design, the fuel handling system design, together with the reactor and cooling system design, is very important from the viewpoint of plant safety and operation. The fuel handling system design flow for Monju is shown in Fig. 10.23.

The fuel handling system design flow includes new fuel delivery to the plant, fuel burning in the core, decay heat removal from fuel, long-term storage in sodium and water, and carrying out of spent fuel from the plant. The most important activity

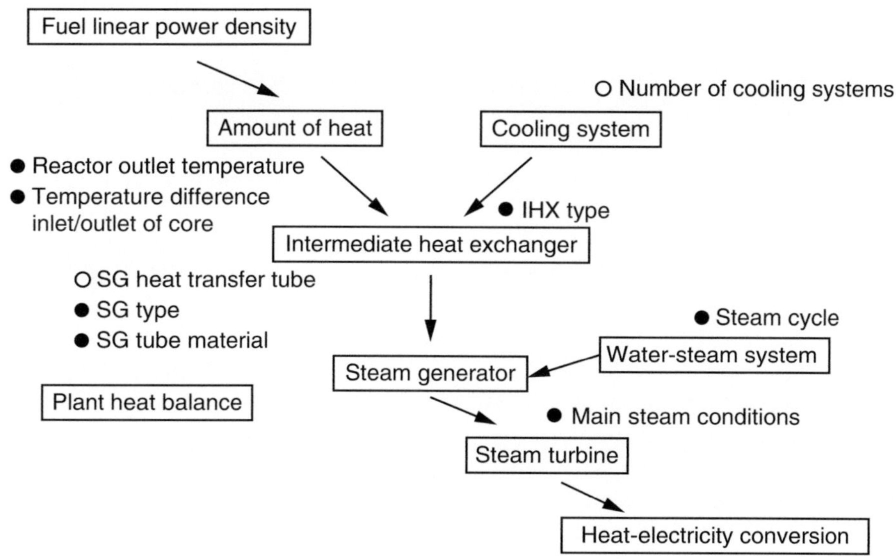

Fig. 10.22 Flow of cooling system design

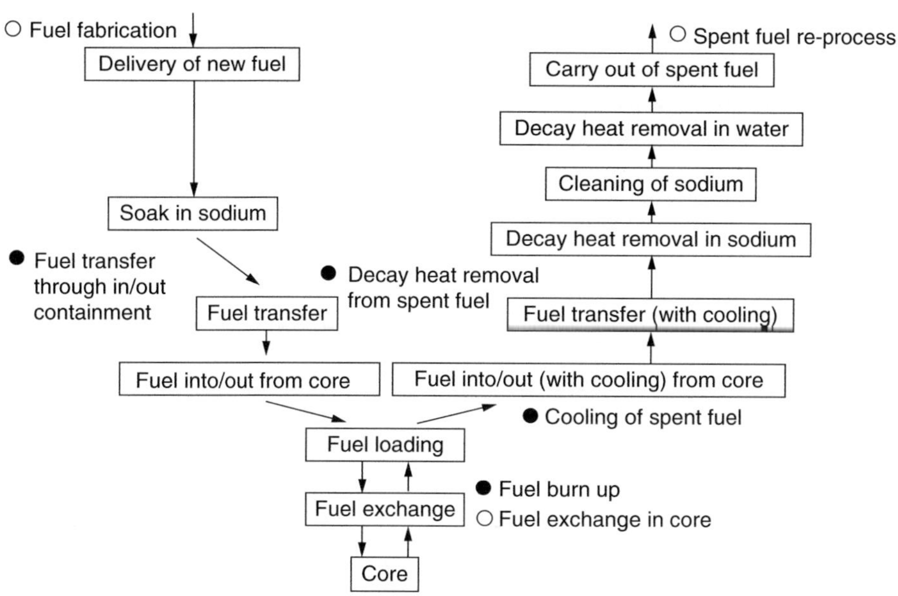

Fig. 10.23 Flow of fuel handling system design

in this flow is the refueling in the core. Refueling requires accurate automatic functions for the insertion of new fuel and drawing out spent fuel from an arbitrary position in the core; however, these functions are difficult to provide.

Accurate and rapid fuel transfer under sodium, angle adjustment by rotating hexagonal fuel subassemblies, and latching and de-latching motions of fuel handling machine are essential functions of the fuel handling system. In addition, the fuel handling machine should be kept outside the core during plant operation. To allow for these functions, the fuel handling machine is required to move in sodium and the cover gas at high temperatures under highly radioactive conditions. Moreover, these functions must be guaranteed continuously from the initial fuel loading into the core to the final state where all spent fuels would be removed from the core for decommissioning. Proven technology and/or R&D test results must be applied to satisfy these functions. In the Monju design, proven technologies were used from the viewpoint of reliability.

Another important subject is where spent fuel should be stored for decay heat removal. In the Monju design (loop-type reactor), the surroundings of the core and a vessel located outside the reactor vessel were compared. The latter location was adopted from the viewpoint of the smaller reactor vessel, the availability of spent fuel transfer, the possibility of handling fuel during plant operation, and the availability of space for installing an ex-core vessel. On the other hand, many pool-type fast reactors adopted the former location because sufficient space can be used in the reactor vessel.

New fuel subassemblies should be cleaned and soaked in liquid sodium before insertion into the core. From these processes to the end of decay heat removal, fuel subassemblies must be kept under sodium. To meet this requirement, the sodium level in the related components must be higher than twice the fuel length.

Cleaning of sodium attached to the fuel surface is performed by a continuous cleaning process using gas, gas-steam mixture, and water. This process was established by R&D.

10.4.2 Features of Design Flow and the Summary

The important features of FR design can be seen in the Monju design flow shown above. The design process and major features are shown in Fig. 10.24.

Criticality in thermal neutron reactors is achieved by appropriate geometrical arrangement of fuel and moderators, while in FR design, criticality is achieved by dense arrangement of fuel, and accordingly, a compact core is feasible in terms of reactor physics design. However, heat removal from the core is a very important problem for satisfying design limits.

Here, the total design flow is shown in Fig. 10.25. It was fortunate that computer technologies had been significantly developed by the time of total conceptual design of Monju was completed so that detailed analyses of structural integrity

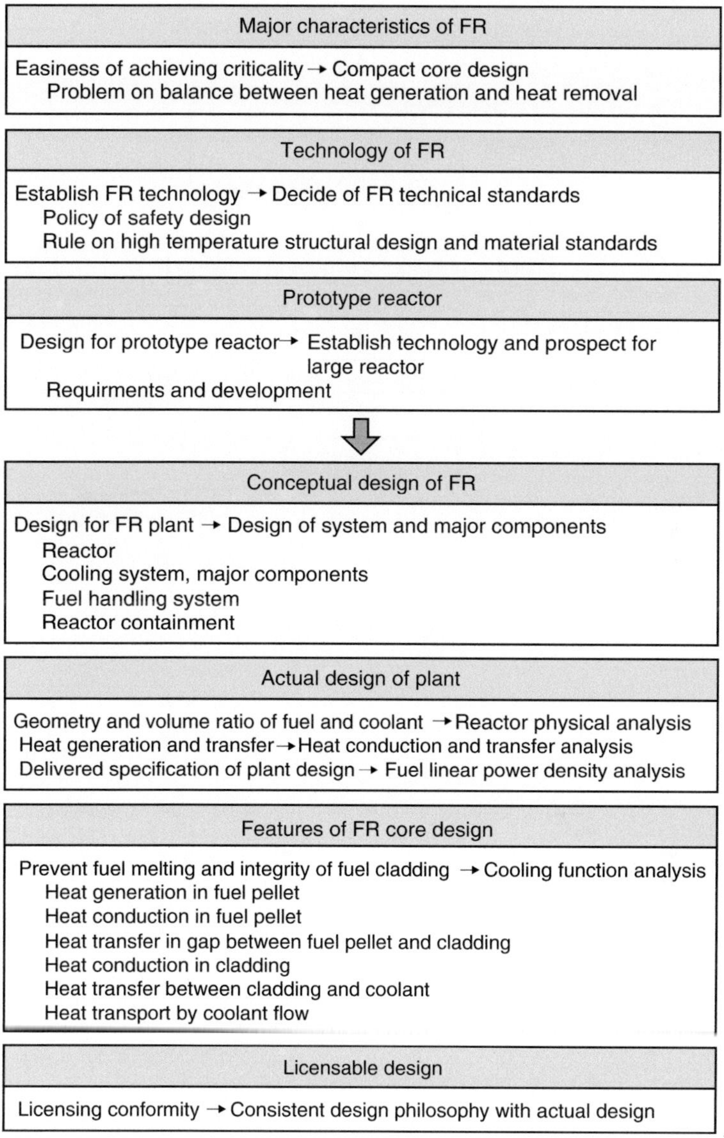

Fig. 10.24 Actual design process and features

and safety evaluation could be carried out. This made it possible for the Monju design to be developed by using new computer technologies.

The check and review shown in the figure was conducted through hearings by the Atomic Energy Commission to evaluate the Monju design. Significant efforts

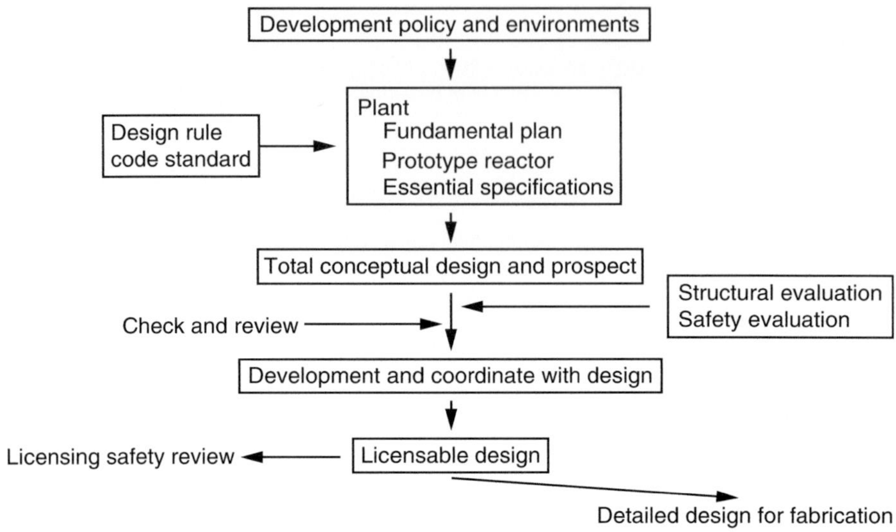

Fig. 10.25 Summary of total design flow

Table 10.6 Major specifications of Monju

Specifications	Value
Electric power	280 MW
Thermal power	714 MW
Outer diameter of fuel pellet	5.4 mm
Number of fuel pins in subassembly	169
Pu enrichment inner/outer	16/21%
Number of subassemblies	108/90
Number of control subassemblies	Fine 3 coarse 10 safety 6
Method of fuel exchange	Single rotating plug and fixed arm
Breeding ratio	Approx. 1.2
Type of reactor	Loop type
Number of main cooling system	3
Type of heat transfer tubes in SG	Helically coiled
Coolant temperature at reactor outlet	529 °C
Steam conditions at turbine inlet	127 kg/cm^2 g 483 °C

were needed to cope with the check and review; however, the successful completion of the review gave great confidence to the main development organization.

Main specifications of the Monju design are shown in Table 10.6, and a main cooling system diagram is shown in Fig. 10.26.

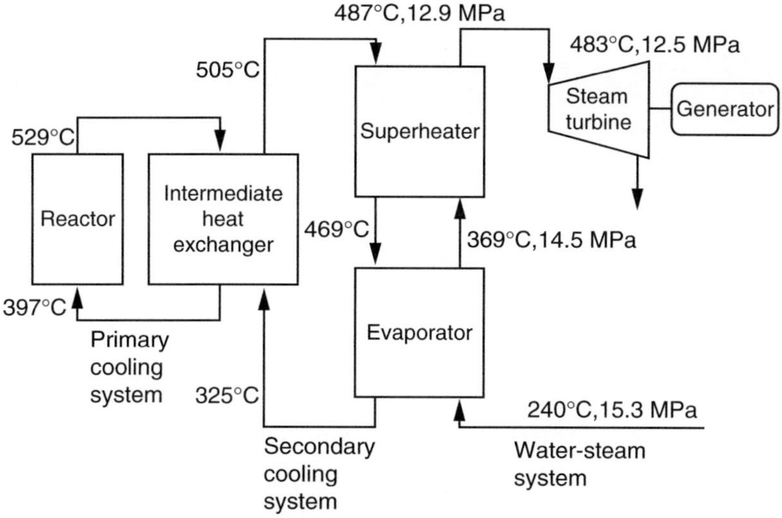

Fig. 10.26 Main cooling system diagram of Monju

10.5 Design Limits of Monju

Design limits are determined from constraints in the design step from mechanism to structure. Here, design limits specified in the Monju system design are described.

First, the meaning of design limits is described. The present concept of design limits is to specify the limit for a state or specification for preventing the possibility of jeopardizing the integrity of components and systems and for maintaining the safety state by avoiding abnormal conditions that cannot be recovered to normal operating conditions. For these definitions, design limits are often considered independent; however, each design limit is actually related to other design limits, and the occurrence of a limited state may often cause other limited states.

For these reasons, it may be difficult to understand the meaning of each design limit and the relationship with other design limits. However, it is important to organize and categorize these design limits from the viewpoint of total plant design.

Main design limits for Monju are shown in Table 10.7; however, many other design limits were used in the actual Monju design.

10.6 Key Items in Monju Design

In this section, the establishment of standards and rules for design, the harmonization of heat generation and removal, and the confirmation of the functions "stop, cool and confine" are described as the typical and important problems in FR design. These key items are essential for comprehensive understanding of the FR design.

Table 10.7 Major design limits of Monju

Specifications	Design limits
1. Diameter of wire spacer (m)	≥0.0005
2. Height of core (m)	1.000≥
3. Number of cooling systems (–)	≥3
4. Coolant velocity at core center (m/s)	9.00≥
5. Steam temperature at turbine inlet (°C)	510≥
6. Evaporator feedwater temperature (°C)	250≥ ≥180
7. Pu enrichment inner and outer (–)	0.35≥
8. Breeding ratio (–)	≥1.05
9. Maximum neutron flux (n/m^2s)	6.07×10^{19}≥
10. Maximum linear power density (W/m)	36,000≥
11. Coolant temperature at reactor outlet (°C)	560≥
12. Secondary coolant temperature at steam generator outlet (°C)	Feedwater temp. + 85≥ ≥ feedwater temp
13. Maximum temperature of fuel (°C)	2350≥
14. Maximum temperature of cladding (°C)	675≥
15. Maximum temperature of coolant (°C)	659≥
16. Coolant temperature at reactor inlet (°C)	≥300

10.6.1 Policy of Safety Design

Although there was a regulatory guide for reviewing the safety assessment of light water power reactor facilities, there was no guide or standard for the safety of FR in those days. Therefore, it was an important task to develop safety design guidelines for FR. Then, investigations were performed for developing new safety guidelines by adding requirements for the special features of FR to the guidelines for light water reactors. The special features considered include the determination of boundaries in FR plants and their integrity, consideration of sodium and plutonium characteristics, the inherent safety of FR, nuclear and thermal characteristics, engineered safety facilities, and the installation of an intermediate cooling system. As a result, the safety guidelines for the liquid metal-cooled fast breeder reactor design were decided, and the conformity of Monju design with the new guidelines was clarified. It should be noted that these guidelines were authorized by the regulatory authority as the standard rule for the present and future FR design in the safety review process.

10.6.2 High-Temperature Structural Design Guide and Material Strength Standards

In the days of Monju design, since there was no guide or standard for FR structural design, designers were required to develop original ones using R&D data, especially in the area of the peculiar conditions for FR. The important parts include the policy to ensure structural integrity against elastic and inelastic creep under high-temperature and thermal transient conditions, ratchet, and creep fatigue due to repeated loading. A lot of effort was put into conducting many tests, analyses, and evaluations, and these results were provided at a meeting at which many specialists examined the results to perform evaluation. However, it should be noted that the developed guide and standards were authorized exclusively for the prototype FR, Monju.

10.6.3 Heat Generation and Removal Capabilities in Reactor

As emphasized above, one of the most important problems in FR design is the balance of heat generation and removal from the core. This problem was considered difficult in the early days of the Monju design. Specifically, the fuel pellet diameter, which followed the experience in Joyo, cladding material properties under high-temperature and high-radiation conditions, sodium hydrodynamics, and the limit of cooling capability were carefully investigated. These problems were examined in normal operating conditions as well as transient conditions in case of accidents. Major design limits were preventing fuel from melting, keeping the integrity of fuel cladding, and preventing coolant from boiling. Moreover the increase in the amount of heat generated due to the accumulation of plutonium in fuel and blanket regions was considered.

Since it was very difficult to examine all these problems in a comprehensive manner, it is thought reasonable that the balance between heat generation and removal capabilities was identified as the most important problem in early design stages.

10.6.4 Reliability of Reactor Shutdown System

Securing the reliability of the reactor shutdown system is one of the most important problems in FR design. In the Monju design, various investigations on this problem were performed to improve the reliability.

First of all, the dual reactor shutdown system was adopted. This system allows rapid insertion of control rods and full insertion of safety rods that are normally withdrawn and kept above the core during normal operation. Second, different structural and action mechanisms of fine, coarse, and safety rods were adopted to prevent the loss of functions by a common cause. Moreover, many performance tests were conducted using various prototypes before the actual components were manufactured to improve the reliability of the reactor shutdown system. Thanks to these activities, sufficient reliability of the system was achieved without using other methods, for example, a function other than inserting absorbers into the core that was only in a conceptual design stage.

10.6.5 Decay Heat Removal Function

Needless to say, it is one of the most important problems for FR design to ensure the decay heat removal function. The basic requirement to ensure this function is to keep coolant level sufficient for core cooling capability using the inherent coolant properties. That means maintaining coolant volume necessary for core cooling even in case of failure of a component or piping. Initially, double-walled piping was considered based on the experience in Joyo; however, the requirement was met by installing the primary system piping at high levels and guard vessels around the relevant components using new guidelines for high-temperature structures. In addition, the reliability of decay heat removal function was further improved by installing an auxiliary air-cooled heat exchanger in each cooling circuit. Actually, detailed investigation was performed to ensure the cooling capability and startup characteristics of the auxiliary heat exchangers, in particular, immediately after the occurrence of an accident.

Since the decay heat removal function is needed for a long time after reactor shutdown, long-term decay heat removal characteristics were investigated in consideration of a failure of cooling system components and external events that may occur in the long term.

10.6.6 Design of Containment Vessel

The type of containment vessel for Monju was determined by adopting a steel fabricated pressure vessel in the early design stage. However, its design conditions were not determined soon because of the delay in selecting the design accidents. Investigations were performed on the pressure change due to coolant leakage or core collapse, the negative pressure caused by weather conditions, and mechanical strength against seismic conditions. As a result, the design conditions were determined considering these phenomena to achieve as robust a structure as possible in a reasonable manner.

10.6.7 Function to Suppress Fire Due to Leaked Sodium

In case of sodium leakage, no fire would occur due to the presence of inert gas in the atmosphere surrounding the primary cooling system. In addition, hydrogen generation due to direct contact between concrete and leaked sodium would be prevented by the steel lining plate on the floor, wall, and ceiling. However, the secondary cooling system is surrounded by air for the rapid identification of sodium leakage and confirmation of sodium fire by white smoke produced by the reaction between sodium and air as well as for simplicity of operation and maintenance. In case of secondary sodium leakage, sodium is rapidly drained to limit the volume of leak. In case of small leakage, leaked sodium falls down on the steel lined floor, while in case of large leakage, leaked sodium flows down through the conducting piping system installed onto the lowest room floor that is covered by steel lining plate. When the leaked volume is significant, leaked sodium drops down to lower floors, and finally, leaked sodium gathers on the bottom floor. To prevent sodium fire on the bottom floor, fire suppression plates are provided above the surface of stored sodium. Moreover, each secondary cooling circuit can be isolated by fire protection doors.

10.7 Verification of Monju Design Technology

One of the remaining subjects in the Monju design is the verification of design technologies using technical information obtained through testing and operation of Monju.

However, it is obvious that design verification has not been sufficiently achieved. This is because the operation time is quite short, the rated power operation has not been achieved, and the economical evaluation of plant construction, operation, and maintenance has not been performed using actual operational information. The accumulation and succession of technical information concerning the design and operation are very important, and these subjects must be performed in the future.

In addition, although the target values for the breeding ratio, fuel burnup, and plant service lifetime were assumed in the early design stage, the validation of these target values will require quite a long time and plenty of operating experience. Since these target values are very important for development of FR, continuous efforts will be required for some time. Moreover, it is necessary to confirm the integrity, safety, and reliability and achieve economic evaluation for Monju to be globally recognized as a valuable FR plant.

Finally, the author earnestly hopes that the Monju design will be evaluated as a typical and secure FR design in the near future.

Further Readings

PNC TN2410 94–023, History of design/construction/test operation for Fast Breeder Prototype Reactor "Monju", Power Reactor and Nuclear Fuel Development Corporation

PNC-TN1340 84–003, Monju Report–Design of Monju, Donen Giho No. 51 (PNC Technical Review), Power Reactor and Nuclear Fuel Development Corporation (present JAEA), (1984)

PNC-TN1340 88–003, State of Fast Breeder Reactor Monju Construction, Donen Giho No. 67 (PNC Technical Review), Power Reactor and Nuclear Fuel Development Corporation (present JAEA), (1988)

PNC-TN1340 90–003, Fabrication of Monju Reactor Vessel, Donen Giho No. 75 (PNC Technical Review), Power Reactor and Nuclear Fuel Development Corporation (present JAEA) (1990)
PNC-TN1340 91–001, Operational Experience and Gained Results of Monju No. 1, Donen Giho No. 77 (PNC Technical Review), Power Reactor and Nuclear Fuel Development Corporation (present JAEA), (1991)
Power Reactor and Nuclear Fuel Development Corporation (present JAEA), Total View of Monju Started its Construction, Genshiryoku kogyo (in Japanese), Vol. 32 No. 5 (1986)
Power Reactor and Nuclear Fuel Development Corporation (present JAEA), Road to construction of Fast Breeder Reactor Monju, Genshiryoku kogyo (in Japanese), Vol. 37 No. 12 (1991)
Power Reactor and Nuclear Fuel Development Corporation (present JAEA), Starting of Monju Operation-Present Status and future development, Genshiryoku kogyo (in Japanese), Vol. 40 No. 6 (1994)
PNC TN2410 87–009, List of R&D Activities of Monju, JAEA, (1987)

References

1. Application for Permission of the Establishment of Monju Reactor, Power Reactor and Nuclear Fuel Development Corporation (present JAEA), (1986)
2. IAEA-TECDOC-1531, Fast Reactor Database 2006 Update, IAEA, (2006)

MIX
Papier aus verantwortungsvollen Quellen
Paper from responsible sources
FSC® C105338

If you have any concerns about our products,
you can contact us on
ProductSafety@springernature.com

In case Publisher is established outside the EU,
the EU authorized representative is:
Springer Nature Customer Service Center GmbH
Europaplatz 3, 69115 Heidelberg, Germany

Printed by Libri Plureos GmbH
in Hamburg, Germany